AN INTRODUCTION TO STATISTICAL PROBLEM SOLVING IN GEOGRAPHY

SECOND EDITION

AN INTRODUCTION TO STATISTICAL PROBLEM SOLVING IN GEOGRAPHY

SECOND EDITION

J. Chapman McGrew, Jr.
Salisbury State University

Charles B. Monroe
The University of Akron

Boston   Burr Ridge, IL   Dubuque, IA   Madison, WI   New York   San Francisco   St. Louis
Bangkok   Bogotá   Caracas   Lisbon   London   Madrid
Mexico City   Milan   New Delhi   Seoul   Singapore   Sydney   Taipei   Toronto

*This text is dedicated to our patient and tolerant wives,
Kathy and Laura. The innumerable hours spent working on this
manuscript meant less time could be devoted to family activities
and to our children, Courtney, Grady, Caroline, and Michael. The
continuous unselfish support of this project by our families is
deeply appreciated.*

McGraw-Hill Higher Education

A Division of The McGraw-Hill Companies

AN INTRODUCTION TO STATISTICAL PROBLEM SOLVING IN GEOGRAPHY,
SECOND EDITION

This book is printed on acid-free paper.

1 2 3 4 5 6 7 8 9 0 KGP/KGP 0 9 8 7 6 5 4 3 2 1 0

ISBN 0–697–22971–8

Vice president and editorial director: *Kevin T. Kane*
Publisher: *Edward E. Bartell*
Sponsoring editor: *Daryl Bruflodt*
Senior marketing manager: *Lisa L. Gottschalk*
Project manager: *Cathy Ford Smith*
Production supervisor: *Laura Fuller*
Coordinator of freelance design: *Michelle D. Whitaker*
Supplement coordinator: *Stacy A. Patch*
Compositor: *Lachina Publishing Services*
Typeface: *10/12 Palatino*
Printer: *Quebecor Printing Book Group/Kingsport*

Freelance cover designer: *Jamie O'Neal*
Cover image: *©Tony Stone*

Library of Congress Cataloging-in-Publication Data

McGrew, J. Chapman.
    An introduction to statistical problem solving in geography/J.
Chapman McGrew, Charles B. Monroe.—2nd ed.
        p. cm.
    Includes bibliographical references and index.
    ISBN 0–697–22971–8
        1. Geography—Statistical methods. I. Monroe, Charles B.
II. Title.
G70.3.M4   2000
910' .2'1—dc21   99-15121
                                                               CIP

www.mhhe.com

CONTENTS

PREFACE

All well-trained geographers need to be proficient in applying statistical techniques to the problems they face. Geography students need a sound introduction to the variety of ways in which statistical procedures are used to explore and solve practical geographic problems. The second edition of this book continues under the principles of the first edition and is designed to provide a comprehensive and understandable introduction to statistical methods in a problem-solving framework. We want students who use this textbook to become well-grounded in and feel comfortable applying statistical techniques in situations they will encounter in their subsequent geographic education and career.

This book is targeted for undergraduate geography majors and entry-level graduate students with limited backgrounds in statistical approaches to geographic problem solving. Because this is an introductory textbook, we assume that students have not taken other courses in statistical analysis and do not have previous experience with statistical methods. However, students with a background in statistics will still find the problem-solving emphasis useful.

In this second edition, many of the features of the first edition are continued and, in many areas, developed further. In this edition, we continue to stress the importance of written narratives that explain each statistical technique fully in ways that undergraduate geography majors and beginning graduate students can easily understand. This is done without compromising or oversimplifying the statistical integrity of the material. As in the first edition, real-world examples and problems are drawn from a variety of topical areas in both human and physical geography, and these examples are fully integrated into the text. Problems are again taken from various spatial levels that range from local to international.

The use of spatial statistics, both for descriptive and inferential purposes, continues to be emphasized. We carry on the practices begun in the first edition of using a flexible, exploratory approach that emphasizes real-world problem solving and the *p*-value method of statistical testing.

Instructors who have used the first edition of *An Introduction to Statistical Problem Solving in Geography* should note these major changes that have been incorporated into the second edition:

- Throughout the book, geography receives increased emphasis. For example, as a technique is introduced, a more detailed, student-friendly discussion about the geographic applications of the technique is included, with more written narrative examples. This change does not mean a decreased role of math and statistics, but rather more supportive narrative material throughout the text to help the geography student master beginning statistical techniques.
- Many new examples are presented throughout the text, particularly in the chapters dealing with inferential techniques. Most of the example problems found in chapters 9, 10, and 11 of the first edition have been replaced by new examples or have been revised. A new geographic example has been added in the second edition for each of the following techniques: two-sample difference of means Z or t test, two-sample difference of proportions test, matched-pairs (dependent-sample) difference test, ANOVA and Kruskal-Wallis test, goodness-of-fit chi-square uniform, goodness-of-fit chi-square proportional, goodness-of-fit Kolmogorov-Smirnov test for normality, contingency analysis, and area pattern analysis.
- A student-directed list of "major goals and objectives" has been added at the end of each chapter.

These goals and objectives will allow students to monitor their own progress and their mastery of geographic statistical materials.

- A new epilogue chapter has been added. This chapter contains a set of more than 150 "geographic situations." The ultimate goal of spatial analysis and the use of statistical techniques in geography is to be able to determine what statistical technique should be used in a given geographic situation. Instructors will find this feature useful to assess student abilities at the completion of the course.

Other narrative and explanatory materials are new to the second edition, including the following:

- The discussion in chapter 1 concerning how geographers use statistics, particularly in the context of a general research process, has been expanded. A new example discussing acid deposition across North America has been added, providing another demonstration of how geographers use statistics in the research process.
- In chapter 5, a new section presents "Basic Probability Terms and Concepts," including a detailed introduction of basic probability rules, laws, and formulas that guide the use of probability. This is accompanied by geographic examples to illustrate the principles discussed. Covered in greater depth are such concepts as statistical independence, complements, the simple and extended multiplication rule of probability, and the addition rules for mutually exclusive and nonmutually exclusive events.
- The material in chapter 7 has been reorganized, and new material has been included. These changes are designed to provide smoother sequencing and more student-friendly explanations of such difficult concepts as the central limit theorem, standard error of the mean, and the finite correction factor. Also, the confidence interval equations associated with tables 7.3 and 7.4 in the first edition have been reorganized and fragmented into smaller tables that are easier for the student to follow.
- The material in chapter 8 has been reorganized to provide a smoother, more logical flow. The one-sample difference of means tests are now separated from the one-sample difference of proportions test. Also, a new table supplies a schematic framework to classify statistical methods. Rows of this table represent different levels of measurement, and the columns group statistical approaches according to the type of question being asked.
- The material grouped together in chapter 9 of the first edition has now been split into two distinct chapters: "Two-Sample and Dependent-Sample

Difference Tests," and "Three-or-More Sample Difference Tests: Analysis of Variance Methods." This reorganization highlights the contrasts that distinguish various multisample difference tests. ANOVA testing schemes receive increased emphasis.

The overall organization of the text remains unchanged from the first edition. Part one (chapters 1 and 2) introduces basic statistical concepts and terminology. Part two (chapters 3 and 4) focuses on descriptive statistics (both nonspatial and spatial) and their use in solving geographic problems. The chapters in part three (5, 6, and 7) make the transition from descriptive analysis to inferential problem solving. Topics covered include probability, sampling, and estimation. Chapters 8 through 12 make up part four of the text and cover a variety of inferential statistics for geographic problem solving. Part five (chapters 13 and 14) examines the related techniques of correlation and regression, procedures that allow measurement of association between variables and prediction of the nature of the relationship between variables. The final epilogue chapter (chapter 15) provides the list of "geographic situations" and matching statistical techniques.

Introductory quantitative methods and spatial analysis courses vary among geography programs. This textbook can be adapted to a one-semester, two-quarter, or two-semester course sequence. However the course is structured, basic instruction in the use of statistical software packages is needed.

In the context of a single semester or two quarters, several approaches to the text are possible. One alternative is to move quickly through the entire text, emphasizing those topics that seem most appropriate. Another alternative is to eliminate certain topics and chapters entirely. Instructors have considerable flexibility in determining the depth of treatment for the topics in the text. The appropriate integration of the text into a specific spatial analysis course may best be determined from the previous statistical and computing background of the students.

In the context of a two-semester (or three-quarter) sequence, the entire text can be covered in considerable detail. One option is to present parts one through three in the first semester. Students would then be well grounded in basic statistical concepts, descriptive statistics, probability, and sampling. Parts four and five would then be presented in the second semester, and students could examine the variety of inferential statistics in considerable depth.

An exercise manual accompanies the textbook. This workbook contains a variety of exercises and problems keyed directly to the material presented in the text. Geographic problems and exercises are available for topics and techniques discussed in every chapter. To address the problems, students are referred to one of several data sets included in the workbook.

The data sets in the workbook are organized to provide maximum flexibility for the instructor and efficiency for the student. Instructors often disagree on the relative merits of requiring students to take samples from a population data set. Although sampling is an integral element of statistical problem solving, it can be a tedious and time-consuming task. The workbook offers the instructor several alternatives by identifying a few "predrawn" random samples within each data set. Different students can use different samples, avoiding the situation of every student analyzing the same set of values, or all students in the class can work with the sample data. The instructor also has the option of requiring students to draw their own sample.

ACKNOWLEDGMENTS

The groundwork for this textbook dates from our experiences together as graduate students at Pennsylvania State University. In the exciting and challenging environment of the Department of Geography, our developing interests in spatial analysis flourished. We again would like to acknowledge the support provided early in our careers by the many fine geography faculty at this institution.

Colleagues at both Salisbury State University and The University of Akron have provided a variety of practical suggestions and ideas for both the first and second editions. At Salisbury State, thanks are extended to Michael Folkoff, Daniel Harris, and Brent Skeeter for suggesting improvements in some of the new examples. Particular thanks goes to Calvin Thomas, chair at Salisbury State, for maintaining a flexible, supportive work environment conducive to the completion of projects such as this. At Akron, Robert B. Kent in Geography and Planning provided much advice and encouragement, and Richard L. Einsporn in Statistics offered pertinent comments on a statistical example.

We would like to thank the many students at both Salisbury State and Akron who were involved with this project. Spatial analysis classes at both institutions field-tested draft examples for the first edition and offered pertinent comments to improve the content in the second edition. Particular thanks are extended to those students who contributed in some way to data collection or statistical analysis of geographic data sets.

The Laboratory for Cartographic and Spatial Analysis at the University of Akron produced the figures and graphic material in the second edition. We especially want to recognize the excellent work and direction of Joseph Stoll, supervisor of the Lab. His patience and dedication have been most important in the successful completion of this book.

Laura Monroe provided excellent editorial and organizational comments throughout the second edition. Her professional "non-geographic" perspective has dramatically improved the written narrative. With her help, we believe a more readable textbook has resulted.

We wish to recognize the following reviewers who made suggestions and helpful comments on our manuscript:

Abu Muhammad Shajaat Ali, Oklahoma State University
Nancy R. Bain, Ohio University
Donald W. Clements, Southern Illinois University
Mark Flaherty, University of Victoria
Joseph R. Oppong, University of North Texas
Yong Wang, East Carolina University

Their constructive criticism contributed greatly to a stronger final product, and we appreciate their efforts.

The editorial and production staff at McGraw-Hill has provided welcome support throughout the development of this second edition. We would particularly like to thank Daryl Bruflodt, Cathy Smith, and Sheila Frank for their guidance in advancing the text through the later stages of production. All members of the McGraw-Hill book team have been supportive of our efforts to complete an improved second edition, and our thanks are extended.

J. Chapman McGrew, Jr.
Charles B. Monroe

PART ONE

BASIC STATISTICAL CONCEPTS IN GEOGRAPHY

CHAPTER 1

Introduction: The Context of Statistical Techniques

1.1 **The Role of Statistics in Geography**

1.2 **Examples of Statistical Problem Solving in Geography**

1.3 **Basic Terms and Concepts in Statistics**

Geography is an integrative spatial science that attempts to explain and to predict the spatial distribution and variation of human activity and physical features on the earth's surface. Geographers study how and why things differ from place to place, as well as how spatial patterns change through time. Of particular interest to geographers are the relationships between human activities and the environment and the connections between people and places. People have been interested in such geographic concerns for thousands of years. Early Greek writers, such as Eratosthenes and Strabo, emphasized the earth's physical structure, its human activity patterns, and the relationships between them.

These traditions remain central to the discipline of geography today. Contemporary geography continues to be an exciting discipline that attempts to solve a variety of problems and issues from spatial and ecological perspectives. The spatial perspective focuses on patterns and processes on the earth's surface, and the ecological perspective focuses on the complex web of relationships between living and nonliving elements on the earth's surface (Geography Education Standards Project, 1994).

The geographer starts by asking *where* questions. Where are things located on the earth's surface? How are features distributed on the physical or cultural landscape? What spatial patterns are observable, and how do phenomena vary from location to location? Historically speaking, geographers have focused on trying to answer these questions. In its popular image, the discipline of geography remains focused almost exclusively on the location of places.

However, in reality professional geographers no longer limit themselves to spatial or locational description. After a spatial pattern has been described and the where questions have been answered adequately, attention shifts to *why* questions. Why does a particular spatial pattern exist? Why does a locational pattern vary in a specific observable way? What spatial or ecological processes have affected a pattern? Why are these processes operating? As such why questions are answered, or as geographers speculate about why a spatial pattern has a particular distribution, we gain a better understanding of the processes that create the pattern. Sometimes different variables are found to be related spatially, and these findings provide insights into underlying spatial processes. In other instances, geographers try to determine if phenomena differ in various locations or regions and seek to understand why such differences exist.

Geographers are increasingly concerned with the practical application of this spatial information. Geographers ask *what-to-do* questions that involve the development of spatial policies and plans. More geographers now want to be active participants in both public and private decision making. Geographers might explore such questions as: What type of policy might best achieve more equal access for urban residents to city services and facilities? and, What sort of government policy would a geographer recommend to balance protection of wetlands and economic development in a fragile environment?

Geography is now a problem-solving discipline, and geographers are concerned with applying their spatial knowledge and understanding to the problems facing the world today. Noted geographer Risa Palm recently stated, "[G]eography involves the study of major problems facing humankind such as environmental degradation, unequal distribution of resources and international conflicts. It prepares one to be a good citizen and educated human being" (Assoc. of American Geographers, n.d.).

1.1 THE ROLE OF STATISTICS IN GEOGRAPHY

Statistics is generally defined as the collection, classification, presentation, and analysis of numerical data. Statistical techniques and procedures are applied in all fields of academic research. In fact, wherever data are collected and summarized or wherever any numerical information is analyzed or research is conducted, statistics are needed for sound analysis and interpretation of results.

Beyond the widespread application of statistics in academic research, use of statistics can also be seen in many aspects of everyday life. Just consider for a moment this sampling of common applications of statistics:

1. In sports, statistics are frequently cited when measuring both individual and team performance. The success of a team (and the job security of the coach!) is often gauged by the team's winning percentage. Baseball fans are well-versed in such individual player statistics as batting average, earned run average, and slugging percentage. Goalies in hockey may be benched if their "goals against average" gets too high. Quarterbacks in professional football are evaluated by their quarterback rating—a multifaceted descriptive statistic with a number of components that few fans can identify.

2. Statistical procedures are used routinely in political polling and opinion gathering. When statistical sampling is applied properly, certain characteristics about a statistical population can be inferred based solely on information obtained from the sample. Political opinion polling is commonplace before any major election, and exit polling of a sample of voters is used to predict the election results as quickly as possible.

3. Statistics are widely used in business market analysis. Businesses constantly scrutinize all aspects of consumer behavior and details of consumer purchasing patterns because a fuller understanding of consumer actions and spending often translates into greater profit. Major business decisions, such as where to advertise and which markets are best to introduce a new product, depend on sample statistics. Many of us are familiar with the ratings systems used to estimate the relative popularity of various television programs. These business decisions, involving literally millions of dollars of television advertising and variable advertising rates, rely on well-designed statistical samples of the viewing population.

4. Most of us use a wide variety of statistical measures and procedures in making personal financial decisions. Whether purchasing a home, buying auto or life insurance, trying to develop a workable budget, or setting up an investment or savings plan, an understanding of different monetary terms and statistical measures is indispensable. Knowledgeable reading of stock tables and financial pages in a newspaper requires an understanding of various statistical measures. Two common examples are price-earnings ratio (the closing price of stock divided by the company's earnings per share for the latest 12-month period) and yield (current annual dividend rate divided by the closing price of a stock, expressed as a percentage).

5. Weather is an everyday practical concern for many people. They want to know which coat they should wear that morning and whether or not they will need an umbrella. Perhaps a winter storm system is approaching, and people want to know the likelihood of getting 2 to 4 inches of snow as the system moves through. A farmer may want to know the probability of getting measurable precipitation over the next two weeks as an aid to scheduling spring planting. Weather forecasting models based on statistics attempt to provide answers to such questions, often using probability estimates from similar historical weather situations.

Geographers use statistics in numerous ways. Statistical analysis benefits geographic investigation by helping answer the where, why, and what-to-do questions posed in the introductory discussion. Among many general applications, the use of statistics allows the geographer to

- describe and summarize spatial data,
- make generalizations concerning complex spatial patterns,
- estimate the likelihood or probability of outcomes for an event at a given location,
- use limited geographic data (sample) to make inferences about a larger set of geographic data (population),
- determine if the magnitude or frequency of some phenomenon differs from one location to another, and
- learn whether an actual spatial pattern matches some expected pattern.

The use of statistics must be placed within the context of a general research process. Most geographers now recognize the overall importance of statistics in research. Some may view the application of statistical methods as an essential element of any scientific geographic research, whereas others view statistical methods as one of many approaches that can be applied in geography. Whatever one's particular perspective, the general methodological shift, or "revolution," in the geographer's view of the world occurred in the late 1950s and early 1960s. During that time, geographers began to move from a qualitative description of the spatial distribution and variation of human and physical features to a quantitative analysis of the same features. As early as 1963, one observer concluded that the quantitative revolution itself was over (Burton, 1963). The application of quantitative methods (including a variety of statistical techniques) across all areas of geographic inquiry became generally accepted by geographers in the 1960s and has continued to serve as a fundamental methodological or procedural approach to much geographic research since that time.

The geographic research process and the roles of statistics in that process are summarized in a general organizational framework (figure 1.1). The left column of the figure lists the series of steps that lead to the formulation of hypotheses. These activities are done early in scientific inquiry. The right column of the figure diagrams the steps involved in scientific research after a hypothesis has been stated. Statistical procedures are involved in geographic research both before and after hypotheses are generated.

The sequence of tasks outlined in figure 1.1 is typical of the process many geographers follow when conducting research. This organizational framework is not the only mode of geographic research, however. Moreover, this framework should not be viewed as a rigid series of steps, but rather as a general, flexible guide used in geographic research.

The research process begins when a geographer identifies a worthwhile geographic problem to investigate. To recognize a productive research problem, the geographer must have background knowledge and experience in the area being studied. There is simply no substitute for having a strong background in the appropriate branch of the discipline.

Formulating a hypothesis is at the center of the research process (figure 1.1). A **hypothesis** is an unproven or unsubstantiated general statement concerning the problem under investigation. The investigator may have sufficient background knowledge or information from previous research (such as a review of the literature) to allow hypotheses to be readily developed. Perhaps a hypothesis can be formulated using a **model,** which is a simplified replication of the real world. A well-known model in geography is the spatial interaction model that predicts the amount of movement expected between two places as a function of their populations and the distance that separates them.

In some cases, a geographer may have identified a possible research area, but is not yet ready to formulate a hypothesis. Perhaps more information is needed, or questions need to be answered about the problem. In these situations, the geographer cannot immediately move down the right column of figure 1.1, but must first address research tasks in the left column. Statistical analysis is central to this process. Geographers often gain spatial insights by collecting data and presenting this information using graphical procedures and maps. During this phase of the investigation, statistical analysis also provides quantitative summaries or numerical descriptions of the data. The information gathered may enable the geographer to draw conclusions about the research questions, develop a model of the spatial situation, and generate suitable hypotheses.

If the geographer has a workable hypothesis, the research process then follows the steps shown in the right column of figure 1.1. Additional data need to be collected and prepared so that the hypothesis can be tested and evaluated. The steps following hypothesis formulation are the core of statistical analysis and are a primary focus of this book. In scientific research, if a hypothesis is repeatedly verified as correct under a variety of circumstances (perhaps at various locations or times), it gradually takes on the stature of a **law.** In other words, a proven hypothesis can eventually become accepted as a law. If various laws are combined, they then constitute a **theory.**

After testing a research hypothesis, the results must be evaluated and conclusions drawn. Several strategies or actions are possible (figure 1.1):

1. The research findings may be incorporated into actual or recommended spatial policies and plans. The applied geographer might be suggesting actions or addressing what-to-do questions.
2. If a hypothesis is verified as correct and valid, the results could be refined further into a spatial model that predicts what is likely to occur under various scenarios. Repeated verification of a hypothesis or model under a variety of circumstances (perhaps at various locations or at different times) might lead to the eventual development of laws and theories concerning the particular geographic problem or issue.
3. If a hypothesis is tested and found to be partially or completely incorrect, the geographer may need to return to an earlier step in the geographic research process. With only a partially validated

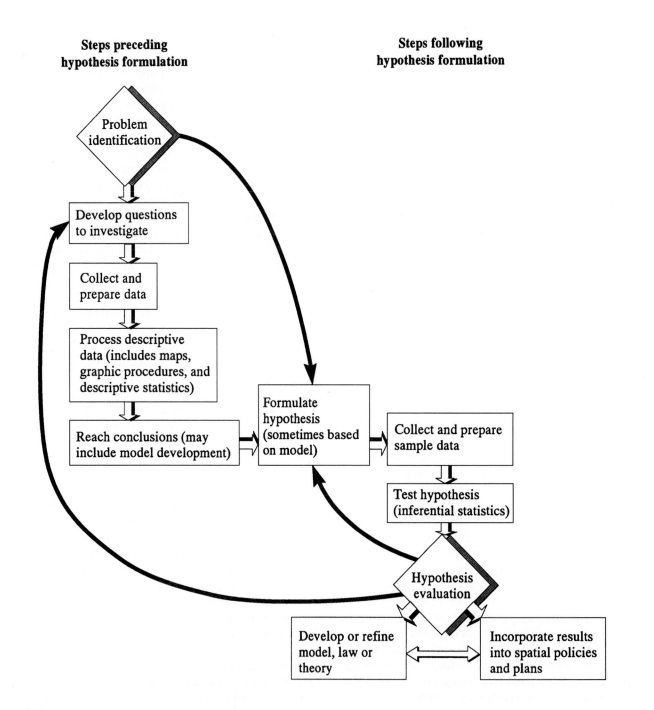

FIGURE 1.1
The Role of Statistics in the Geographic Research Process

hypothesis, one could return to hypothesis formulation to restate or refine the original hypothesis. If a hypothesis is proven totally wrong, it could be necessary to return to question development, the initial step.

The general research process outlined in figure 1.1 takes on many operational forms in geography. However, geographic research centers on the investigation of spatial geographic patterns and processes through descriptive analysis, hypothesis generation, and inferential statistical tests, with a possible goal of developing laws and theories. Collection, presentation, and processing of data all play central roles in this research process.

1.2 EXAMPLES OF STATISTICAL PROBLEM SOLVING IN GEOGRAPHY

How can geographers use statistics in the research process to approach locational issues and solve spatial problems? This will depend partly on the nature of the issue or problem under consideration and partly on the purposes and objectives of the research itself. To illustrate the diversity of approaches, two different spatial patterns are examined:

- Population change in the United States during the 1980s
- Acid deposition across North America

Both of these examples show interesting locational patterns, which are the result of complex spatial processes that are not completely understood. Figures show *where* population growth occurs in the United States and *where* acid deposition occurs in North America, but they do not explain *why* these particular spatial patterns exist.

These two examples emphasize different research needs. The first example focuses on the state-level growth rate pattern from 1980 to 1990, and discussion is closely connected to the steps in the research process (figure 1.1). The goal is to formulate a set of hypotheses based on survey responses from a typical or representative sample of interstate migrants. The second example uses an isoline map of acid deposition in North America to initiate the research process, and discussion is not tied as closely to the research process. Instead, the discussion leads to formulation of hypotheses about acid deposition patterning around a single, isolated emission source.

Population Change in the United States (1980–1990)

Suppose a geographer is interested in analyzing growth trends in the United States during the 1980s

(figure 1.2). The fastest growing states are generally located in the South and West, with much of the national growth occurring in California, Florida, and Texas. Nearly all of the states losing population or experiencing slow growth are located in the Northeast and Midwest.

Given this map of state-level population growth, the geographer may want to explore further and ask why this spatial pattern exists. Why does the growth rate pattern vary in this way? What factors can be suggested to help explain the nature of this spatial distribution? What spatial process or processes might have been operating?

Various factors may be at work to produce the spatial pattern shown in figure 1.2. Examination of previous studies may help the geographer identify potential relationships or explanations that might be relevant. For example, many geographers feel that climatic amenities and environmental considerations influence the spatial pattern of population change in the United States. The southern and western growth regions, including Florida, California, Arizona, and Hawaii, are known as the Sunbelt because of their warm and sunny climates. Conversely, the northeastern and midwestern states that are either losing population or experiencing slow growth are in regions called the Snowbelt or Frostbelt. The implication is that people are fleeing the cold winters and snow for year-round warm weather and outdoor recreation opportunities.

Economic factors undoubtedly influence growth in a number of ways. In surveys examining why people change residence, respondents frequently cite job opportunities and related economic reasons. Over the last few decades, many more new jobs have been created in the southern and western states than in the Northeast and Midwest. The region containing the states of Pennsylvania, Ohio, Michigan, Indiana, and Illinois is sometimes called the Rust Belt because of its traditional economic reliance on heavy industry and manufacturing—sectors of the U.S. economy that suffered a relative decline in the 1980s.

Patterns of immigration also influence population growth. A large number of migrants from Asia and Latin America have settled in California, Florida, and Texas, contributing to the high growth rates of these states in the 1980s. However, many immigrants have also settled in Illinois and New York, states that have experienced little recent change in population size.

Geographers studying recent migration trends suggest that low-density residential areas, such as rural regions and small towns, are increasingly attractive locations for many Americans. If this trend toward nonmetropolitan growth or decentralization

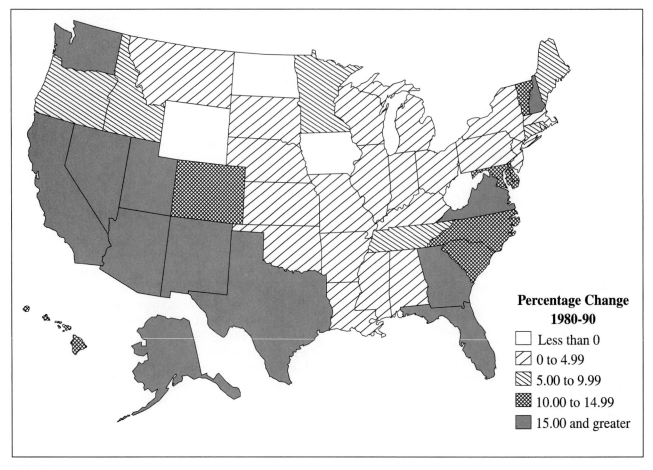

FIGURE 1.2
Population Change by State, United States, 1980–1990 (*Source:* Data from Bureau of the Census, U.S. Dept. of Commerce.)

of the population is true, then states with smaller populations and low population densities should have high growth rates.

Other investigators cite the importance of the oceans for water-based recreation and economic activity as a growth factor. Rapidly growing coastal (peripheral) communities contrast with stagnant or declining interior (heartland) locations.

The spatial pattern of population change is clearly the result of a set of complex spatial processes. A number of questions are worthy of geographic investigation: Does a relationship exist between climate and population change? Are Sunbelt states in the South and West growing faster than the Snowbelt states in the Northeast and Midwest? How important are economic structure and employment in "explaining" population growth? Are states with economies that depend on heavy industry and manufacturing losing population faster than other states? What is the relationship between the residential choices of recent immigrants and state population growth rates? Are coastal states growing faster than interior states?

Once these questions have been posed, the geographer must collect and prepare data that can help generate answers (figure 1.1). Some of the research steps are easily accomplished. It is a straightforward matter to classify each state as coastal or noncoastal, allocate states to the Sunbelt or Snowbelt, and collect data on the number of immigrants settling in each state during the 1980s. Other data collection and preparation tasks may not be quite as direct. For example, some form of operational definition is needed to measure the economic structure of a state. No single clear definition of "economic structure" exists, and it is not immediately obvious how this term should be defined for this problem.

After all data have been collected and prepared, they are "processed" using various graphic procedures and statistical measures (figure 1.1). A series of choropleth maps could be constructed, such as the map showing state-level population change (figure 1.2). Displaying information with a histogram, ogive (cumulative frequency diagram), or scattergram may also be useful. For example, it would be informative to construct a graph showing the relationship between

1980 to 1990 population change for each state and the number of 1980 to 1990 immigrants who chose to reside in each state. These graphic procedures are discussed in detail in section 2.5.

Summary measures or numerical descriptions of data may also help the geographer answer the research questions being studied. Statistical procedures provide the average or typical value of a variable and determine the amount of variability within the set of values. For example, calculating and comparing the average growth rate of coastal states to noncoastal states might provide some insights into the spatial pattern. Various descriptive statistics are presented in chapter 3.

After the data have been processed, the geographer can draw some preliminary conclusions about the geographic research questions. These questions can be understood more fully by using graphic evidence, map summaries, and statistical summary descriptions. The initial investigation of population change in the United States could reveal general answers to some of the questions posed: Are Sunbelt states growing faster than Snowbelt states? Are coastal states growing faster than noncoastal states?

Some geographic studies may be complete at this point, the geographer content with a descriptive analysis of a geographic pattern and some of its relationships. However, geographers often continue the research process down the right column of figure 1.1 through an expanded inferential approach.

In addition to offering factual insights, an initial investigation may allow the geographer to propose a model that describes the situation and to formulate potential hypotheses for testing. For example, analysis of state-level population change permits one to develop hypotheses relating this spatial pattern to factors that produced the change. By testing these hypotheses with inferential procedures, the geographer will be able to probe the underlying spatial processes causing the pattern in figure 1.2.

However, spatially aggregate data (like the state-level data used for the choropleth map) are often not adequate for developing hypotheses in research problems. The researcher may need to obtain individual and family responses to questions regarding personal patterns of migration to understand the underlying reasons for moves. Since the researcher cannot possibly survey all migrants, information from a sample of movers needs to be collected. If done properly, this sample information will adequately reflect what is occurring in the overall population of movers from which the sample is drawn.

How might a geographer investigating the patterns and processes of population change generate hypotheses for testing sample data? That is, how might one move from the creation of general con-cepts and ideas to hypotheses specifically targeted and "testable"? The details of this transition will vary from problem to problem, but the procedure can be clarified by further discussion of observations from figure 1.2. Note that Michigan's population grew little during the 1980s, whereas Florida's population increased substantially. Michigan is a noncoastal, Snowbelt state with an economic and employment structure considered less favorable for growth than Florida, a coastal, Sunbelt state offering jobs in many new service (tertiary) and information-processing (quaternary) fields. What factors influenced the population change? What reasons do former Michigan residents give when asked why they moved? What reasons do newly arrived Florida residents give when asked why they decided to relocate? A preliminary survey of out-migrants from Michigan and in-migrants to Florida would provide information to help answer these questions. Analysis of survey data could allow one to reach certain new conclusions and construct research hypotheses for the problem. Remember that a hypothesis is an unproven general statement regarding a research question. The following are possible hypotheses for this problem:

- The reasons for leaving a Sunbelt state differ greatly from reasons given for leaving a Snowbelt state.
- Most people moving from a Snowbelt state to a Sunbelt state say they have done so to live in a warmer climate.
- People who move to coastal communities from noncoastal communities say they have done so more for water-based recreation than for employment reasons.
- Recent immigrants select a community that already has a large number of residents from their country.
- Most currently employed people relocating from one state to another have a higher income in their new state of residence.
- Most unemployed people who move from one state to another join the workforce less than a month after moving.
- Most people moving from a county in a metropolitan area to a small town or rural area cite the rural lifestyle as more attractive than the urban lifestyle.

Acid Deposition across North America

Another problem or issue of general relevance is the spatial distribution of acid deposition across North America (figure 1.3). Acid deposition refers to any precipitation—rain, snow, sleet, or fog—that is more acidic than what is considered normal. It also refers to the fallout of dry acidic particles. As measured on the pH scale, unpolluted precipitation is normally

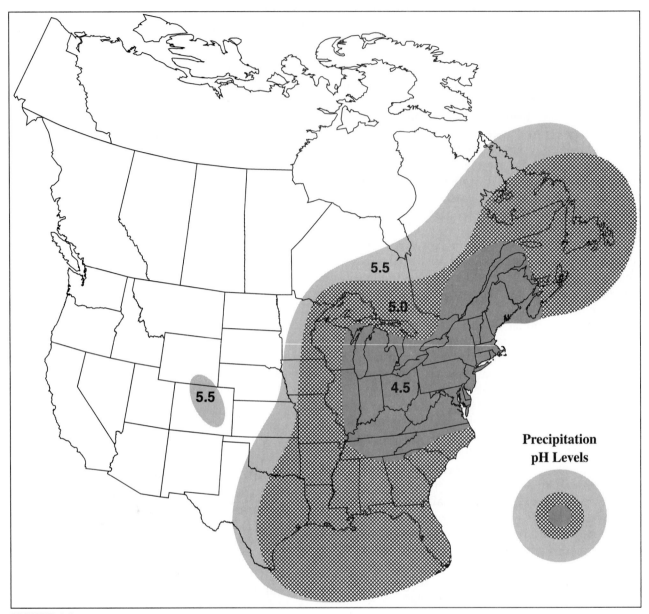

FIGURE 1.3
Acid Deposition Pattern, Eastern North America (*Source:* Modified from Nebel, B.J. and R. Wright. 1993. *Environmental Science: The Way the World Works* (4th edition). Englewood Cliffs, NJ: Prentice-Hall.)

slightly acidic (pH = 5.6), because carbon dioxide in the air readily combines with water to produce carbonic acid. Acid precipitation is therefore any precipitation with a pH of 5.5 or less. As shown in figure 1.3, most of eastern North America is subjected to weakly acidic or moderately acidic precipitation.

At the most fundamental level, the chemical and atmospheric processes are well understood. Sulfur dioxide and nitrogen oxides enter the atmosphere from a variety of natural and anthropogenic (human-generated) sources. Once in the troposphere (the lowest level of the atmosphere), these emissions react with sunlight, water, and other chemicals to form sulfuric and nitric acids. These acids dissolve readily in water or adsorb on particles and return to earth as acid deposition. *Natural sources* contribute substantial quantities of pollutants to the atmosphere:

sulfur from volcanic activity and nitrogen oxides from lightning, biomass burning, and microbial processes. The primary *anthropogenic sources* of sulfur dioxide in North America are the burning of fossil fuels (especially coal-burning power plants) and related industrial processes. Nitrogen oxide emissions are traced primarily to combustion of fossil fuels and transportation.

In the eastern United States and Canada, the Ohio River Valley region has been identified as an important source of acid deposition. Although a variety of industrial activities and processes contribute to the development of pollutants, a set of coal-burning power plants, located in the states of Pennsylvania, Ohio, Indiana, Illinois, West Virginia, and Kentucky, has been identified as particularly heavy pollutant emitters (Nebel and Wright, 1993).

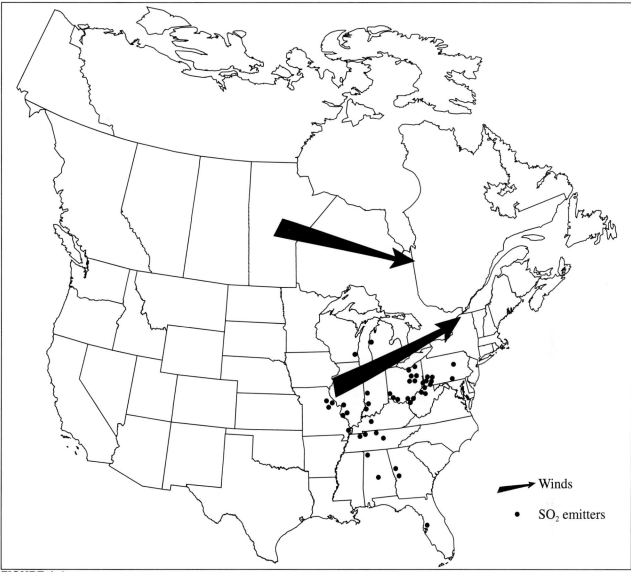

FIGURE 1.4
Wind Pattern Map, Eastern North America (*Source:* Modified from Nebel, B. J. and R. Wright. 1993. *Environmental Science: The Way the World Works* (4th edition). Englewood Cliffs, NJ: Prentice-Hall.)

Given the prevalent wind patterns from these sources of pollution (figure 1.4), it is not surprising to see most acidic deposition taking place in the northeastern United States and southeastern Canada.

Acid deposition creates many problems that produce environmental and human concern. It adversely affects aquatic ecosystems in a variety of complex ways, including well-documented dramatic declines in fish populations in lakes in Ontario and the Adirondack Mountains of upstate New York. Forested areas may also be dramatically affected by acid deposition, with tree die-offs, nutrients leached from soils, and ancillary effects on wildlife, soil erosion, waterway sedimentation, and possible flooding. The effects of acid deposition on such human artifacts as limestone and marble buildings and monuments are also readily noticeable. The corrosion of build-

ings, monuments, and outdoor equipment caused by acid deposition costs billions of dollars for replacement and repair (Nebel and Wright, 1993).

Given the complicated and synergistic relationships associated with the problem of acid deposition, it would seem to be difficult to develop questions for investigation that would lead to the collection and preparation of data and the subsequent formulation of hypotheses (figure 1.1). Many research strategies are possible, and literally thousands of scientists are involved with this important problem in some way. The complexities appear almost insurmountable. For example, air circulation patterns are constantly shifting; it takes different amounts of time for the chemical reactions to take place (depending on a variety of dynamic atmospheric conditions); and the collective actions of multiple sources of pollutants are not easily predictable.

One possible way to proceed, however, is with an "isolated location" strategy. That is, identify a single, isolated emission source, such as a large coal-burning power plant that emits sulfur dioxide. Finding such an isolated source of emissions might be problematic. If feasible, however, this would allow us to begin to measure and describe the contribution of that particular source to acid formation and resultant precipitation. It is possible that some initial insights will be gained by looking for relationships and formulating, testing, and evaluating hypotheses in such an isolated, or "controlled," setting. Then, if some hypothesis is repeatedly verified in this controlled experimental situation, it may be tested elsewhere in more complex settings to lead eventually to the development of a more generalized model, law, or theory.

Consider this simple experimental design: Suppose a regularly spaced (systematic) set of 24 pH-collection stations is placed around a single pollution source (figure 1.5). It is now possible to formulate a number of testable hypotheses:

1. It is hypothesized that pH levels of precipitation are related to distance from the emission source, with lower average recorded pH values (e.g., more acidic precipitation) expected at collection stations closer to the emission source. Thus it is hypothesized that point 13 will have more acidic

precipitation than point 14, that point 17 will have more acidic precipitation than point 22, and so on.

2. It is hypothesized that pH levels of precipitation are related to wind direction from the emission source, with lower average recorded pH values expected at collection stations directly downwind from the emission source and higher average pH values expected directly upwind from the emission source. If the prevalent wind direction is from west to east, it is hypothesized that points 13 and 14 (directly downwind) will have more acidic precipitation than points 5 and 24 (not as directly or consistently downwind); that points 11 and 12 (directly upwind) will have less acidic precipitation than points 1 and 20; and so on.

3. It is hypothesized that the spatial diffusion of acid precipitation is related to wind speed. Higher wind speeds are generally associated with more intense winter cyclonic systems (as contrasted with less intense systems during other times of the year). If lower pH values are diffused more widely when wind speeds are higher, then it may be hypothesized that sample points at a substantial distance from the emission source (e.g., points 5, 14, and 24) will have more acidic precipitation in the winter months than during the rest of the year.

4. Hypothesis #2 should be refined if wind direction varies by elevation. For example, when precipitation is the result of an active cyclonic system moving through the area, wind direction at the surface is likely to differ from wind direction aloft, with surface winds generally having an easterly component and winds aloft moving from southwest to northeast. If it is hypothesized that surface wind flows dominate acid precipitation diffusion in the immediate area, then the lowest pH values should be west of the emission source (e.g., point 12 will have more acidic precipitation than points 9, 13, and 18). On the other hand, if it is hypothesized that winds aloft dominate the acid precipitation diffusion pattern, then the lowest pH values should be northeast of the emission source (e.g., point 5 will have more acidic precipitation than points 1, 20, and 24).

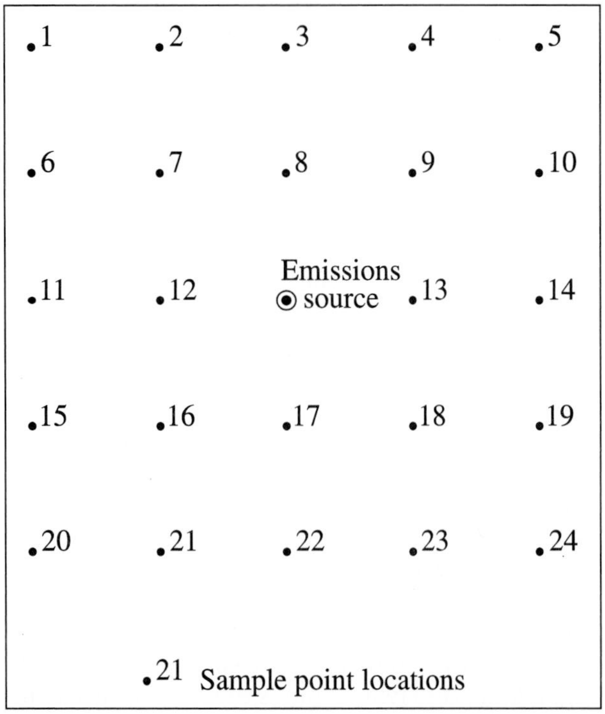

These hypotheses are not overly complex, but may provide some preliminary insights about the pattern of acid precipitation in an isolated (controlled) research setting. As these hypotheses are tested and evaluated, they may serve as effective building blocks for more complex models, laws, and theories. It might eventually be possible to advance toward the development of effective models that accurately predict expected pH levels under a variety of climatic conditions or under circumstances when multiple emission sources interact in different ways.

FIGURE 1.5
Proposed Placement of Data Collection Stations Around Emission Source

1.3 BASIC TERMS AND CONCEPTS IN STATISTICS

The most basic element in statistics is **data** or numerical information. Geographers often use groups of data, which are referred to as a **data set** and presented in tabular format (table 1.1). A data set consists of observations, variables, and data values. The elements or phenomena under study for which information (data) is obtained or assigned are often referred to as **observations.** Observations are sometimes called *individuals* or *cases.* Geographers use many types of observations in their research. Some are spatial locations such as cities or states, and others are nonspatial items such as people or households. A property or characteristic of each observation that can be measured, classified, or counted is called a **variable** because its values vary among the set of observations. The resulting measurement, code, or count of a variable for each observation is a **data value.**

The data set in table 1.1 shows a two-dimensional geographic data array with several variables chosen from a larger set of observations. In this matrix, variables are presented across the top of the table and observations (countries) are listed down the left side. A value or unit of data inside the table represents the magnitude of a single variable for a particular observation. Examples of data values are Thailand's 1994 population of 58 million, New Zealand's area of 271,000 square kilometers, and Chile's 1984–94 average annual rate of inflation of 18.5 percent.

An alternative type of geographic data matrix is shown in table 1.2. This table depicts multiple data values for a single variable—the number of Canadian interprovincial migrants during 1991. Geographers are often interested in the description and analysis of such spatial interaction (origin-destination) data. It is interesting to note that the single largest data value is the number of people moving from Alberta to British Columbia (with 30,654 migrants), that not a single person moved from Yukon to Prince Edward Island in 1991, that the province with the largest positive net migration value is British Columbia (net migration of 33,447), and that the province with the largest negative net migration value is Quebec (net migration of −12,259). If interprovincial migration patterns were evaluated over time (e.g., 1971, 1981, and 1991), additional dimensionality would be included in the analysis.

TABLE 1.1

Two-Dimensional Geographic Data Matrix

Country	Population*	Area**	Fertilizer consumption†	Inflation rate‡
Chile	14.0	757	849	18.5
Iran	62.5	1,648	755	23.4
Madagascar	13.1	587	25	15.8
New Zealand	3.5	271	12,745	4.6
Switzerland	7.0	41	3,340	3.7
Thailand	58.0	513	5,441	5.0
United States	260.6	9,364	1,011	3.3

* millions, mid-1994
** in thousands of square kilometers
† hundreds of grams of nutrients per hectare of arable land, 1992–93
‡ average annual, in percent, 1984–94

Source: Data compiled by World Bank and published in *World Development Report 1996.* New York: Oxford University Press.

TABLE 1.2

Geographic Flow Matrix: Number of Interprovincial Migrants, Canada, 1991

Province of origin	NF	PE	NS	NB	QC	ON	MB	SK	AB	BC	YT	NT
Newfoundland		228	2,175	766	393	6,026	284	117	1,413	1,112	39	175
Prince Edward Island	182		1,469	627	159	1,107	116	73	575	393	—	21
Nova Scotia	1,564	681		2,959	1,170	8,971	688	413	1,951	2,461	47	256
New Brunswick	617	551	3,818		2,588	5,658	560	201	1,358	1,349	58	96
Quebec	376	153	1,348	2,473		26,723	761	314	3,138	5,021	21	222
Ontario	6,333	980	8,602	5,297	18,223		6,044	2,654	16,921	23,636	183	726
Manitoba	109	82	584	363	876	7,412		2,994	6,392	7,290	34	285
Saskatchewan	124	46	356	215	519	2,793	2,991		15,250	6,539	171	478
Alberta	910	268	1,735	1,030	1,882	12,136	3,778	9,167		30,654	512	1,360
British Columbia	411	130	1,857	396	2,228	11,489	3,279	3,534	21,509		1,097	628
Yukon	43	—	30	7	31	178	54	25	401	1,029		94
Northwest Territories	98	50	174	47	222	502	203	161	1,788	818	219	
Total in-migration	10,767	3,169	22,148	14,477	28,291	82,995	18,758	19,653	70,696	80,302	2,381	4,341
Total out-migration	12,728	4,722	21,161	16,854	40,550	89,599	26,421	29,482	63,432	46,855	1,892	4,282
Net migration	−1961	−1553	987	−2377	−12259	−6604	−7663	−9829	7,264	33,447	489	59

Column group header above NF–NT: **Province of destination**

Source: Statistics Canada

In the discussion and application of the geographic research process, two forms of statistical analysis are briefly described. The early steps of the process focus on the descriptive processing of data, and later stages involve testing hypotheses using inferential methods. This fundamental distinction between descriptive and inferential statistics requires further explanation.

Descriptive statistics provide a concise numerical or quantitative summary of the characteristics of a variable or data set. Descriptive statistics describe—usually with a single number—some important aspect of the data, such as the "center" or the amount of spread or dispersion. For most geographic problems, using such descriptive summary measures is superior to working directly with a large group of values. Descriptive statistics allow geographers to work efficiently and communicate effectively.

Replacing a set of numbers with a summary measure necessarily involves the loss of information. Various descriptive statistics are available, each with different advantages and limitations. It is important to select the descriptive measure whose characteristics seem appropriate to the geographic problem being analyzed. In many geographic situations, the objective is to minimize the loss of relevant information when moving from unsummarized data to descriptive measures. Descriptive statistics are discussed in chapters 3 and 4.

The purpose of **inferential statistics** is to make generalizations about a statistical population based on information obtained from a sample of that population. In the context of inferential statistics, a **statistical population** is the total set of information or data under investigation in a geographic study. In the examination of state-level population change, a geographer might be interested in the statistical population of *all* U.S. residents and their migration patterns. In the acid deposition example, the goal might be to develop a model that predicts average pH values of deposition for the statistical population of *all* locations around an emission source.

It is not practical to question *all* U.S. citizens concerning their migration behaviors, nor is it possible to place pH-monitoring systems at *all* locations surrounding an emission source. However, collecting information from a sample of people or a sample of locations in these statistical populations may be feasible. A **sample** is a clearly identified subset of the observations in a statistical population. If one is to make proper inferences about the population based on the collection of sample data, then the sampled subset must be typical or representative of the entire statistical population from which it is drawn.

Inferential statistics use descriptive measures obtained from samples and link this descriptive information to probability theory. General statements can then be made about the nature or characteristics of the population from which the sample has been drawn. For example, using a representative (unbiased) sample of recent U.S. migrants, general statements can be made about the movement patterns of all U.S. migrants. Similarly, from the pH values of a representative sample of locations around an emission source, general statements are possible regarding pH values at all locations in that area.

Estimation and **hypothesis testing** are the two basic types of statistical inference. In some instances, sample statistics are used to estimate a population characteristic. For example, a sample proportion of migrants indicating that a higher income job was the most important reason for moving could be used to estimate the proportion of all migrants moving for that reason. In a similar manner, the pattern of a sample of locations around an emission source having moderately acidic precipitation (pH values between 4.0 and 5.0) could be used to estimate the pattern of all locations having moderately acidic precipitation. Sampling procedures and estimating population characteristics from samples are discussed in chapters 6 and 7.

In inferential hypothesis testing, sample data are used to reach conclusions about population characteristics. For instance, an inferential hypothesis test could be applied to determine if the reasons for moving differ significantly between a sample of interstate migrants from the Midwest and another sample of interstate migrants from the South. If the responses from the two sample groups are significantly different, it can be inferred that the samples were drawn from two different statistical populations. It could also be concluded that the reasons for moving are different among the populations of all midwesterners and southerners. In the example of acid deposition patterning around an emission source, an inferential hypothesis test could be used to determine if the typical pattern of highly acidic precipitation from a sample of winter cyclonic systems differs from the pattern of highly acidic precipitation from a sample of summer convective systems. If the samples of winter cyclonic and summer convective patterns are significantly different, it could be concluded (inferred) that the samples were drawn from two different statistical populations and that the patterns around the emission source differ locationally among all winter cyclonic and summer convective systems. Parts IV and V (chapters 8 through 14) consider a variety of inferential hypothesis tests.

KEY TERMS

MAJOR GOALS AND OBJECTIVES

If you have mastered the material in this chapter, you should now be able to do the following.

1. Identify the types of questions geographers ask.
2. Understand the general importance and uses of statistics.
3. List the potential applications of statistics in geography.
4. Understand the role of statistics in the geographic research process.
5. Explain the role of statistics both before and after hypothesis formulation.
6. Formulate possible hypotheses when presented with a spatial pattern or locational data set.
7. Distinguish between questions, hypotheses, laws, theories, and models in geography.
8. Explain the typical organization or relationship of observations, variables, and data values in geographic data sets (such as tables 1.1 and 1.2).
9. Explain the basic difference between descriptive statistics and inferential statistics.
10. Distinguish between a statistical population and a sample.

REFERENCES AND ADDITIONAL READING

Abler, R. F., J. S. Adams, and P. R. Gould. 1971. *Spatial Organization: The Geographer's View of the World.* Englewood Cliffs, NJ: Prentice-Hall.

Amedeo, D. and R. G. Golledge. 1975. *An Introduction to Scientific Reasoning in Geography.* New York: John Wiley and Sons.

American Statistical Association. "What is Statistics?" "What Do Statisticians Do?" "Careers in Statistics." www. amstat .org/education.

Association of American Geographers. No date. *Geography: Today's Career for Tomorrow* (informational brochure). Washington, DC: Assoc. of Amer. Geographers.

Burton, I. 1963. "The Quantitative Revolution and Theoretical Geography." *The Canadian Geographer* 7:151–62.

Cobb, G. 1993. "Reconsidering Statistics Education: A National Science Foundation Conference." *Journal of Statistics Education.* www.stat.ncsu.edu/info/jse.

DASL, in StatLib, Department of Statistics, Carnegie Mellon University. 1999. http://libstat.cmu.edu/dasl. (The Data and Story Library, an online library of data files and stories that illustrate the use of basic statistics methods, with several closely related to geography.)

Gaile, G. L. and C. J. Willmott (editors). 1989. *Geography in America.* Columbus, OH: Merrill.

Geography Education Standards Project. 1994. *Geography for Life: National Geography Standards 1994.* Washington, DC: National Geographic Research and Exploration.

Haring, L. L. and J. F. Lounsbury. 1992. *Introduction to Scientific Geographic Research* (4th edition). Dubuque, IA: Wm. C. Brown.

Hubbard, R. 1997. "Assessment and the Process of Learning Statistics." *Journal of Statistics Education.* www.stat .ncsu.edu/info/jse.

Journal of Statistics Education. 1999. www.stat.ncsu.edu/ info/jse. (A refereed electronic journal of postsecondary teaching of statistics. Many articles are of interest to geographers.)

Nebel, B. J. and R. Wright. 1993. *Environmental Science: The Way the World Works* (4th edition). Englewood Cliffs, NJ: Prentice-Hall.

CHAPTER 2

Geographic Data: Characteristics and Preparation

Before performing statistical processing and analysis, one must know a number of characteristics of spatial data. One needs to understand how variables are organized as well as how data are arranged within this organization. In this chapter, basic concepts are introduced that provide the background information needed to characterize data before statistical techniques are applied.

Questions about data and variables arise early in the scientific research process. When identifying an appropriate geographic research problem and formulating meaningful hypotheses, one must decide the sources of available data, the method of collecting data, and the variables to be included in the analysis. In section 2.1, these dimensions of geographic decision making are discussed.

Several measurement issues must be considered before any statistical analysis. Variables may be organized and displayed in various ways, and different levels of measurement are used depending on the geographic problem. The characteristics of nominal, ordinal, and interval/ratio measurement scales are reviewed in section 2.2. Section 2.3 discusses potential measurement errors and addresses the issues of precision, accuracy, validity, and reliability.

Basic methods for data classification are reviewed in section 2.4. Goals and purposes of classification are emphasized, including its importance in geographic research. The basic classification strategies of subdivision and agglomeration are reviewed, and several specific operational methods or rules of classification are applied to a set of spatial data.

A few of the simpler graphic procedures commonly used by geographers to summarize, classify, or display spatial data are reviewed in section 2.5. This review includes discussions of frequency distributions, histograms, and ogives. Also, scattergrams are introduced as a simple graphic mechanism to display relationships between geographic variables.

2.1 SELECTED DIMENSIONS OF GEOGRAPHIC DATA

In the scientific research process, questions about data arise almost immediately. When trying to identify an appropriate geographic research problem or to formulate a hypothesis, questions like these usually emerge: What sources of data are available? Which method(s) of data collection should be used? What type of data will be collected and then analyzed statistically? The various aspects of a geographic problem must be carefully considered to ensure that when data are collected and analyzed, it will be possible to answer research questions effectively and reach meaningful conclusions. The dimensions of geographic data discussed here include sources of information, methods of data collection, and selected characteristics of data that distinguish geographic research problems.

A simple distinction can be made between primary and secondary data sources. **Primary data** are acquired directly from the original source. The geographer conducting the study usually collects this information "in the field." Primary data collection is often quite time-consuming and generally involves making decisions about sample design so a set of representative data may be acquired.

Secondary (or **archival**) **data** are generally collected by some organization or government agency and can be used by the geographer. Because the data have already been collected and are probably organized in an accessible, convenient form—such as a written report, compact disk, or computer data file—secondary sources are generally less expensive and time-consuming to use than primary sources of data. In addition, many of the problems associated with sampling and survey design may not be experienced with archival sources. Secondary sources are often very comprehensive, including a census or total enumeration from a very large population. Duplicating these efforts with primary data collection would be virtually impossible for the researcher.

However, difficulties can also occur with secondary data sources. The data may have been collected, organized, or summarized improperly. Errors may have occurred in the editing and collating of data, especially if the information was obtained from a number of different original sources. Information from secondary sources is not always measured properly, resulting in other potential problems.

Several basic procedures are used to collect geographic data. If primary data are necessary for the study, sampling will almost certainly be part of the data collection process, and survey questionnaires may need to be designed. Among the options for primary data collection are direct observation, field measurement (especially in physical geography research), mail questionnaires, personal interviews, and telephone interviews.

To select the appropriate method of data collection, the nature of the research problem must be evaluated carefully. Even if a suitable method has been selected, problems in survey design are common. When using a survey to collect data, each question must be properly worded, all possible responses to a question must be considered in advance, and the sequence of questions must be determined. In fieldwork, logistic problems often occur, or special arrangements are necessary. Preliminary site reconnaissance may not reveal all the difficulties. A more detailed discussion of data collection methods is found in chapter 6.

Other dimensions or characteristics of data help distinguish geographic research problems. Some studies are considered **explicitly spatial** because the locations or placement of the observations or units of data are themselves directly analyzed. For example, a geographer responsible for selecting potentially profitable locations for a new retail store might calculate the "center of gravity" or average location of

certain "target" households in the area. In another application, a biogeographer might analyze the spatial pattern of a sample of diseased trees in a national forest to determine whether these trees are randomly distributed throughout the area or clustered in certain places. In both of these examples, the data are spatially explicit because the locations of the observations or units of data are analyzed directly. An important set of spatial statistics is used to investigate these problems. Later in the book, a variety of descriptive spatial statistics are discussed (chapter 4), and several inferential spatial statistics are applied to point and area patterns to test for randomness (chapter 12).

Other geographic studies are **implicitly spatial.** An implicitly spatial situation exists when the observations or units of data represent locations or places, but the locations themselves are not analyzed directly. For example, a geomorphologist might wish to determine if significant differences occur in alluvial fan development in two different basins in the basin-and-range region of the American Southwest. A random sample of alluvial fans from each of the basins could be taken and relevant aspects of alluvial fan development compared. In a suburban neighborhood, a geographer may be investigating the relationship between the assessed valuation of homes and their ages. In both of these examples, the observations (alluvial fans in a basin or homes in a suburban neighborhood) obviously have locations on the earth's surface, *but the locational pattern itself is not under scrutiny.* Looking ahead, a two-sample difference test might be appropriate in the alluvial fan study (chapter 9), while a correlation analysis may be needed to study the suburban housing question (chapter 13).

Another important dimension of geographic research problems is the contrast between **individual-level** and **spatially aggregated** data sets. In some geographic problems, each data value represents an individual element or unit of the phenomenon under study. In other problems, each value entered into the statistical analysis is a summary or spatial aggregation of individual units of information for a particular place or area. The best way to see this distinction is by discussing an example. Suppose a population geographer is researching current fertility patterns in Nigeria. One approach would be to collect a set of individual-level data, perhaps through personal interviews of a random sample of Nigerian women. Another possible approach would be to obtain birthrate estimates from officials in each of Nigeria's administrative divisions (21 states and 1 territory) and use these 22 spatially aggregated values as data units to estimate the nationwide fertility pattern.

Using spatially aggregated data raises special issues. Geographers must always be extremely cautious when trying to transfer results or apply conclusions "down" from larger areas to smaller areas or from smaller areas to individuals. If conclusions are derived from the analysis of data spatially aggregated for large areas, it may not be valid to reach conclusions about smaller areas or individuals. In the study of Nigerian fertility patterns, for example, valid, spatially aggregated conclusions can be drawn about the degree of acceptance or rejection of family-planning programs in each of these states by using birthrate estimates from the administrative divisions. However, taking these aggregate conclusions down to the level of the individual family will probably result in deductive errors. Even in the Nigerian state with the lowest birthrate, a number of families will not be practicing any form of birth control. This invalid transfer of conclusions from spatially aggregated analysis to smaller areas or to the individual level is known as the **ecological fallacy.**

Conversely, taking individual-level data and aggregating it to larger spatial units is generally not a problem. In fact, Nigerian officials probably collected data at the individual and village level, then aggregated that information to obtain state-level estimates. The effects of level of spatial aggregation on descriptive statistics is discussed in chapter 3.

Variables in a data set can be characterized as either discrete or continuous. A **discrete variable** has some restriction placed on the values the variable can assume. A **continuous variable** has an infinite number of possible values along some interval of a real number line. In general, discrete data are the result of counting or tabulating the number of items, and potential values are limited to whole integers. Continuous data are the result of measurement, and values can be expressed as decimals. The following are examples of discrete variables: the number of households in a county with videocassette recorders; the number of immigrants currently living in a city; the number of survey respondents in favor of a local bond issue; and the number of active volcanoes in a country. Continuous variables could include inches of precipitation at a weather station collected over a period of time; total area under irrigation in a country; distance traveled by a family on its annual vacation; and average wind speed at the summit of a mountain.

The rounding off of data values must not be confused with the distinction between discrete and continuous data. For example, the elevation of a mountain is almost always expressed to the nearest foot or meter. Representation of elevation as a whole number may give the impression that this variable is discrete. However, since elevation can be measured more precisely than the nearest foot or meter, it is considered a continuous variable.

When discussing probability distributions, the distinction between discrete and continuous data is important. Geographic problems with discrete variables often require the application of different probability distributions than problems with continuous variables. Several practical geographic examples of both discrete and continuous distributions are presented in chapter 5.

Variables in a set of data are either quantitative or qualitative. If a variable is **quantitative,** the observations or responses are expressed numerically; that is, units of data are assigned numerical values. On the other hand, if a variable is **qualitative,** each observation or response is assigned to one of two or more categories. Suppose an agricultural geographer asked 80 farmers to identify their primary cash crop and received the following responses: 43 corn, 28 wheat, and 9 barley. It might be tempting to conclude that these are quantitative data, since 43, 28, and 9 are clearly numerical values. However, the *variable responses* (corn, wheat, barley) are nonnumeric. The values of 43, 28, and 9 are not the raw data, but rather *frequency counts* of observations assigned to the nonnumerical categories, making this an example of qualitative (or **categorical**) data. Other examples of qualitative variables are type of land use, sex (male or female), political party affiliation, religious preference, and climate type. Special types of statistical tests have been developed to handle qualitative (categorical) data, and these are discussed in chapter 11.

2.2 LEVELS OF MEASUREMENT

Just as the organization of variables determines how data can be analyzed statistically, so too does the related issue of variable measurement. That is, the level of measurement of values or units of data must be considered when selecting an appropriate statistical technique to solve a geographic problem. Several different levels of measurement can be distinguished (table 2.1), and different statistical procedures are appropriate to each.

Nominal Scale

The simplest scale of measurement for variables is the assignment of each value or unit of data to one of at least two qualitative classes or categories. In **nominal scale** classification of variables, each category is given some name or title, but no assumptions are made about any relationships between categories—only that they are different. Values are "different" if they are assigned to different categories, or "similar" if assigned to the same category. Thus, problems using variables placed on a nominal scale are considered categorical (qualitative).

TABLE 2.1

Summary of Levels of Measurement

Level of measurement	Brief description
Nominal	Each value or unit of data is assigned to one of at least two categories or qualitative classes; no assumptions are made about relationships between categories—only that they are "different."
Ordinal	Values themselves are placed in some rank order.
Strongly ordered	Each value or unit of data is given a particular position in a rank-order sequence; that is, each value is assigned its own particular rank.
Weakly ordered	Each value or unit of data is assigned to a category, and the categories are then rank ordered.
Interval	Each value or unit of data is placed on a measurement scale, and the interval between any two units of data on this scale can be measured; origin or zero starting point is assigned arbitrarily (i.e., origin does not have a "natural" or "real" meaning).
Ratio	Each value or unit of data is placed on a measurement scale, and the interval between any two units of data on this scale can be measured; origin or zero starting point is "natural" or non-arbitrary, making it possible to determine the ratio between values.

An urban planner could assign each parcel of land in a city to one of several nominal land-use categories (residential, commercial-retail, industrial, recreation-open space, etc.) without inferring that residential land use is "greater than" recreation-open space or "less than" industrial. In fact, the only necessary conditions for a proper nominal scale classification of variables are that the categories are **exhaustive** (every value or unit of data can be assigned to a category) and **mutually exclusive** (it is not possible to assign a value to more than one category because the categories do not overlap).

Geographers create nominal variables in many ways. Individuals can be classified by religious affiliation (Baptist, Catholic, Methodist, Presbyterian, etc.) or political party (Democrat, Republican, Independent); cities can be classified by primary economic function (manufacturing, retail, mining, transportation, tourism, etc.); counties can be organized by primary type of home heating fuel (fuel oil, utility gas, coal, wood, electricity, etc.); and countries can be distinguished by predominant language family (Indo-European, Afro-Asiatic, Sino-Tibetan, Ural-Altaic, etc.).

If a variable has only two categories, a special subset of nominal classification is used. This sort of dichotomous (binary) assignment is used when no greater degree of qualification is necessary or possible. For example, each person could be assigned a 1 if he or she attended a private elementary school, and a 0 if not. Although this type of information is clearly limited, statistical analysis on the frequency counts of the number of values or individuals in the categories can be conducted. Many geographic problems have only "yes–no" or "presence–absence" data available.

When variables are organized nominally, geographers are limited in the types of numerical analysis that can be applied. Nevertheless, appropriate statistical techniques exist and are presented in chapter 11.

Ordinal Scale

The next higher level of measurement involves placement of values in rank order to create an **ordinal scale** variable. The relationship between observations takes on a form of "greater than" and "less than." With data in rank order, more quantitative distinctions are possible than with nominal (qualitative) scale variables.

Geographers can easily identify examples of ordinal scale variables. An important distinction needs to be made, however, between a strongly ordered ordinal variable and a weakly ordered ordinal variable. When each value or unit of data is given a particular position in a rank-order sequence, the variable is considered **strongly ordered.** The city ranking schemes that occasionally appear in newspapers or magazines, such as the "ten best places to live" or "fifty best American cities," are typical examples of this sort of survey. A popular publication is the *Places-Rated Almanac,* which provides a ranking of U.S. cities, based on the aggregate compilation of many variables. Since each city is assigned its own particular rank, these "preference rankings" are examples of strongly ordered ordinal variables. Other examples of strongly ordered variables include the ranking of countries by gross national product per capita and the ranking of states in terms of dollars spent per resident on higher education.

By contrast, in a **weakly ordered** variable, the values are placed in categories, and the categories themselves are rank ordered. Suppose one is constructing a choropleth map showing the percent of population change in each county of the United States from 1990 to 1995. To depict the population change cartographically, six ordinal categories are selected (greater than 10 percent increase, 5–9.9 percent increase, and so on), and each of the more than 3,000 counties nationwide is assigned to one of these six categories. When frequency counts of counties are made in each category, the variable is weakly rather than strongly ordered. It is weak or "incomplete" in the sense that two counties assigned to the same category on the map cannot be distinguished, even though in reality the counties almost certainly have different population change values.

As with nominally scaled variables, both strongly and weakly ordered ordinal variables have specific statistical tests. Some of these techniques will be discussed in chapters 9, 10, and 11.

Interval and Ratio Scales

With variables measured on either an **interval** or **ratio scale,** the magnitude of difference between values can be determined. That is, the interval between any two units of data can be measured on the scale. This means that not only is the relative position of each value known (a value is above or below another), but also how different each unit of data is from all other values on the measurement scale.

Interval and ratio measurement scales can be distinguished by the way in which the origin or zero starting point is determined. With interval-scale measurement, the origin or zero starting point is assigned arbitrarily. The Fahrenheit and Celsius scales used in the measurement of temperature are two widely known interval scales. The placement of the zero degree point on both of these measurement scales is arbitrary. With the Fahrenheit scale, zero degrees is the lowest temperature attained with a mixture of ice, water, and common salt, whereas with the Celsius scale zero degrees corresponds to the melting point of ice.

In ratio-scale measurement, by contrast, a natural or nonarbitrary zero is used, making it possible to determine the ratio between values. If Montreal, Canada, receives 40 inches of annual precipitation and Chihuahua, Mexico, receives only 10 inches, the ratio between these two measures is easily calculated (40/10 = 4). Furthermore, it is correct to conclude that Montreal has received four times as much precipitation as Chihuahua. Zero inches of precipitation is a natural or nonarbitrary zero.

It should be noted that ratio-type statements cannot be made with interval-scale variables. For example, since zero degrees is an arbitrary value on the Fahrenheit scale, 60° Fahrenheit cannot be considered twice as warm as 30° Fahrenheit. Many other variables of interest to geographers are measured on a ratio scale, including distance, area, and such demographic and socioeconomic variables as infant mortality rate and median family income.

Values or units of data from the same variable can be expressed at different measurement scales, depending on how they are collected, organized, and displayed. For example, a state resource planner interested in the type of energy used in homes across the state could use data collected and measured in several different ways. Data could be collected at the individual household level and organized nominally by type of energy used (number of homes in the state using coal, utility gas, fuel oil, etc.). Alternatively, data might be available as county-level summaries. Counties in the state could then be arranged in a strongly ordered sequence by the percent of households in each county using coal (with the county having the highest percentage of households using coal assigned rank one, and so on). A choropleth map of the county values could display a weakly ordered graphic representation of the percentage of households using coal. As yet another alternative, the number of households per county using coal could be organized and displayed on a ratio scale.

2.3 MEASUREMENT CONCEPTS

As proficiency improves when working with descriptive and inferential statistics, it becomes tempting to believe that the analysis is truly error-free and that the geographic research problem has been solved. Results from statistical analysis often seem very exact and definitive. If the data are analyzed on the computer, the results are nicely displayed, and the same result is obtained if the data are resubmitted. It may seem appropriate to accept the answers as error-free automatically. However, an error-free result cannot be guaranteed just because a set of data is submitted correctly into some statistical analysis. In fact, several interrelated sources of measurement error can operate separately or in combination to produce problems for the geographer.

Precision

Precision refers to the level of exactness associated with measurement. Precision is often associated with the calibration of a measuring instrument, such as a rain gauge. As a frontal system moves through an area, suppose that the amount of rainfall is recorded by two standard rain gauges having different calibration systems. On the coarsely calibrated gauge, the amount of rainfall might be estimated as somewhere between 1.2 and 1.3 inches. However, the more finely calibrated gauge provides a more precise estimate of between 1.26 and 1.27 inches.

In many geographic problems, the issue of **spurious precision** must be considered. The computer (or calculator) output will often provide statistics with six or more decimal places, even when the data are in integer form. Reporting seemingly precise statistics based on less precise input is a deceptive but relatively commonplace occurrence. Unless confidence in such a level of measurement precision is warranted, it should be avoided.

Accuracy

The concept of **accuracy** refers to the extent of systemwide bias in the measurement process. It is quite possible for measurement to be very precise, yet inaccurate. Return to the rain gauge example for a moment. Suppose another more finely calibrated (more precise) rain gauge is used, but the gauge has not been calibrated properly. The person reading the gauge estimates the amount of rainfall to be 1.19 inches rather than the actual rainfall total of about 1.26 or 1.27 inches, resulting in an inaccurate reading. Unfortunately, discovering systematic bias in a measurement instrument is often quite difficult.

To understand the relationship between precision and accuracy, consider this "target analogy" showing results from successive firings of a gun at a target (figure 2.1). In case 1, the five bullet holes are closely clustered (precise) and centered on the middle of the target (accurate), making this the best of the four alternatives presented. Cases 2 and 3 are both flawed: the inaccuracy of the bullet holes in case 2 and the imprecision of the holes in case 3 result in different types of errors. Case 4 appears to have the severest problems, with an inaccurate systematic bias toward the upper-left corner of the target as well as a considerable scatter of bullet holes that also make the results imprecise. Geographers must take care to distinguish between the interrelated concepts of precision and accuracy when working with locational data.

Validity

In many geographic problems, the spatial distribution or locational pattern being analyzed is the result of complex processes. With such complex processes, it is understandably difficult to express the "true" or "appropriate" meaning of that concept through the measurement of any simple variable or set of variables. **Validity** addresses the measurement issues on the nature, meaning, or definition of a concept or variable. The discipline of geography is saturated with multifaceted, complex variables, such as "level of poverty," "environmental quality," "economic well-being," "quality of education," "level of pollution," and "quality of life." To express the true meaning of such concepts is often not possible, so geographers find it necessary to create **operational definitions** that can serve as indirect or surrogate measures. The question then becomes whether the operational definition is valid. For example, a geographer studying the spatial pattern of "quality of education" in a metropolitan area might evaluate elementary schools on the basis of "average student score on the California Achievement Test (CAT)" and evaluate high schools by "percent of graduates who subsequently go to college." Clearly, the concept "quality of education" involves much more than what is reflected by these operational definitions, and their validity must be questioned. In this case, it should be asked whether "average CAT score" is a fully valid, somewhat valid, or invalid measure of "quality of education."

Admittedly, the degree of validity in a geographic problem may prove difficult (even impossible) to determine. Consequently, this question is often ignored, and problems with validity are sometimes assumed away as being inconsequential. A good geographic study involving complex variables will discuss the degree of validity of any operational definitions used in the analysis.

Reliability

A final measurement concept of concern in many geographic problems is **reliability.** When data are collected over time or when changes in spatial patterns are analyzed over time, the geographer must

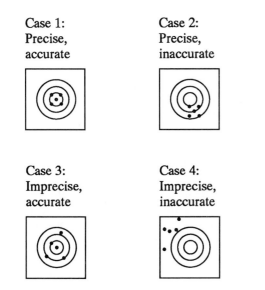

Case 1:
Precise,
accurate

Case 2:
Precise,
inaccurate

Case 3:
Imprecise,
accurate

Case 4:
Imprecise,
inaccurate

FIGURE 2.1
The Measurement Concepts of Precision and Accuracy: The Target Analogy

question the consistency and stability of the data. For example, if repeated or replicate samples of water are taken from the same set of locations over time, a consistent, well-defined method of water sampling is necessary, or the results of subsequent statistical analysis may be unreliable. Suppose, for example, a geographer wanted to examine whether the spatial pattern of poverty across the United States varies more today than it did 50 or 100 years ago. A reliable and consistent measure of poverty is needed to answer this question.

Reliability problems often occur when using international data. Fully comparable and totally consistent methods of collecting data rarely exist from country to country. A developing nation has fewer resources (e.g., personnel, money) than one that is more developed, and sources of measurement error inevitably affect the data collection process. International comparative statistics are often unreliable. Even within the same country or region, locational variations in data collection and processing methods can render data unreliable. Data collection procedures can change drastically from one time period to the next, again resulting in unreliable data.

One way to assess the degree of reliability of a measurement instrument is to compare at least two applications of the data collection method used at different times. This "test-retest" procedure is one way to evaluate the reliability of IQ test scores, medical diagnoses, and SAT scores, for example. Reliability checks need to be used more frequently in behavioral geography, particularly when analyzing the results from a survey or questionnaire that contains individual attitudes and opinions, which may be uncertain and subject to frequent change.

2.4 BASIC CLASSIFICATION METHODS

Although categories and classification have already been introduced, the methods describing *how* to classify data into categories and the reasons *why* classification is necessary have not been examined. In this section, the purposes and importance of classification in geographic research are explained, and basic classification methods are reviewed.

Geographers regularly face the problem of deciding how to classify or group spatial data. Classification is used for several important reasons. Classification schemes organize, simplify, and generalize large amounts of information into effective or meaningful categories, bringing relative order and simplicity to complexity. As a result, communication is enhanced, detailed spatial information is better understood, and complex spatial patterns are represented more clearly. Maps created with properly classified data result in more effective graphic communication. Classi-

fication is also an integral part of the scientific research process, helping in the formation of hypotheses and guiding further investigations.

In classification, values are organized according to their degree of similarity. That is, similar values should be generally placed in the same category, and dissimilar values should be generally placed in different categories. The result is a grouping of data that seems to minimize the amount of fluctuation or dispersion of values within the same category and to maximize the dispersion of values between different categories. Whatever the specific method of classification, the resulting categories must be both mutually exclusive and exhaustive.

Many specific classification methods are available, each approaching these general goals from a slightly different perspective. No matter what specific method of classification is used, some information is lost when large amounts of information are simplified and generalized. Information is lost if individual-level values have been spatially aggregated and only the aggregated data are available. Similarly, information is lost if values have been classified, and only the classified data are available. In fact, spatially aggregated data are simply individual-level data classified by location.

At the most fundamental level, classification uses one of two *conceptual strategies*. The first of these is **subdivision** (sometimes called **logical subdivision**). At the start of the subdivision process, all units of data in a population are grouped together. Then, through a series of steps or iterations, individual values are allocated to an appropriate subdivision using carefully defined criteria. This strategy "works down" by disaggregating all values into logically subdivided classes. Most practical geographic examples of subdivision are hierarchical, with multiple levels of subdivision, depending on the problem or situation. A clear and consistent set of rules is always needed to assign values to the proper category at each stage of the subdivision procedure. The characteristics or values associated with each category are defined before the classification procedure begins.

The subdivision strategy of classification is illustrated with two examples. The Soil Conservation Service has developed a basic system of soil classification that is used for making and interpreting soil surveys. Soils are first subdivided into orders, then further divided into suborders, great groups, subgroups, families, and specific soil series. This subdivision system is an hierarchical taxonomy, concerned with relationships among many different soil properties, the assemblage of soil horizons, and soil-forming processes. The sorting process in this classification is complex: properties relevant to the sorting in one soil order may have little meaning in another.

Ten soil orders have been determined for the United States, with 47 suborders, more than 200 great groups, and about 9,500 soil series at the lowest level of the hierarchy. A small portion of the soil classification system is shown in table 2.2, including examples of the descriptive nomenclature used to identify different soil classes.

The standard industrial classification (SIC) system used to categorize manufacturing products can also be used to illustrate subdivision. In the United States, data on manufacturing activity by product grouping is reported in the *Census of Manufactures,* which is published every five years by the Bureau of the Census. The SIC classification scheme subdivides all industrial activity according to numerical code, ranging from a one-digit general product grouping at the top level of the hierarchy to a more specific four-digit breakdown of activity at the lowest level. A small portion of the SIC classification is shown in table 2.3.

The second conceptual classification strategy is **agglomeration.** With this general approach, each unit of data or value in a population or data set is separate and distinct from others at the start of the classification process. The agglomeration procedure then "works up" by allocating values into classes according to well-defined grouping criteria. Agglomeration is accomplished when similar values are combined into the same category, and dissimilar values are placed in different categories. The agglomeration method of grouping is an opposite concept to subdivision. The agglomeration strategy of classification is very important in the geographic research process, and data for many geographic problems are summarized numerically or graphically using agglomeration.

Beyond these general conceptual strategies of logical subdivision and agglomeration, a variety of specific *operational procedures* or *rules* is applied in practical classification. These operational methods are not necessarily pure logical subdivision or agglomeration, but may contain elements of both. The remainder of this section focuses on four simple operational methods or rules of classification that can be applied in geography (table 2.4).

1. **Equal intervals based on range.** The **range** is simply the difference in magnitude between the largest and smallest values in an interval/ratio set of data. To determine class breaks (the values that separate one class from another), the range is divided into the desired number of equal-width class intervals. The procedure is easy to use and results in class intervals of equal width, which may be an advantage for some applications. However, because all class breaks are derived from the two units of data with the most extreme values, results can sometimes be misleading. To maintain the equal width of each class, the numbers in class breaks are not usually rounded off. The number of values in each category may also vary considerably. This may be an advantage or a disadvantage, depending on the purposes of the classification and the goals of the analysis.

2. **Equal intervals not based on range.** This method of classification also designates class breaks to create equal-interval classes, but the exact range is not used to select the class breaks. Instead, a convenient or practical interval width is selected arbitrarily, based on rounded-off class-break values. Units of data are then assigned to the categories.

TABLE 2.2

A Portion of the Soil Classification System Developed by the Soil Conservation Service

Order	Suborder	Great group	Subgroup	Family	Series
Alfisols					
Aridisols					
Entisols	Aquents				
Arents					
	Fluvents	Cryofluvents	Typic cryofluvents	Coarse-loamy, mixed, acid	Susitna
		Torrifluvents	Typic torrifluvents	Fine-loamy, mixed (calcareous), mesic	Jocity and Youngston
			Vertic torrifluvents	Clayey over loamy, mixed (calcareous), hyperthermic	Glamis
	Orthents	Cryorthents	Typic cryorthents	Loamy-skeletal, carbonatic	Swift Creek
			Pergelic cryorthents	Loamy-skeletal, mixed (calcareous)	Durelle
Histosols					

Source: Soil Conservation Service, U.S. Dept. of Agriculture.

TABLE 2.3

A Portion of the Standard Industrial Classification (SIC) System

1 Digit	2 Digit	3 Digit	4 Digit
0 Agriculture, forestry, fisheries			
1 Mining and construction			
2 Manufacturing (nondurable)			
3 Manufacturing (durable)--------	30 Rubber and plastics		
	31 Leather		
	32 Stone, clay, glass, concrete		
	33 Primary metals		
	34 Fabricated metals		
	35 Machinery, except electrical		
	36 Electrical and electronic machinery		
	37 Transportation equipment --------	371 Motor vehicles------- and equipment	3711 Motor vehicles and passenger car bodies
	38 Instruments, photographic goods, optical goods, watches, and clocks		3713 Truck and bus bodies
	39 Miscellaneous		3714 Motor vehicle parts and accessories
			3715 Truck trailers
4 Transportation, communication, utilities (electrical, gas, sanitary services)		372 Aircraft and parts	
5 Wholesale trade		373 Ship and boat building	
6 Finance, insurance, and real estate		374 Railroad equipment	
7 Services (personal, business)		375 Motorcycles, bicycles, and parts	
8 Services (professional, education)		376 Guided missiles, space vehicles and parts	
9 Public administration (federal, state, and local government)		379 Miscellaneous transportation equipment	

TABLE 2.4

Summary of Operational Classification Methods

Classification method	Brief description
Equal intervals based on range	Class breaks determined by dividing range (difference between the lowest- and highest-valued units of data) into desired number of equal-width class intervals.
Equal intervals not based on range	Based on convenience or practical considerations; rounded-off class breaks and class interval widths arbitrarily selected.
Quantile breaks	Equally divide the total number of values into the desired number of classes; two commonly used divisions are quartiles (4 categories) and quintiles (5 categories).
Natural breaks	Place units of data in rank order, identify "natural breaks" or separations between adjacent ranked values, and locate class breaks in the largest of these natural breaks. Iterative process if single-linkage version used, with largest natural break selected as first class break location, next largest natural break selected second, and so on until desired number of classes created.

This method of classification is preferred for constructing a frequency distribution, histogram, or ogive to represent the data graphically (see section 2.5). Many institutions and government agencies use this method to map complex spatial patterns. The convenient class-break values generally result in maps that are easy to understand and interpret. The number of values in each cate-gory could vary widely, which may be either an advantage or disadvantage, depending on the goals of the classification.

3. **Quantile breaks.** This method approaches classi-fication from a somewhat different perspective. The total number of values is divided as equally as possible into the desired number of classes. Two frequently used alternatives are the division

of data into quartiles (four categories) or quintiles (five categories). The allocation of an equal number of values to each category is often an advantage in choropleth mapping, particularly if an approximately equal area on the map is desired for each category. However, the possible disadvantages of quantile breaks classification should be evaluated before deciding to use this method. Class breaks are frequently not convenient rounded-off values, and class interval widths are usually not equal for different categories. If a large number of data units are clustered relatively close together—a frequent occurrence with geographic data—these similarly sized values are likely to be split unnaturally by a class break to keep an equal number of values in each category.

4. **Natural breaks.** Yet another perspective is available with the natural-breaks method of classification. The most elementary natural-breaks method is known as the single-linkage approach (Abler, Adams, and Gould, 1971). The logic is to identify natural breaks in the data and separate values into different classes based on these breaks. The process is done iteratively, with the largest gap or separation between adjacent values on a number line selected as the first class-break location. The next largest gap is selected second, and so on until the desired number of classes has been created. With this classification process, similar values are kept together in the same category, dissimilar values are separated into different categories, and gaps in the data are incorporated directly in the grouping procedure. This method will highlight extreme values, placing unusual outliers of data into their own unique categories. Depending on the research problem, highlighting extreme values may (or may not) be a primary goal of the classification. Another common consequence of natural breaks is the clustering of large numbers of values into one or two categories. Again, it must be decided whether this is an advantage or disadvantage.

To illustrate how each of these methods works, 1996 unemployment rates by state are classified. The set of 50 ranked unemployment rates (table 2.5) is too detailed and cumbersome for convenient study and interpretation. This is particularly true if one goal is to illustrate the spatial pattern of unemployment in a choropleth map, with each state allocated to one of several unemployment classes.

Deciding on the *number of classes* is very important in choropleth mapping. If data are allocated into too few categories, key details in the spatial pattern are likely to be lost. If, on the other hand, too many categories are used, the map reader becomes over-

TABLE 2.5

Ranked Unemployment Rates by State, United States, 1996 (in Percent)

State	Labor force unemployed	State	Labor force unemployed
Nebraska	2.9	Rhode Island	5.1
North Dakota	3.1	Delaware	5.2
South Dakota	3.2	Idaho	5.2
Utah	3.5	Tennessee	5.2
Wisconsin	3.5	Illinois	5.3
Iowa	3.8	Montana	5.3
Minnesota	4.0	Pennsylvania	5.3
Indiana	4.1	Arkansas	5.4
Oklahoma	4.1	Nevada	5.4
Colorado	4.2	Arizona	5.5
New Hampshire	4.2	Kentucky	5.6
Massachusetts	4.3	Texas	5.6
North Carolina	4.3	Connecticut	5.7
Virginia	4.4	Oregon	5.9
Kansas	4.5	South Carolina	6.0
Georgia	4.6	Mississippi	6.1
Missouri	4.6	New Jersey	6.2
Vermont	4.6	New York	6.2
Maryland	4.9	Hawaii	6.4
Michigan	4.9	Washington	6.5
Ohio	4.9	Louisiana	6.7
Wyoming	5.0	California	7.2
Alabama	5.1	West Virginia	7.5
Florida	5.1	Alaska	7.8
Maine	5.1	New Mexico	8.1

Source: Bureau of Labor Statistics, U.S. Dept. of Labor.

whelmed with detail and could miss certain important generalizations of the spatial pattern. The decision on the number of classes to use always involves the trade-off between effective generalization and communication of sufficient detail. Five classes are used in this unemployment example—a number many cartographers consider reasonable for choropleth mapping of geographic patterns.

Each simple method of classification is applied to the state-level unemployment rates, and for consistency, exactly five classes and four class breaks are used with each method. The results are shown in two ways—with number lines (figure 2.2) and with choropleth maps (figure 2.3, cases 1–4). On the number lines, a short vertical bar indicates each class break. On each choropleth map, the class intervals derived from application of each classification scheme are shown in the legend.

Application of the various classification methods results in different class breaks along the number lines as well as dramatically different unemployment pattern maps. It is important to realize that such startling visual differences on choropleth maps are to be expected from one method of classification to another, even when all the maps are created from the same data set. These sharp visual contrasts do

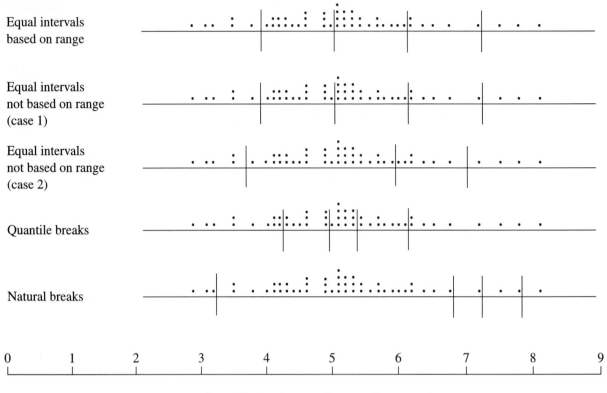

Equal intervals based on range

Equal intervals not based on range (case 1)

Equal intervals not based on range (case 2)

Quantile breaks

Natural breaks

0   1   2   3   4   5   6   7   8   9

Rate of unemployment by state (in percent)

FIGURE 2.2
Number Lines and Class Break Values for Each Method of Classification: Rate of Unemployment by State, 1996 (*Source:* Data from Bureau of Labor Statistics, U.S. Dept. of Labor)

not mean that any of the classification methods are inaccurate or biased. Rather, this situation illustrates the distinctiveness of the goals in classification. It seems that different impressions regarding the reality of the situation emerge with application of the different classification methods.

Perhaps the greatest visual contrast exists between the maps of quantile breaks and natural breaks (figure 2.3, cases 3 and 4). The objective in using quantile breaks is to allocate an equal number of values to each category. With the quantile method, New Mexico—the state with the highest unemployment rate (8.1 percent)—is grouped with many other states. The overall impression from the quantile map is that high unemployment states are scattered throughout the country, and there is no particular visual focus on New Mexico.

By contrast, the goal in natural breaks is to separate values at places on the number line where large "natural breaks" occur. The mapped result is sharply different from quantiles, as New Mexico's high unemployment is highlighted graphically. In fact, the dominant impression from the natural breaks map is that most of the United States has comparable un-

employment rates, with the exception of a few outliers, such as New Mexico.

In this example problem, minor operational difficulties are encountered with each of the simple methods of classification. One limitation is that the employment data are expressed in tenths of a percent. This is a reasonably precise degree of measurement, but it causes application problems with some methods of classification. As a result of this precision level, many ties exist in the data set. For example, four states are reported with a 5.1 percent unemployment rate. Obviously, those four states do not have an *exact* rate of 5.1 percent, so they are not *really* tied, but the full reality of the situation is not known. The Bureau of Labor Statistics recognizes that more precise estimates of unemployment rates (say, to the nearest hundredth) could be calculated, but they would create meaningless, spurious precision.

The first method (figure 2.3, case 1) is equal intervals based on range. The range from lowest unemployment in Nebraska to highest unemployment in New Mexico is 5.2 percent, and five *exactly equal* interval categories cannot be created at this precision level over a range of 5.2 units of data without modifying the data in some way. The closest one can get to equal intervals is the classification shown in the map legend, where the first four categories have an interval of 1.1 units, and the last category has an interval of 0.9 units (see the legend in figure 2.3, case 1).

The second method (case 2) is equal intervals not based on range. Again, it is not possible to create five exactly equal interval categories where the lower bound of the first category is 2.9 *and* the upper bound of the fifth category is 8.1. The solution shown on the map uses a class interval of 1.1 unemployment units, with the first category "anchored" on 2.9. If a class interval of exactly 1.1 units is to be maintained throughout the classification, a flaw occurs in the final category, where the upper bound of the category is 8.3 (see the legend of figure 2.3, case 2). While these class breaks are convenient, another disconcerting factor is that the class-break values are hardly rounded off. Nevertheless, this is a fully valid and objective application of equal intervals not based on range, working within the constraints of the data.

The third method (case 3) is quantile breaks. Tied observations create a minor problem here as well. It is not possible to allocate *exactly* ten states to each of the five categories without splitting multiple states that are tied. A valid operational rule would be to minimize the disparities between groups, creating a classification that is as close as possible to having an equal number of observations in each category. One such allocation is shown on the map, where the maximum number of states in any category is 11 and the minimum number is 9 (figure 2.3, case 3).

The final method (case 4) is the natural-breaks, single-linkage method. Here, the lack of precision generates a major problem. The single largest natural break in the data set is 0.5 unemployment units, separating Louisiana (6.7 percent) from California (7.2 percent). If the classification process were to stop at this juncture, there would be no problem, but the resultant map would have only two categories: (1) a category of size 46 containing all states with unemployment rates of 6.7 or less, and (2) a category of size 4 containing all states with unemployment rates of 7.2 or more. A two-category choropleth map will not reveal much of the actual spatial pattern and is clearly not the best strategy. Also, a condition in this problem was to create classification schemes having exactly five categories; so this condition would also be violated.

Proceeding iteratively using the single-linkage approach, the next largest natural break in the data set is 0.3 units. This causes a classification dilemma because this magnitude of interval occurs six times. If all of these natural breaks are implemented simultaneously, the choropleth map will have a total of

seven categories, too many to meet the condition of exactly five categories.

How should the natural-ranks method proceed? No solution is fully satisfactory, and natural breaks should probably be rejected as a method of classification for this data set. However, a compromise strategy is now implemented, *for illustrative purposes only.* You be the judge as to whether this approach is valid—it is offered here to generate discussion.

The Bureau of Labor Statistics has the original data from which the unemployment rates were calculated. Accessing these data, the unemployment rates were recalculated, but with a greater level of precision—to the nearest hundredth. For example, Nebraska's rate is 2.94, North Dakota's rate is 3.11, and so on. With these more precise (and perhaps spurious) estimates, it is possible to break all of the ties in the data. Proceeding with the single-linkage method, the second largest natural break is 0.32 units, separating California from West Virginia. Continuing, the third iteration separates Alaska from New Mexico (at 0.31 units), and the final iteration separates South Dakota from Utah (at 0.28 units). The result is a natural-breaks choropleth map having exactly five categories, similar to all the other classification methods (see the legend of figure 2.3, case 4).

Subjectivity can enter into the classification process in other ways. For example, when applying the equal intervals not based on range method, different results can occur, depending on which "convenient" class-break values are arbitrarily selected. In figure 2.4, case 1, the class intervals are equal throughout (at 1.1 percent unemployment), and the lower bound of the first class interval is conveniently "anchored" at 2.9 (Nebraska). Note that the map in figure 2.4, case 1, is the same as figure 2.3, case 2. Suppose now that another set of class breaks is selected, using the same equal-interval width and same number of categories, but this time conveniently anchored at 8.1 (New Mexico). The resultant map pattern (figure 2.4, case 2) has a different visual appearance.

What can be concluded about these disparities among classification methods? Depending on the method of classification used, outcomes can be quite different, even though the same data set is used and the same number of classes created. The different class-break values in figure 2.2 and visually distinctive choropleth maps in figures 2.3 and 2.4 illustrate the substantial effects of various classification decisions on spatial patterning and the conclusions that can be drawn about these spatial patterns. The logical conclusion is to recognize that any observed spatial pattern (map) is a function of the specific classification method applied and that using a different

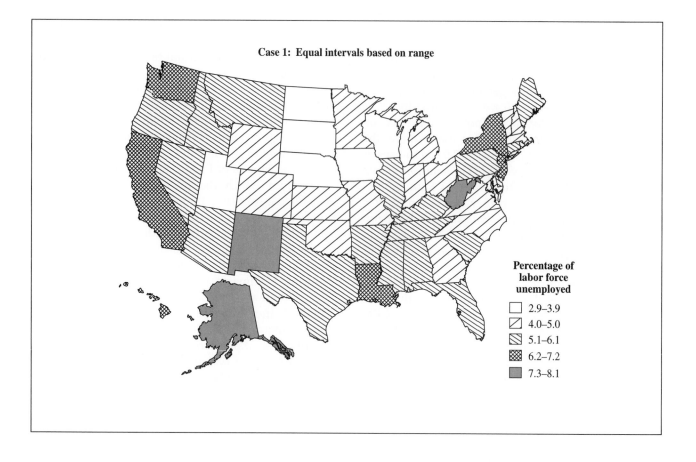

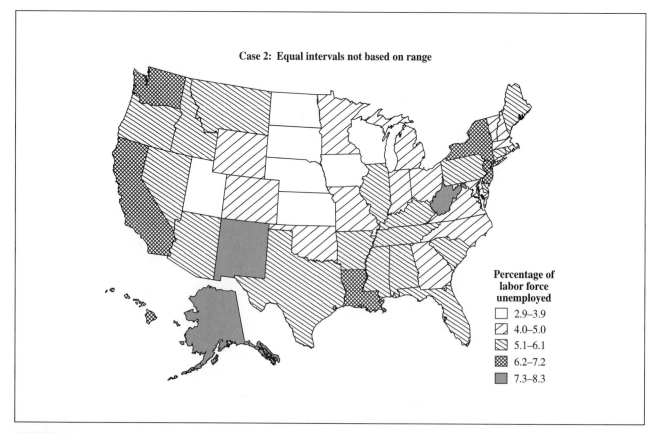

FIGURE 2.3
Choropleth Maps for Each Model of Classification: Rate of Unemployment by State, 1996

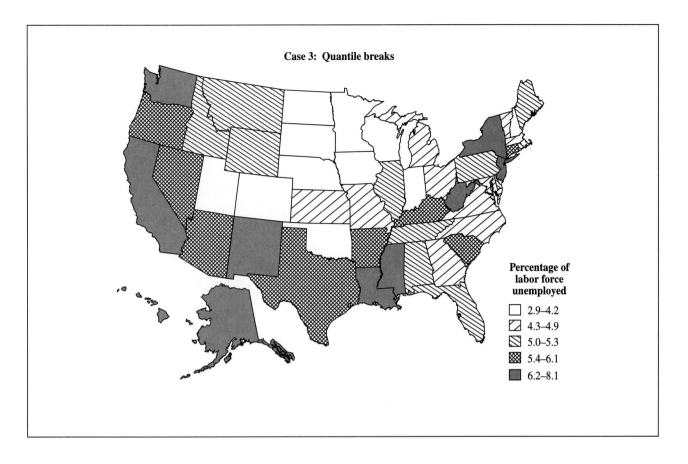

Case 3: Quantile breaks

Percentage of
labor force
unemployed

- 2.9–4.2
- 4.3–4.9
- 5.0–5.3
- 5.4–6.1
- 6.2–8.1

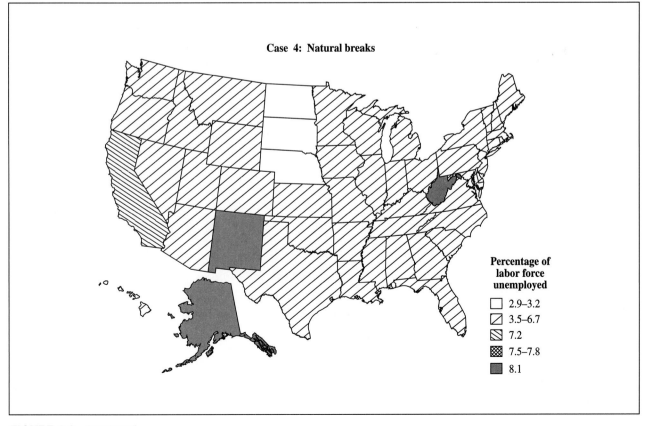

Case 4: Natural breaks

Percentage of
labor force
unemployed

- 2.9–3.2
- 3.5–6.7
- 7.2
- 7.5–7.8
- 8.1

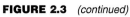

FIGURE 2.3 *(continued)*

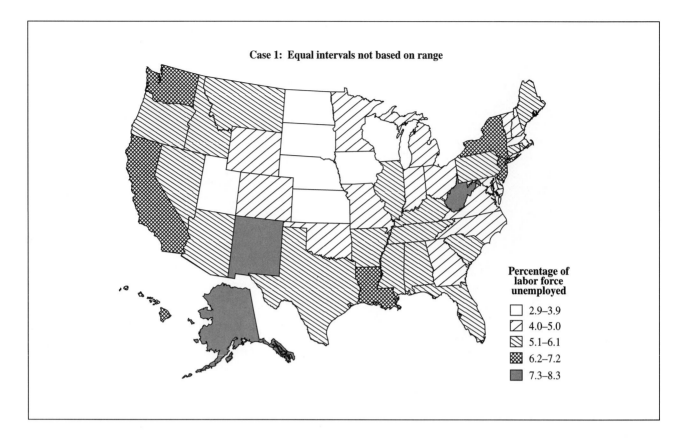

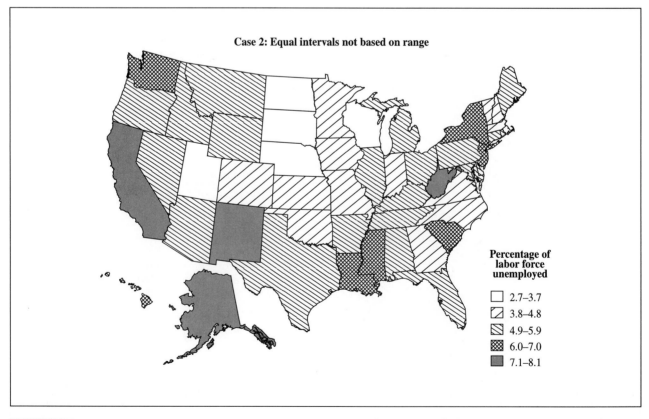

FIGURE 2.4
Choropleth Maps for Alternate Versions of Equal Interval Not Based on Range: Rate of Unemployment by State, 1996

method of classification will likely result in a visually distinctive map.

In addition to these simple classification methods, other more powerful classification methods exist, including sophisticated clustering procedures that classify multiple variables simultaneously. These operational methods are beyond the scope of this introductory test, however. Several references that discuss cluster analysis, more advanced methods of numerical taxonomy, and other multivariate classification procedures in more detail are cited at the end of the chapter.

2.5 GRAPHIC PROCEDURES

When geographers classify, they are attempting to simplify and generalize information to understand it more fully. Representing data graphically is another way to simplify complex data and increase understanding. In addition to working with a data set as a group of numerical values, geographers can also describe or summarize data graphically to show the same information visually. Clearly, the unemployment number lines in the previous section are one type of graphic display.

Geographers often work with the frequency counts of individual data values or data grouped into categories. In the previous section, for example, the frequency of states grouped into unemployment categories by each classification method can easily be determined and presented in a table. However, the same results can often be shown more effectively with a frequency diagram. Figures can be constructed using several different formats, but they all share some common characteristics. Usually the frequency of values is shown on the vertical axis and the range of data (the minimum value to the maximum value) is presented on the horizontal axis. Information can be shown as absolute frequencies (actual frequency counts) or as relative frequencies (percentages or probabilities). Annual precipitation data from Washington, D.C., collected over a 40-year period are used to illustrate frequency diagrams (table 2.6).

The frequency counts of values in a distribution can be shown with a histogram or a polygon, the two most common frequency diagrams. In a **histogram,** the frequency of values is shown as a series of vertical bars, one for each value or class of values. If the data are discrete, the height of each bar represents the frequency of values in a particular category or integer value. To show the frequency of different family sizes for a sample of families, the horizontal scale could be divided into integer values starting at one and continuing to six or more. The height of each bar would represent the number of families having the specified number of persons.

TABLE 2.6			
Annual Precipitation for Washington, D.C.: A Ranked 40-Year Record (in Inches)			
26.87	35.20	39.86	45.62
26.94	35.38	40.21	46.02
28.28	35.96	40.54	47.73
29.48	36.02	41.11	47.90
31.56	36.65	41.34	48.02
32.78	36.83	41.44	51.17
33.62	38.15	41.94	51.97
34.98	39.34	43.30	54.29
35.09	39.62	43.53	57.54

Source: National Climatic Data Center, U.S. Dept. of Commerce.

However, if data have continuous values or the range of data is large, the histogram is usually constructed to show the frequency of each class or group derived from an appropriate classification system (table 2.7). When using categories instead of actual values along the horizontal scale of a histogram, classification by *equal intervals not based on range* is usually the best technique. With this procedure, the class breaks occur at convenient, rounded-off positions, and widths of class intervals are uniform. Since the precipitation data for Washington, D.C., are continuous and extend from a low of 26.87 inches to a high of 57.54 inches, the histogram of the data is constructed using a series of five-inch intervals from 25 inches to 60 inches (figure 2.5).

A **frequency polygon** is very similar to a histogram, except that the vertical position of each data value or class is shown as a point rather than a bar. If the values have been categorized into groups or classes, the single point for displaying the frequency is usually placed at the midpoint of the class interval. The points are then connected by straight lines to produce the frequency polygon. In figure 2.5, the frequency polygon for the Washington, D.C., precipitation example is superimposed on the histogram.

Instead of displaying the absolute frequency count on the vertical axis, a histogram or frequency polygon can easily be converted to show relative frequency. For example, by dividing the individual frequency values by the sum of all frequencies for the data set, the diagrams would display the frequency percentages for each value or class. This change would not affect the general shape of the graphic, only the scale of values along the vertical axis. These graphic tools are especially useful for displaying descriptive statistics and probability.

Another useful method for displaying data in a relative frequency format is a **cumulative frequency diagram,** also known as an **ogive.** Instead of showing actual frequencies for each value or class, this graphic aggregates frequencies from value to value

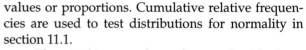

TABLE 2.7

Classification of Washington, D.C., Precipitation Data Using Equal Intervals Not Based on Range Method

Class	Interval	Absolute frequency	Cumulative absolute frequency	Relative frequency	Cumulative relative frequency
1	25–29.99	4	4	0.100	0.100
2	30–34.99	5	9	0.125	0.225
3	35–39.99	12	21	0.300	0.525
4	40–44.99	9	30	0.225	0.750
5	45–49.99	5	35	0.125	0.875
6	50–54.99	4	39	0.100	0.975
7	55–59.99	1	40	0.025	1.000
		40		1.000	

or class to class and displays the cumulative frequencies at each position. By starting at the lowest value and cumulating higher values, this technique is equivalent to presenting the number of values that are "equal to or less than" each value or class along the horizontal axis. When the cumulative absolute frequency values from table 2.7 are plotted on the vertical axis for each data category or class on the horizontal axis, the result produces a typical pattern for the ogive that has an S shape (figure 2.6). Such a diagram is useful for comparing the value of a particular observation with all other values in the distribution. The researcher can tell how many observations are "less than or equal to" a particular unit of data. In addition to displaying cumulative absolute frequencies, the data can also be shown as cumulative relative frequencies (or cumulative proportions)—see the right vertical axis label on figure 2.6. Following the procedure discussed earlier, the cumulative absolute frequencies can be divided by the sum of all frequencies to obtain cumulative relative

values or proportions. Cumulative relative frequencies are used to test distributions for normality in section 11.1.

The graphic procedures discussed so far have focused on ways of displaying the distribution or frequency for one variable (a **univariate distribution**). Another type of graph, termed a **scattergram**, or scatterplot, shows the pattern of association or relationship between two variables (a **bivariate relationship**). A scattergram requires that both variables be measured on an interval/ratio scale for a common set of observations. The most familiar organizational framework for showing a bivariate relationship is the Cartesian coordinate system (named for the French mathematician and philosopher, René Descartes). When one variable is represented on the vertical or Y-axis and the other on the horizontal or X-axis, this system allows the precise positioning of each observation as a point in two-dimensional space. So for example, the location of coordinate pair ($X = 5$, $Y = 4$) is shown (figure 2.7).

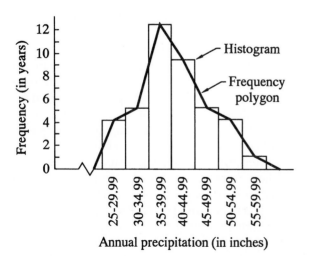

FIGURE 2.5

Histogram and Frequency Polygon for 40-Year Annual Precipitation Data in Washington, D.C.

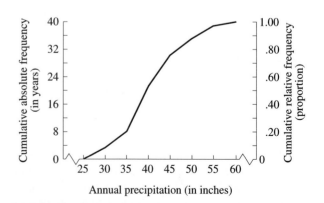

FIGURE 2.6

Cumulative Frequency Polygon (Ogive) for 40-year Annual Precipitation Data in Washington, D.C.

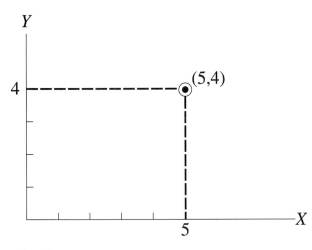

FIGURE 2.7
Location of the Cartesian Coordinate Pair (5,4)

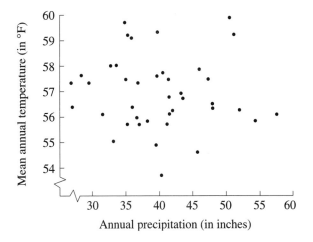

FIGURE 2.8
Scatterplot Showing Association between Precipitation and Temperature for 40-year Period in Washington, D.C.

If a set of observations is plotted, analysis of the scatter of points suggests the amount and nature of association or relationship that exists between the two graphed variables. For example, the 40 years of precipitation data in Washington, D.C., can be matched with the corresponding yearly data for mean annual temperature to construct a scattergram of the two variables (figure 2.8). The circular-shared arrangement of points on this graph seems generally random, suggesting the lack of meaningful association between precipitation and temperature. In chapters 13 and 14, scattergrams are used with correlation and regression analyses to investigate bivariate relationships.

KEY TERMS

categorical (qualitative) variable, 18
classification strategies: subdivision
   (logical subdivision) and agglomeration, 22
discrete and continuous variables, 18
ecological fallacy, 18
exhaustive and mutually exclusive categories, 19
explicitly spatial data, 17
frequency polygon, 31
histogram, 31
implicitly spatial data, 17
individual-level and spatially aggregated
   data sets, 17
measurement concepts: precision, accuracy,
   validity, reliability, 20
measurement scales: nominal, ordinal,
   interval, ratio, 18
methods of classification: equal intervals
   based on range, equal intervals not based
   on range, quantile breaks, natural breaks, 23
ogive (cumulative frequency diagram), 31
operational definitions, 21

primary data, 16
quantitative and qualitative variables, 18
range, 23
secondary (or archival) data, 17
scattergram (scatterplot), 32
spurious precision, 21
strongly ordered and weakly ordered
   ordinal variable, 19
univariate distribution and bivariate relationship, 32

MAJOR GOALS AND OBJECTIVES

If you have mastered the material in this chapter, you should now be able to do the following.

1. Categorize geographic variables and data sets on a variety of dimensions (e.g., primary or secondary [archival], explicitly or implicitly spatial, individual-level or spatially aggregated, discrete or continuous, categorical [qualitative] or quantitative variables).

2. Identify the measurement scale (nominal, ordinal, interval, or ratio) of a variable and understand the characteristics of each measurement scale.

3. Describe the measurement concepts of precision, accuracy, validity, and reliability. When presented with data from any particular geographic problem or situation, evaluate the variable in terms of these potential sources of measurement error.

4. Explain the general logic associated with each of the two conceptual classification strategies (subdivision and agglomeration).

5. Understand the specific procedures, as well as both the advantages and disadvantages associated with each of the operational classification methods (equal intervals based on range, equal intervals not based on range, quantile breaks, and natural breaks). When given a set of geographic data, select an appropriate classification method.

6. Describe or summarize a given set of data graphically with a histogram, frequency polygon, ogive, and scattergram.

REFERENCES AND ADDITIONAL READING

Abler, R. F., J. S. Adams, and P. R. Gould. 1971. *Spatial Organization: The Geographer's View of the World.* Englewood Cliffs, NJ: Prentice-Hall.

Blalock, H. M., Jr. 1982. *Conceptualization and Measurement in the Social Sciences.* Beverly Hills, CA: Sage.

Jenks, G. F. and R. C. Coulson. 1963. "Class Intervals for Statistical Maps." *International Yearbook of Cartography* 3:119–34.

Monmonier, M. S. 1975. *Maps, Distortion, and Meaning* (Resource Paper No. 75-4). Washington, D. C.: Assoc. of Amer. Geographers.

Monmonier, M. S. and H. J. DeBlij. 1996. *How to Lie with Maps* (2nd edition). Chicago: University of Chicago Press.

Robinson, A. H, J. L. Morrison, and P. C. Muehrcke (contributor). 1995. *Elements of Cartography* (6th edition). New York: John Wiley and Sons.

Sokal, R. R. 1966. "Numerical Taxonomy." *Scientific American* 215, No. 6:106–16.

PART TWO

DESCRIPTIVE PROBLEM SOLVING IN GEOGRAPHY

CHAPTER 3

Descriptive Statistics

3.1 Measures of Central Tendency

3.2 Measures of Dispersion and Variability

3.3 Measures of Shape or Relative Position

3.4 Spatial Data and Descriptive Statistics

A basic distinction between descriptive and inferential statistics is discussed in chapter 1. Recall that the overall goal of descriptive statistics is to provide a concise, easily understood summary of the characteristics of a particular data set. For most geographic problems, such quantitative or numerical summary measures are clearly superior to working with unsummarized raw data. With these easily understood and widely used descriptive statistics, geographers can communicate effectively.

Some of the advantages of summarizing spatial information have already been demonstrated in the discussion of basic classification methods (section 2.4) and through the use of such visual procedures as histograms and ogives (section 2.5), which provide *graphic* summary descriptions of a data set. In this chapter, through the use of basic descriptive statistics, we will show how complementary *numerical* or *quantitative* summary measures "describe" a data set. The descriptive statistics discussed in this chapter are illustrated with the same set of precipitation data used to construct the graphics in section 2.5.

A data set can be summarized in several different ways:

- *Measures of central tendency*—numbers that represent the center or typical value of a frequency distribution, such as mode, median, and mean (section 3.1)
- *Measures of dispersion*—numbers that depict the amount of spread or variability in a data set, such as range, interquartile range, standard deviation, variance, and coefficient of variation (section 3.2)

- *Measures of shape or relative position*—numbers that further describe the nature or shape of a frequency distribution, such as skewness, which indicates the amount of symmetry of a distribution, or kurtosis, which describes the degree of flatness or peakedness in a distribution (section 3.3).

Choosing the proper descriptive statistic for a particular geographic problem depends partly on the level of measurement—whether the data are nominal, ordinal, or interval/ratio. In addition, different calculation procedures are applied if the data are grouped (weighted) or ungrouped (unweighted).

Geographers must be cautious when applying descriptive statistics to spatial or locational data. The way in which a geographic problem is structured can affect the resulting descriptive statistics. These issues related to the structure of problems are discussed in section 3.4: (1) the effects of boundary line delineation and study area location on descriptive measures; (2) the effect of altering internal subarea boundaries within the same overall study area—the so-called modifiable areal units problem; and (3) the impact of using different levels of spatial aggregation or different scales on descriptive statistics.

3.1 MEASURES OF CENTRAL TENDENCY

The central or typical value of a set of data can be described numerically in several different ways. Each of these measures of central tendency has advantages and disadvantages, and the logic underlying the calculation procedure for each measure is different. To select the most appropriate measure in a particular geographic situation requires an understanding of each measure and its characteristics. The discussion in this section is limited to three widely used measures of central tendency: mode, median, and mean.

Mode

The **mode** is simply the value that occurs most frequently in a set of ungrouped data values. When nominal data are used, the mode would be the category containing the largest number of observations. With ordinal or interval/ratio data grouped into classes, the category with the largest number of observations is defined as the **modal class.** The midpoint of the modal class interval is the **crude mode.** A mode can be calculated for data at all levels of measurement. For nominal data, however, the mode is the only available descriptive measure of central tendency.

Although it is a useful measure of central tendency in many data sets, the mode may not always provide a practical result. For example, the mode would not be an appropriate measure for the annual precipitation data from Washington, D.C. (table 2.6). In this particular data set, no annual precipitation figure occurs more than once over the 40-year sample period, making the mode ineffective. In fact, a large number of tied values is not likely to occur in most geographic situations where data are interval or ratio scale. However, if the precipitation data are grouped (table 2.7), the modal class is 35–39.99 (with 12 values), and the crude mode of 37.5 is the midpoint of this modal class interval.

Median

The median is the middle value from a set of ranked observations and is therefore the value with an equal number of data units both above it and below it. With an odd number of observations, the middle value is unique and defines the median. With an even number of observations, the median is defined as the midpoint of the values of the two "middle ranks." The Washington, D.C., data has 40 values, so the two middle values are rank 20 (39.62 inches of precipitation) and rank 21 (39.86 inches), and the median is their midpoint (39.74 inches).

Mean

The **mean** (also called the arithmetic mean or the average) is the most widely used measure of central tendency. It is usually the most appropriate measure when using interval or ratio data. The arithmetic mean ($\overline{X}$) is the sum of a set of values divided by the number of observations in the set. In standard statistic notation, the mean is defined as follows:

$$\overline{X} = \frac{\sum_{i=1}^{n} X_i}{n} = \frac{X_1 + X_2 + \ldots + X_n}{n} \qquad (3.1)$$

where
$\overline{X}$ = mean of variable X
X_i = value of observation i
Σ = summation symbol (uppercase sigma)
n = number of observations

It is generally understood that summation is over all n observations, so the symbols above and below sigma are usually omitted:

$$\overline{X} = \frac{\Sigma X_i}{n} \qquad (3.2)$$

The calculation of mean annual precipitation for the Washington, D.C., example data is shown in table 3.1.

In many geographic problems, a sample mean must be differentiated from a population mean. Using conventional notation, lowercase n refers to sample size and uppercase N to population size. Population characteristics are customarily defined using Greek letters, but sample measures are not. The formula for a population mean (μ = lowercase Greek mu) with N values is

$$\mu = \frac{\Sigma X_i}{N} \qquad (3.3)$$

Because it is possible to draw many samples of size n from a large population of size N, a different-magnitude sample mean may result from each sample drawn. The population mean is a fixed value,

TABLE 3.1

Worktable for Calculating Arithmetic Mean of Washington, D.C., Precipitation Data

Observation i	Precipitation X_i
1	41.11
2	54.29
3	35.09
. . .	. . .
38	34.98
39	35.96
40	50.50
Total	1598.00

$$\overline{X} = \frac{\Sigma X_i}{n} = \frac{41.11 + 54.29 + \ldots + 50.50}{40} =$$

$$\frac{1598.00}{40} = 39.95$$

however. In chapter 7, the distinction between samples and populations is explored further in the discussion of applying sample information to estimate population characteristics.

The mean can also be calculated for grouped data. In some practical situations, computing a grouped or weighted mean is the only viable option. Suppose, for example, the only information available is in summary form, like the histogram and frequency polygon for the Washington, D.C., annual precipitation data shown in figure 2.5. Perhaps the original data or source from which the histogram has been constructed (like the National Climatic Data Center data in table 2.6) is not available. In other practical situations, it may be most effective first to classify a set of data, then use the classified (grouped) information to calculate the weighted mean. Generally, if a graphic representation such as a histogram is part of the analysis, the most conventional or commonly used method of classification is equal intervals not based on range, which is how the data in figure 2.5 have been classified.

Whatever the specific circumstance, a **weighted mean** is calculated from only the class intervals and class frequencies presented in figure 2.5, using this formula:

$$\overline{X}_w = \frac{\sum\limits_{j=1}^{k} X_j f_j}{n} \qquad (3.4)$$

where $\overline{X}_w$ = weighted mean
X_j = midpoint of class interval j
f_j = frequency of class interval j
k = number of class intervals
n = total number of values = $\sum\limits_{j=1}^{k} f_j$

To permit this calculation, two related assumptions are made about the distribution of values within each category or class interval: (1) without any information to the contrary, the data are assumed to be distributed evenly within each class interval; therefore, (2) the best summary representation of the values in each interval is the **class midpoint.** Quite simply, the class midpoint is located exactly midway between the extreme values that identify the class interval. So, for example, the class midpoint of the interval from 25 to 29.99 is 27.5, which also happens to be the mean of those two values (table 3.2).

A quick comparison of the unweighted and weighted means for the Washington, D.C., precipitation data reveals that the unweighted mean is only slightly smaller than the weighted mean (the unweighted mean is 39.95, whereas the weighted mean is 40.25). In general, weighted and unweighted means

TABLE 3.2

Worktable for Calculating Weighted Mean of Washington, D.C., Precipitation Data

Class interval j	Class midpoint X_j	Class frequency f_j	$X_j f_j$
25–29.99	27.5	4	110.0
30–34.99	32.5	5	162.5
35–39.99	37.5	12	450.0
40–44.99	42.5	9	382.5
45–49.99	47.5	5	237.5
50–54.99	52.5	4	210.0
55–59.99	57.5	1	57.5
Total		40	1610.0

$$\overline{X}_w = \frac{\sum X_j f_j}{n} = \frac{1610.0}{40} = 40.25$$

should be expected to differ only slightly, depending on the overall effect of using class midpoints as representative summary values for each category. For example, a weighted mean will be higher than the corresponding unweighted mean if the majority of original values in most of the categories happen to fall below the class midpoints.

Selecting the Proper Measure of Central Tendency

Deciding which measure of central tendency to use depends both on the geographic application and certain key characteristics of the data set. On the surface, the mean seems to have certain advantages. It is the most widely applied and generally understood measure of central tendency. In addition, it is always affected to some extent by a change or modification of any data set value. This is not always true of either the mode or the median. The mean is also an element in many tests of statistical inference when estimates of population parameters are made from sample information. When working with interval/ratio data, the mean is the measure of choice.

Selection of the "best" measure of central tendency often depends on certain characteristics of the distribution of the data, as the following situations illustrate:

- If a distribution is **unimodal** (having a single distinct mode) and **symmetric** (evenly balanced around a single distinct mode or central vertical line), then all three "centers" are similarly located and no advantage accrues to any statistic (figure 3.1, case 1).
- If a frequency distribution has some degree of **skewness** (is not completely symmetric about a central line), the various measures of central

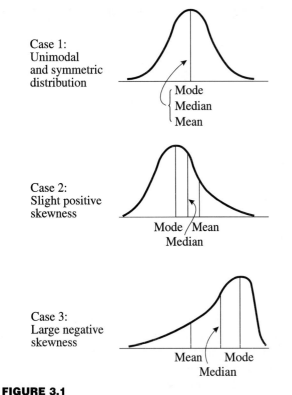

FIGURE 3.1
Measure of Central Tendency Placed on Symmetric and Skewed Frequency Distributions

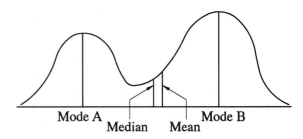

FIGURE 3.2
Measures of Central Tendency Placed on a Bimodal Frequency Distribution

tendency will be positioned at different places along the distribution. With slight positive skew (figure 3.1, case 2), the median is pulled slightly to the right toward the positive tail of the distribution. The location of the mean is affected even more significantly, positioned even further out the positive tail than the median. With large negative skew (figure 3.1, case 3), the magnitude of displacement of the median and, especially the mean, is even more pronounced.

- If a frequency distribution is **bimodal** (having two distinct modes) or **multimodal** (with more than two modes), the mean may be located on the distribution at a place that would not be considered typical. In figure 3.2, neither the mean nor median depicts useful central locations; providing the values of mode A and mode B is probably more informative.
- If a frequency distribution contains one or more extreme or atypical values (**outliers**), the mean will be heavily influenced by these values, and its effectiveness as a measure of centrality may be reduced. The existence of an extreme value or values is actually a form of skewness.

A simple example using the incomes of seven families illustrates the sensitivity of the mean to a

single outlier (table 3.3). The mean is ineffective as a measure of central tendency in this example because it is greatly affected by the income of a single atypical wealthy family. The mode is too low an income to represent a useful central value from this frequency distribution. The median provides the most suitable measure in this situation. In fact, with variables such as income, which often contain a large amount of skewness, the median is generally the descriptive statistic of choice.

3.2 MEASURES OF DISPERSION AND VARIABILITY

Simple Measures of Dispersion

The three measures of central tendency—mode, median, and mean—specify the center of a distribution, but provide no indication of the amount of spread or variability in a set of values. Dispersion can be calculated in several different ways. The level of measurement and nature of the frequency distribution determine the most appropriate dispersion statistic.

The simplest measure of variability is the **range,** already defined in the previous chapter as the difference between the largest and smallest values in an interval/ratio set of data. Because it is derived

TABLE 3.3

Sensitivity of the Mean to a Single Outlier

Values	Statistics
$21,000	Total = $500,000
21,000	
22,000	Mode = $21,000
26,000	
27,500	Median = $26,000
32,500	
349,000	Mean = $500,000/7
	= $71,428.57

solely from the two most extreme or atypical values and ignores all other values, the range can be a misleading measure. The range measured for the 40 precipitation values in the Washington, D.C., example is $(57.54 - 26.87) = 30.67$.

If data are grouped, the range is defined as the difference between the upper value in the highest numbered class interval and the lower value in the lowest numbered class interval. With the grouped precipitation data, these values are $(59.99 - 25) = 34.99$. As with the ungrouped range, the grouped range can be a deceptive measure of dispersion. Given only the range, the degree of clustering or dispersion of the values between these two extremes is unknown.

To provide more information, one can examine certain intervals, portions, or percentiles within a frequency distribution. If data are divided into equal portions or percentiles (**quantiles**), then the range of values within any quantile can be calculated and graphed. Although any logical subdivision is possible, data are often classified into quartiles (fourths), quintiles (fifths), or deciles (tenths). The median, which is the 50th percentile, may also be one of these subdivisions.

When data are divided into four portions, or quartiles, the **interquartile range** is defined as the difference between the 25th percentile value (the lower quartile) and the 75th percentile value (the upper quartile). The interquartile range thus encompasses the "middle half" of the data. In the Washington, D.C., precipitation example, the interquartile range is $(43.53 - 35.20) = 8.33$.

A **dispersion diagram** can be constructed to display graphically the entire frequency distribution within lines that box in the interquartile range and show the upper quartile, median, and lower quartile. Figure 3.3 shows dispersion diagrams of annual precipitation for three American cities—San Diego, California, St. Louis, Missouri, and Buffalo, New York. This graphic representation of data offers useful information for comparing the variability of several distributions.

Other measures of dispersion are based on an examination of individual deviations from the mean value. The difference between each value and the mean is calculated, and these deviations are used as the building blocks to measure dispersion. An individual deviation (d_i) is calculated as follows:

$$d_i = (X_i - \overline{X}) \tag{3.5}$$

where d_i = deviation of value i from the mean
       X_i = value of observation i
       $\overline{X}$ = mean of variable X

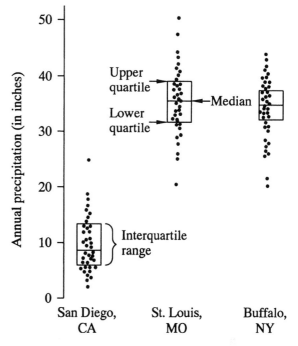

FIGURE 3.3
Dispersion Diagrams of Annual Precipitation for Three Cities (*Source:* National Climatic Data Center, U.S. Dept. of Commerce.)

The **average deviation**, or **mean deviation**, is based on the mean of the set of individual deviations. That is, the mean deviation (m) for a sample of values is

$$m = \frac{\Sigma |X_i - \overline{X}|}{n} \tag{3.6}$$

where $|X_i - \overline{X}|$ = the absolute value of the difference between X_i and $\overline{X}$

Some units of data are larger than the mean, making the difference $(X_i - \overline{X})$ a positive number. Conversely, values below the mean make $(X_i - \overline{X})$ a negative number. However, the sum of the deviations about a mean is always zero. To avoid the problem of negative differences offsetting positive differences, the absolute value of each individual deviation is calculated. With absolute values, negative deviations are converted to positive ones—for example: $|-2| = |2| = 2$.

Another important property associated with the mean is its "least-squares" character. Not only is the sum of deviations (i.e., unsquared deviations) of all observations about a mean always zero, but the sum of *squared* deviations of all observations about a mean is the minimum sum possible. That is, the sum of squared deviations about a mean is less than the sum of squared deviations about any other number.

This important attribute is called the **least squares property of the mean,** and can be summarized as

$$\overline{X}: \min \Sigma \, (X_i - \overline{X})^2 \qquad (3.7)$$

Standard Deviation and Variance

The least squares property of the mean carries over into the most common measure of variability or dispersion—**standard deviation**. Given a sample of values, the standard deviation (s) is defined as

$$s = \sqrt{\frac{\Sigma \, (X_i - \overline{X})^2}{n - 1}} \qquad (3.8)$$

In the numerator of equation 3.8, the deviation of each value from the mean is squared, then all of the squared deviations are summed, incorporating the least-squares property into this measure of dispersion. Note also that this squaring process removes the problem of negative deviations offsetting positive deviations in a more effective way than absolute values. This least-squares sum is then divided by (n − 1) to reduce sampling bias. Finally, to reverse the effect of squaring, the square root of this quotient is taken.

For a population, the standard deviation formula is written as

$$\sigma = \sqrt{\frac{\Sigma(X_i - \mu)^2}{N}} \qquad (3.9)$$

where σ = lowercase Greek sigma

Note the different denominators of the two standard deviation formulas. For a sample, the denominator is n − 1, but is N for a population. If the standard deviation is being calculated for a large sample, (n > 30), the difference between n and (n − 1) is small, and the difference between s and σ will also be small. However, with a small sample (n < 30), the true population standard deviation (σ) is underestimated if division is by n. When dividing by (n − 1), s becomes larger, correcting the underestimation problem. The incorporation of this sample size correction provides a better estimate of the true population standard deviation for smaller samples.

The **variance** of a data set, defined as the square of the standard deviation, provides a measure of the average squared deviation of a set of values around the mean. Variance is seldom used directly as a summary descriptive statistic, since the calculated variance is generally a very large number and difficult to interpret. Nevertheless, variance is an extremely important descriptive statistic that has many applications in inferential procedures. The for-

mulas for sample variance (s^2) and population variance (σ^2) are as follows:

$$s^2 = \frac{\Sigma \, (X_i - \overline{X})^2}{n - 1} \qquad (3.10)$$

$$\sigma^2 = \frac{\Sigma \, (X_i - \mu)^2}{N} \qquad (3.11)$$

Both standard deviation and variance are fundamental building blocks in statistical analysis. When discussing sampling procedures in chapter 7, these descriptive measures are used as the theoretical basis for making probabilistic statements. Confidence intervals are placed around estimates derived from samples to predict population parameters. These two statistics are also integral components in inferential hypothesis testing. For example, variance is used in analysis of variance (ANOVA), a statistical technique that determines whether multiple samples differ significantly from one another (chapter 10).

A final issue regarding standard deviation and variance requires clarification—the procedural question of which to use, a more convenient computational formula or a more cumbersome and time-consuming definitional formula. Formulas based on the definition of standard deviation are valuable because they show how variability is a direct function of individual deviations ($X_i - \overline{X}$) or ($X_i - \mu$). However, if it is necessary to calculate standard deviation manually without computer assistance, more efficient computational formulas are available. These alternative formulas, based on algebraic manipulations of the definitional formulas, are summarized in table 3.4. Since variance is the square of the standard deviation, it is calculated by simply removing the square root symbol from each formula in table 3.4.

The calculation procedures for both the definitional and computational formulas of sample standard deviation are demonstrated in table 3.5 using

TABLE 3.4		
Definitional and Computational Formulas for Standard Deviation of Sample and Population		
	Definitional formula	Computational formula
Sample (s)	$\sqrt{\dfrac{\Sigma \, (X_i - \overline{X})^2}{n - 1}}$	$\sqrt{\dfrac{\Sigma \, X_i^2 - n\overline{X}^2}{n - 1}}$
Population (σ)	$\sqrt{\dfrac{\Sigma \, (X_i - \mu)^2}{N}}$	$\sqrt{\dfrac{\Sigma \, X_i^2}{N} - \mu^2}$

TABLE 3.5

Worktable for Calculating Sample Standard Deviation of Washington, D.C., Precipitation Data

Observation i	X_i	$X_i - \overline{X}$	$(X_i - \overline{X})^2$	X_i^2
1	41.11	1.16	1.35	1690.03
2	54.29	14.34	205.64	2947.40
3	35.09	−4.86	23.62	1231.31
. . .	. . .	. . .	. . .	. . .
38	34.98	−4.97	24.70	1223.60
39	35.96	−3.99	15.92	1293.12
40	50.50	10.55	111.30	2550.25
Total	1598.00		2192.76	66,032.86

$$n = 40 \quad n - 1 = 39 \qquad \overline{X} = \frac{\Sigma X_i}{n} = \frac{1598.00}{40} = 39.95$$

Using the definitional formula for sample standard deviation:

$$s = \sqrt{\frac{\Sigma (X_i - \overline{X})^2}{n - 1}} = \sqrt{\frac{2192.76}{39}} = 7.50$$

Using the computational formula for sample standard deviation:

$$s = \sqrt{\frac{\Sigma X_i^2 - n\overline{X}^2}{n - 1}} = \sqrt{\frac{66032.86 - 40(39.95)^2}{39}} = 7.50$$

the Washington, D.C., precipitation data. The statistical parity of the definitional and computational formulas can be easily seen.

For calculating standard deviation and variance from grouped data, the same assumptions are made as when calculating a weighted mean. That is, the values are assumed to be evenly distributed within each class interval, making the midpoint the best representation of the middle of that class interval. As with the weighted average, the weighted standard deviation (s_w) and variance (s_w^2) formulas use these class midpoints (X_j) and class frequency counts (f_j). Although a definitional formula for weighted standard deviation exists, table 3.6 uses the simpler computational formula for calculating this index with the Washington, D.C., precipitation data.

The weighted standard deviation ($s_w = 7.59$) is similar in magnitude to the unweighted standard deviation ($s = 7.50$). A slight difference between the two descriptive statistics is expected given the contrasting assumptions that underlie the calculation procedures.

TABLE 3.6

Worktable for Calculating Weighted Standard Deviation of Washington, D.C., Precipitation Data

Class interval j	Class midpoint X_j	Class frequency f_j	$X_j f_j$	$(X_j)^2 f_j$
25–29.99	27.5	4	110.0	3025.00
30–34.99	32.5	5	162.5	5281.25
35–39.99	37.5	12	450.0	16875.00
40–44.99	42.5	9	382.5	16256.25
45–49.99	47.5	5	237.5	11281.25
50–54.99	52.5	4	210.0	11025.00
55–59.99	57.5	1	57.5	3306.25
Total		40	1610.0	67,050.00

Using the computational formula for sample standard deviation:

$$s_w = \sqrt{\frac{\Sigma (X_j)^2 f_j - ((\Sigma X_j f_j)^2 / n)}{n - 1}}$$

$$= \sqrt{\frac{67050 - ((1610)^2 / 40)}{39}} = 7.59$$

Coefficient of Variation

For many types of geographic research, it is extremely valuable to compare the amount of variability in different spatial patterns directly to see which has the greatest spatial variation. In other problems, it may be useful to compare variability in some phenomenon as it changes over time. For example, a climatologist might want to compare the variability in annual rainfall at different meteorological stations to learn which locations have the greatest variation from year to year. An economic development planner might be interested in comparing family income in several different counties to learn which region has the greatest internal variation. Asking these investigative, comparative questions allows geographers to explore practical problems in a highly productive manner.

Using standard deviation or variance to compare locations or regions directly is inappropriate because they are both **absolute measures.** That is, their values depend on the size or magnitude of the units from which they are calculated. A data set with large numbers (i.e., in the thousands) is described with a large average, standard deviation, and variance. Conversely, analysis of a data set containing single-digit numbers results in small absolute descriptive measures. Clearly, direct comparison of averages, standard deviations, and variances is limited across data sets with different magnitudes.

To resolve this problem and compare the spatial variation of two or more geographic patterns, a relative measure of dispersion has been developed. Called the **coefficient of variation (CV)**, or **coefficient of variability,** this index is simply the standard deviation expressed relative to the magnitude of the mean:

$$CV = \frac{s}{\overline{X}} \text{ or } CV = \frac{s}{\overline{X}}(100) \qquad (3.12)$$

The coefficient of variation is usually expressed as a proportion or percentage of the mean and may be used with either sample or population data. Dividing the standard deviation by the mean removes the influence of the magnitude of the data and allows direct comparison of relative variability in different data sets.

In the following example, variability for locational data whose magnitude varies from place to place is directly compared using the coefficient of variation. Sharp contrasts are evident when comparing 40 years of annual precipitation data for three sample locations—Buffalo, St. Louis, and San Diego (table 3.7).

While the average amounts of precipitation for Buffalo and St. Louis are similar (35.47 versus 35.56), St. Louis has a large absolute amount of variability in precipitation from year to year ($s = 6.62$ vs.

TABLE 3.7

Descriptive Statistics Using 40 Years of Annual Precipitation Data for Three Cities

City	Mean	Standard deviation (absolute)	Coefficient of variation (relative)
Buffalo, N.Y.	35.47	4.70	13.25
St. Louis, Mo.	35.56	6.62	18.62
San Diego, Calif.	9.62	4.42	45.95

$s = 4.70$). Location can explain this result. St. Louis is found in the middle of the continent, whereas Buffalo is close to a consistent source of moisture, Lake Erie. Of the three cities, however, San Diego has by far the greatest degree of relative variability in precipitation from year to year. Although the standard deviation for San Diego is only 4.42 (the smallest of the three cities), that magnitude of variability is relatively large when compared to San Diego's rather low average precipitation of 9.62. The higher coefficient of variation for San Diego provides a clear statistical measure of the relatively large fluctuations in precipitation that can occur from year to year in a semiarid climate. This type of investigative comparison of coefficient of variation values could be taken further and has been used to explore practical problems dealing with spatial patterns of climatic variability.

Only ratio-scale data should be used to calculate the coefficient of variation. Data measured at the interval scale are not appropriate because the interval metric has an arbitrary zero. As a result, numeric manipulations involving multiplication or division are meaningless. Since the coefficient of variation is a ratio of the standard deviation to the mean, it follows that a coefficient of variation value has no meaning when calculated from interval-scale data.

Geographers should use the coefficient of variation more frequently in their research. This relative index of dispersion is easily calculated from simple descriptive statistics and has many potential applications. As a numeric measure of relative variability, geographers can use the coefficient of variation to summarize maps and other spatial patterns quantitatively. Given multiple sets of locational or explicitly spatial data, geographers can use the coefficient of variation to measure and compare the relative variability in the data. Changes in a regional or areal pattern over time can be numerically summarized and regional trends toward dispersal or clustering observed. With easy access to computers and computer databases, such direct comparisons between locations or regions can be made with relative ease.

3.3 MEASURES OF SHAPE OR RELATIVE POSITION

Skewness and Kurtosis

In addition to the coefficient of variation, two other relative measures—skewness and kurtosis—can further describe the nature or character of a frequency distribution. As already mentioned, **skewness** measures the degree of symmetry in a frequency distribution by determining the extent to which the values are evenly or unevenly distributed on either side of the mean. **Kurtosis** measures the flatness or peakedness of a data set. Like the coefficient of variation, geographers underutilize these indices, yet they provide important descriptive insights about a frequency distribution and offer considerable potential in spatial research.

Introducing the concept of moments of a distribution about the mean provides a more complete understanding of skewness and kurtosis. The first moment is the sum of individual deviations about the mean and must equal zero:

$$\text{First moment} = \Sigma d_i = \Sigma (X_i - \overline{X})^1 = 0 \qquad (3.13)$$

The second moment of a frequency distribution is the numerator in the expression that defines variance:

$$\text{Second moment} = \Sigma (X_i - \overline{X})^2 \qquad (3.14)$$

Skewness involves use of the third moment of a frequency distribution. One commonly used measure of relative skewness contains the third moment in the numerator:

$$\text{Skewness} = \frac{\Sigma (X_i - \overline{X})^3}{ns^3} \qquad (3.15)$$

The denominator of this expression contains the cubed standard deviation, which effectively standardizes the third moment. This allows geographers to compare the amount of relative skewness in different frequency distributions directly.

If a frequency distribution is symmetric, with an equal number of values on either side of the mean, the distribution has little or no skewness. If a value in a distribution is greater than the mean, its cubed deviation will be positive. However, if a value is less than the mean, it will produce a negative cubed deviation. In a symmetric distribution, these positive and negative cubed deviations will counterbalance each other, and the sum (the third moment) will be zero. In a distribution with a tail to the left, large negative cubed deviations will cause the sum of all deviations (the third moment) to be negative. In this case, the resultant distribution is said to be **negatively skewed.** In a distribution with a tail to

the right, on the other hand, large positive cubed deviations will dominate the sum, resulting in a **positively skewed** distribution.

Another measure of skewness, known as Pearson's coefficient, is based on a comparison of the mean and median:

$$\text{Pearson's skewness} = \frac{3(\overline{X} - Median)}{s} \qquad (3.16)$$

With a unimodel and symmetric distribution, the mean and median have the same value, and the skewness coefficient is zero (figure 3.1, case 1). When the mean is greater than the median (as in case 2), positive skewness results, and when the mean is less than the median (as in case 3), the skewness measure is negative. Division by the standard deviation provides a standardization of the Pearson's skewness values, allowing direct comparisons.

Kurtosis measures the fourth moment of a frequency distribution using the following formula:

$$\text{Kurtosis} = \frac{\Sigma (X_i - \overline{X})^4}{ns^4} \qquad (3.17)$$

If a large proportion of all values is clustered in one part of the distribution, it will have a pointed or peaked appearance, a high level of kurtosis, and be considered **leptokurtic** (figure 3.4). In a data set having low kurtosis (a "flat" or **platykurtic** distribution), values are dispersed more evenly over many different portions of the distribution. Some distributions are considered **mesokurtic,** or "moderate," having a bell-shaped appearance, which is neither very peaked nor very flat.

The interpretation of kurtosis is enhanced by comparing the peakedness of a distribution to that of a normal probability distribution. Although the importance of the normal curve is discussed in more detail in chapter 5, it is worth mentioning here because kurtosis formulas assign characteristic values to a normal distribution. In equation 3.17, a leptokurtic distribution has a kurtosis greater than 3.0, a normal distribution is mesokurtic, with a kurtosis

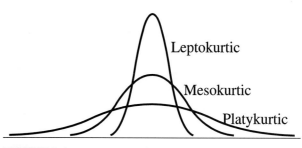

FIGURE 3.4
Different Levels of Kurtosis

value of 3, and a platykurtic distribution has a value less than 3. Some computer software packages use equation 3.18, subtracting 3 from the quotient, so that a normal distribution has zero kurtosis. With this alternative, platykurtic distributions are negative, mesokurtic distributions are close to zero, and leptokurtic distributions are positive.

$$\text{Kurtosis} = \frac{\Sigma (X_i - \overline{X})^4}{ns^4} - 3 \qquad (3.18)$$

Exploratory Comparison of Skewness and Kurtosis Values

Examination of skewness and kurtosis values can provide geographers with useful insights about the comparative nature of spatial patterns. Geographers may want to compare descriptive statistics for a particular variable at different locations during a single time period. They may ask such questions as "During this period of time, which locations had the greatest relative variability? . . . the greatest amount of skewness? . . . the most leptokurtic distribution?" After these questions have been examined and the data analyzed, geographers can continue the investigation by asking *why* these observed skewness or kurtosis values occurred. By using these descriptive statistics more fully, geographers gain a better understanding of their data.

In the previous section, annual precipitation data from Buffalo, St. Louis, and San Diego were compared over the same 40-year period using mean, standard deviation, and coefficient of variation. In table 3.8, intercity comparisons are made for skewness and kurtosis. Note both the high positive level of skewness and large positive (leptokurtic) kurtosis value for San Diego relative to Buffalo and St. Louis. Because skewness and kurtosis are standardized third and fourth moments, respectively, they are quite sensitive to values that have a large relative difference from the mean. This sensitivity is particularly notable with kurtosis, where the deviations between individual values and the mean are taken to the fourth power.

The precipitation data for San Diego clearly demonstrate the sensitivity of skewness and kurto-

sis to extreme values. With a low mean annual precipitation of 9.62 inches and a semiarid climate, San Diego's 40-year rainfall record shows several very wet years. For example, the 40 years of data include annual totals of 24.93 inches, 17.74 inches, and 19.03 inches, all of which are several standard deviations above average ($s = 4.42$). These unusually high annual precipitation values are shown on the dispersion diagram (figure 3.3), and their existence dominates the subsequent descriptive statistics. The result is a large positive skewness value, with a few very wet years extending the tail of the distribution to the right.

San Diego also has a peaked kurtosis value. Even more than skewness, kurtosis will be influenced by one or two extreme outliers. When contrasted with the mean and taken to the fourth power (see equations 3.17 and 3.18), the unusually wet years result in large positive kurtosis. Nearly all of the 40 precipitation values for San Diego are grouped in the range between 6 and 12 inches. In contrast, the skewness value for Buffalo is slightly negative, with more of a tail to the left (toward relatively dry years). In Buffalo, the highest precipitation total over the 40-year record is 44.78 inches, only about 9 inches over the 40-year mean. Although Buffalo receives a considerable amount of precipitation each year, the total seldom deviates much from the mean, especially in the direction of increased precipitation.

3.4 SPATIAL DATA AND DESCRIPTIVE STATISTICS

Geographers need to recognize potential difficulties associated with the analysis of spatial or location-based data. Problems addressed here include **boundary delineation, modifiable areal units,** and level of **spatial aggregation** or **scale.** Although these issues are rarely discussed outside the discipline, an understanding of the nature of these problems is essential for conducting geographic research.

Impact of Boundary Delineation

The location of an external study area boundary and the consequent positioning of internal subarea boundaries affect various descriptive statistics. Consider first the possible impact of study area size on absolute measures, such as the mean or standard deviation. Suppose a social geographer is conducting a study on the number and spatial pattern of families below the poverty level in census tracts of a large urban area. If the geographer chooses a smaller study area boundary, corresponding to older inner-city tracts (boundary A in figure 3.5), the average number of families below the poverty level will likely be higher than if the average is calculated using a larger

TABLE 3.8

Additional Descriptive Statistics Using 40 Years of Annual Precipitation Data for Three Cities

City	Mean	Standard deviation	Skewness	Kurtosis
Buffalo, N.Y.	35.47	4.70	−0.389	0.672
St. Louis, Mo.	35.56	6.62	0.080	−0.096
San Diego, Calif.	9.62	4.42	1.393	2.684

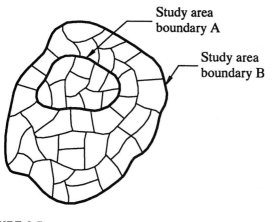

FIGURE 3.5
Alternative Study Area Size

study area boundary that includes suburban tracts (boundary B). Other absolute measures, such as standard deviation and variance, whose values are partially a function of the mean, will also be affected. It is clear, therefore, that absolute descriptive statistics should be evaluated comparatively only in relation to a particular study area.

Location of boundaries and the placement of internal subarea boundaries may also influence statistical analysis. Suppose the shaded area in figure 3.6 represents the location of a high concentration of a particular demographic group. Notice how the positioning of the study area and subarea boundaries seem to determine whether the group is segregated (case 1) or integrated (case 2) within the large region. In each case, the location of the demographic group is the same; only the location of the study area and subareas has been changed. Again, it is advisable to conclude that summary descriptive statistics can have valid interpretations only for the area and subarea configuration over which they are calculated.

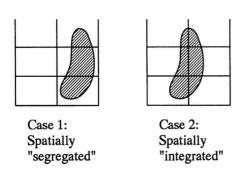

FIGURE 3.6
Effect of Alternative Study Area Boundaries on Subarea Representation of a Demographic Group

Modifiable Area Units

Descriptive statistics can also be influenced significantly by using alternative subdivision or regionalization schemes within the same overall study area. In many cases, geographers are given little or no choice in the way data are subdivided. For example, certain demographic data within a state may be available only at the county level. As a result, county-level areal units and boundaries must be used for any substate analysis. However, for studies where alternative subdivision or regionalization schemes are possible, geographers should realize that each subdivision scheme may produce a different result.

Consider the impact of the alternative areal subdivision schemes shown in figure 3.7. The descriptive measures calculated from each scheme convey completely different statistical summary impressions, even though the same 12 data points are used in each case. Regionalization scheme A suggests spatial separation of different magnitude values—low values (and a low mean) in region A_1, intermediate-sized values in region A_2, and high values in region A_3. Only slight intraregional variation occurs with scheme A, as indicated by the low standard deviation values. By contrast, the means in regionalization scheme B show little difference in magnitude, but considerable intraregional variation, as denoted by the relatively large standard deviation values. The inclusion of large, intermediate, and small values within each subdivision in scheme B results in larger coefficients of variation figures than those derived from scheme A.

The effects of subarea boundary modification are often not as predictable as in this example. When constructing a regionalization scheme, the geographer must realize that the resultant summary statistics are a direct function of the subarea boundary configuration. Summary statistics can be obtained to convey different impressions (e.g., slight internal variation in one configuration and considerable internal variation in another). One must ask whether a particular regionalization scheme contains too much "spatial bias." All subarea boundary schemes have bias, however, since the resultant statistics are affected by whatever configuration is chosen. Descriptive statistics should be interpreted and evaluated carefully, keeping in mind the particular boundary scheme used in the study.

Spatial Aggregation or Scale Problem

In data analysis, geographers can often choose from many different levels of spatial aggregation or scale. Socioeconomic variables, for example, are often available at the block, census tract, enumeration district, election district, county, planning region, and state levels. When the same data are aggregated at different

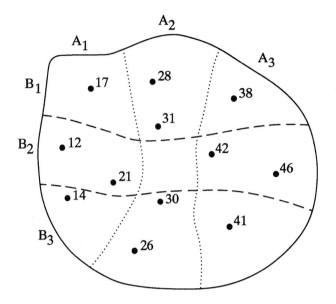

.... Subdivision scheme A
— — — Subdivision scheme B

Region	Mean	Standard deviation	Coefficient of variation
A_1	16.00	3.9	0.244
A_2	28.75	2.2	0.077
A_3	41.75	3.3	0.079
B_1	28.50	8.7	0.305
B_2	30.25	16.4	0.542
B_3	27.75	11.1	0.400

FIGURE 3.7
Impact of Alternative Subdivision Schemes on Descriptive Statistics

spatial levels or scales, the resultant descriptive statistics will vary, sometimes in a systematic, predictable fashion and sometimes in an uncertain way.

The following example uses data on farm population from the last eight decades, analyzed at two different levels of spatial aggregation: census division and state. The 50 states of the United States are aggregated by the Bureau of the Census into nine census divisions (New England, Middle Atlantic, East North Central, West North Central, South Atlantic, East South Central, West South Central, Mountain, and Pacific) (figure 3.8). Various descriptive statistics regarding farm population can be compared both at different spatial scales and at different times (table 3.9).

For each of the time periods from 1920 to 1990, both the mean and standard deviation values are higher at the census division level (where the greater level of aggregation exists) and smaller at the state

level (where the lower level of aggregation exists). This result is expected because absolute descriptive measures are directly influenced by the magnitude of the data from which they are calculated. The absolute measures decrease in size at a particular level of aggregation over time. For example, the mean farm population by state steadily decreases from over 658,000 in 1920 to barely 77,000 in 1990. This trend reflects a dramatic and continuous reduction in farm population nationally.

In comparison to the absolute measures, the columns of table 3.9 showing relative measures are more difficult and challenging to interpret geographically. Figure 3.9 summarizes graphically the differences in coefficient of variation, skewness, and kurtosis values at both levels of spatial aggregation. Each of the graphs shows substantial differences by spatial scale over all time periods. In figure 3.9 (case 1), coefficient of variation values at both the census

FIGURE 3.8
United States Census Divisions

TABLE 3.9

Descriptive Statistics of Farm Population Data at Various Dates and Levels of Aggregation

Date and level of aggregation		Absolute measures (in 1000s)		Relative measures		
		Mean	Standard deviation	Coefficient of variation	Skewness	Kurtosis
1920	Census Division	3512.33	2279.36	64.90	−0.207	−2.147
	State	658.56	533.76	81.05	0.713	0.092
1930	Census Division	3382.67	2174.05	64.27	−0.233	−2.260
	State	634.25	519.56	81.92	0.875	0.860
1940	Census Division	3394.00	2141.33	63.09	−0.202	−2.161
	State	636.38	512.88	80.59	0.712	0.024
1950	Census Division	2560.67	1610.47	62.89	−0.167	−2.025
	State	480.13	376.98	78.52	0.482	−0.804
1960	Census Division	1736.67	1090.55	62.80	−0.100	−2.032
	State	312.60	255.15	81.62	0.498	−0.858
1970	Census Division	1079.22	738.02	68.38	0.457	−0.998
	State	194.26	162.63	83.72	0.585	−0.908
1980	Census Division	624.44	499.73	80.03	1.111	0.279
	State	112.40	103.62	92.18	0.895	−0.159
1990	Census Division	430.22	323.88	75.28	1.032	0.096
	State	77.44	69.76	90.08	0.863	−0.314

Source: Calculated from data published by Economic Research Service, U.S. Dept. of Agriculture.

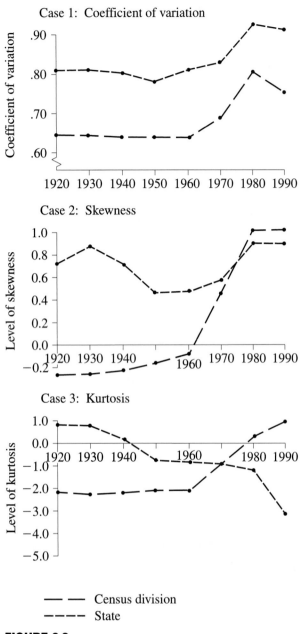

Case 1: Coefficient of variation

Case 2: Skewness

Case 3: Kurtosis

——— Census division
----- State

FIGURE 3.9
Relative Descriptive Statistics of Farm Population Data at
Different Spatial Levels

nois, Ohio, and Wisconsin) contained nearly 28 percent of the nation's total farm population.

Skewness values also differ significantly at various levels of spatial aggregation over time (figure 3.9, case 2). In recent decades, skewness figures have become strongly positive, with the greatest skewness in both 1980 and 1990 at the level of census division. This recent trend toward positive skewness results from the concentration of farm population in the two midwestern census divisions—East North Central and West North Central. In fact, the five states with the largest farm populations in 1990 are all located in these two census divisions, accounting for the increasing positive skewness at the state level as well.

Kurtosis values differ by spatial level and vary over time (figure 3.9, case 3). The kurtosis statistic tends to be less stable for data sets with smaller numbers of values. This instability is evident in the greater fluctuation of kurtosis values at the census division level over time. In contrast, the variation in kurtosis values over time is less dramatic at the state level. Generally, kurtosis values at both the state and census division level have become less platykurtic in the last few decades. In fact, kurtosis values have become more leptokurtic (peaked) at the census division level in 1980 and 1990, as contrasted with earlier dates. This trend would seem to reflect the increasing concentration of farming in the Midwest. If this spatial concentration intensifies in the future, increasingly leptokurtic distributions should be expected, as nearly all states will have virtually no farm population.

KEY TERMS

division and state levels were fairly stable from 1920 to 1960, but then climbed to somewhat higher values in later decades (1970 to 1990). These increases in relative dispersion indicate a spatial concentration of farm population in fewer places. The concentration is especially noticeable at the state level, where coefficient of variation values are higher. The five states with the largest farm populations in 1960 (North Carolina, Texas, Iowa, Minnesota, and Tennessee) accounted for less than 25 percent of the total farm population nationally. By 1990, farm population had concentrated to the degree that the five states with the largest farm populations (Iowa, Minnesota, Illi-

MAJOR GOALS AND OBJECTIVES

If you have mastered the material in this chapter, you should now be able to do the following.

1. Define the basic descriptive measures of central tendency (mode, median, and mean), explain the advantages and disadvantages of each, and select the most appropriate measure given a particular frequency distribution or particular geographic situation.

2. Define the basic descriptive measures of dispersion (range, quantiles, mean deviation, standard deviation, variance) and explain the characteristics of each.

3. Recognize whether a geographic problem or situation requires weighted or unweighted descriptive statistics.

4. Understand the concept of relative variability and its value in comparing different spatial patterns directly by using the coefficient of variation.

5. Define the measures of shape or relative position (skewness and kurtosis) and recognize their potential value in descriptive analysis in geography.

6. Recognize the potential effects of locational data on descriptive statistics. Possible influences include (a) boundary line delineation and study area location, (b) placement of internal subarea boundaries within the same overall study area, and (c) impact of using different levels of spatial aggregation.

REFERENCES AND ADDITIONAL READING

Looking at other introductory textbooks in quantitative geography and spatial analysis can be a valuable experience. Other texts may be simpler, more difficult, or roughly comparable, but whatever their level of difficulty, they will provide *different* insights. Expect some "symbol shock," as each statistics text will likely use somewhat different statistical notation. A few introductory textbooks are listed here.

Burt, J. E. and G. M. Barber. 1996. *Elementary Statistics for Geographers* (2nd edition). New York: Guilford Press.

Clark, W. A. V. and P. L. Hosking. 1986. *Statistical Methods for Geographers.* New York: John Wiley and Sons.

Earickson, R. and J. Harlin. 1994. *Geographic Measurement and Quantitative Analysis.* New York: Macmillan.

Griffith, D. A. and C. G. Amrhein. 1991. *Statistical Analysis for Geographers.* Englewood Cliffs, NJ: Prentice-Hall.

Taylor, P. J. 1983. *Quantitative Methods in Geography: An Introduction to Spatial Analysis.* Prospect Heights, IL: Waveland Press.

In addition to textbooks on quantitative methods written primarily for geographers, dozens of introductory statistics textbooks are currently in print. Most are traditional in the way they handle such topics as probability and sampling. A few recent books, however, present more activity-based ("constructivist") and data-oriented problem-solving frameworks. Many include exercises. A few introductory statistics textbooks are listed here.

Freedman, D., R. Pisani, and R. Purves. 1997. *Statistics.* (3rd edition). New York: W. W. Norton and Co.

Moore, D. S. 1995. *The Basic Practice of Statistics.* New York: W. H. Freeman.

Rossman, A. J. 1996. *Workshop Statistics: Discovery with Data.* New York: Springer-Verlag.

Scheaffer, R. L., M. Gnanadesikan, A. Watkins, and J. A. Witmer. 1996. *Activity-Based Statistics: Instructor Resources.* New York: Springer-Verlag.

Descriptive Spatial Statistics

4.1 Spatial Measures of Central Tendency

4.2 Spatial Measures of Dispersion

4.3 Locational Issues and Descriptive Spatial Statistics

In the preceding chapter, a variety of basic descriptive statistics were examined, including the mean, standard deviation, and coefficient of variation. To summarize point patterns, a set of descriptive spatial statistics has been developed that are areal or locational equivalents to these nonspatial measures (table 4.1). Since geographers are particularly concerned with the analysis of locational data, these descriptive spatial statistics, appropriately referred to as **geostatistics**, are often applied to summarize point patterns and to describe the degree of spatial variability of some phenomena. Geostatistics can also provide a useful summary of an areal pattern on a choropleth map, if each area on the map can be operationally represented by a point.

Spatial measures of central tendency like mean center and median center are examined in section 4.1. Each of these measures has characteristic properties and a set of practical geographic applications. The most important absolute measure of spatial dispersion is standard distance, the spatial equivalent to standard deviation. Standard distance is discussed in section 4.2, in addition to relative distance, a measure

of relative spatial dispersion. Finally, selected locational issues related to the use of descriptive spatial statistics are examined in section 4.3.

4.1 SPATIAL MEASURES OF CENTRAL TENDENCY

Mean Center

The mean was discussed as an important measure of central tendency in the previous chapter. If this concept of central tendency is now extended to a set of points located on a Cartesian coordinate system, the average location, known as the **mean center**, can be determined.

Consider the scattergram shown in figure 4.1. These points might represent any spatial distribution of interest to geographers—the only stipulation is

TABLE 4.1

Nonspatial and Spatial Descriptive Statistics

Statistic	Central tendency	Absolute dispersion	Relative dispersion
Nonspatial	Mean	Standard deviation	Coefficient of variation
Spatial	Mean center or median center or Euclidean median	Standard distance	Relative distance

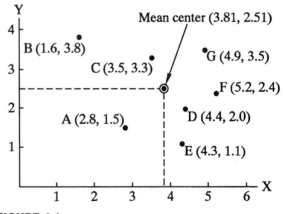

FIGURE 4.1

Graph of Locational Coordinates and Mean Center

that the phenomenon can be displayed graphically as a set of points in a two-dimensional coordinate system. The directional orientation of the coordinate axes and location of the origin are both arbitrary.

Given a coordinate system with the X and Y coordinates of each point determined, the mean center can be calculated by separately averaging those X and Y values, as follows:

$$\overline{X}_c = \frac{\Sigma\, X_i}{n} \text{ and } \overline{Y}_c = \frac{\Sigma\, Y_i}{n} \qquad (4.1)$$

where $\overline{X}_c$ = mean center of X

$\overline{Y}_c$ = mean center of Y

X_i = X coordinate of point i

Y_i = Y coordinate of point i

n = number of points in the distribution

For the point pattern shown in figure 4.1, the mean center coordinates are $\overline{X}_c$ = 3.81 and $\overline{Y}_c$ = 2.51 (table 4.2).

We saw that the nonspatial mean is strongly affected by an outlier or small number of extreme values (section 3.1); the mean center is influenced in a similar way. Suppose, for example, that one additional point with coordinates (15, 13) is included in the previous example (figure 4.2). The mean center location would shift dramatically from (3.81, 2.51) to (5.21, 3.82), the latter being a coordinate position having larger X and Y coordinates than any of the other seven points. Thus, while the mean center represents an average location, it may not represent a "typical" or "central" location.

The mean center may be considered the **center of gravity** of a point pattern or spatial distribution.

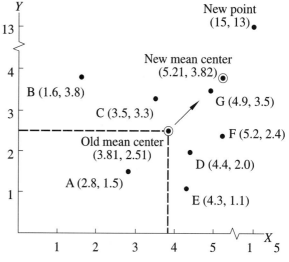

FIGURE 4.2
Affect of an Outlier on Mean Center Location

TABLE 4.2

Worktable for Calculating Mean Center

Point	Locational coordinates*	
	X_i	Y_i
A	2.8	1.5
B	1.6	3.8
C	3.5	3.3
D	4.4	2.0
E	4.3	1.1
F	5.2	2.4
G	4.9	3.5

$n = 7 \quad \Sigma\, X_i = 26.7 \qquad \Sigma\, Y_i = 17.6$

$$\overline{X}_c = \frac{\Sigma\, X_i}{n} = \frac{26.7}{7} = 3.81 \qquad \overline{Y}_c = \frac{\Sigma\, Y_i}{n} = \frac{17.6}{7} = 2.51$$

Mean center coordinates: (3.81, 2.51)*

*See figure 4.1 for graph of locational coordinates and mean center.

Perhaps the most widely known application of the mean center is the decennial calculation of the **geographic "center of population"** by the U.S. Bureau of the Census. This is the point where a rigid map of the country would balance if equal weights (each representing the location of one person) were situated on it. Over the last two centuries, the westward movement of the U.S. population has continued without significant interruption, as reflected in the concomitant westward shift of the center of population (figure 4.3).

In the previous seven-point example, each point is given an equal weight in the mean center calculation—that is, each point is equally important statistically. In many geographic applications, however, points of a spatial distribution should be assigned differential weights. These weights are analogous to frequencies in the calculation of grouped statistics, such as the weighted mean. The points might represent cities and the frequencies the number of people, or the points could be retail store locations and the frequencies could be volume of sales per store. The **weighted mean center** is defined as follows:

$$\overline{X}_{wc} = \frac{\Sigma\, f_i\, X_i}{\Sigma\, f_i} \text{ and } \overline{Y}_{wc} = \frac{\Sigma\, f_i\, Y_i}{\Sigma\, f_i} \qquad (4.2)$$

where $\overline{X}_{wc}$ = weighted mean center of X

$\overline{Y}_{wc}$ = weighted mean center of Y

f_i = frequency (weight) of point i

Each point in figure 4.1 is now assigned a weight (figure 4.4), and the locational coordinates of the weighted mean center are calculated (table 4.3).

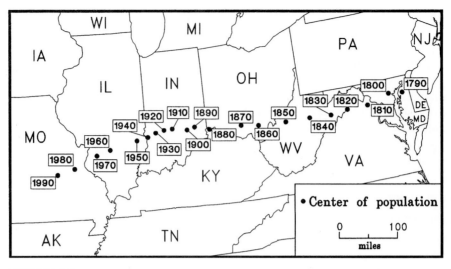

FIGURE 4.3
Geographic Center of U.S. Population, 1790–1990 (Source: Bureau of the Census, U.S. Dept. of Commerce.)

These locational coordinates are somewhat different from the coordinates of the comparable unweighted mean center. The weighted mean center is heavily affected by the relatively large frequency (20) associated with point B. Gravitation of the center toward a point with an unusually heavy weight will occur even if that point is located peripherally within the spatial distribution.

The mean and mean center share an important characteristic that has locational ramifications. Recall from the discussion on mean deviation (section 3.2) that the sum of squared deviations of all observations about a mean is zero. In addition, the sum of squared deviations of all observations about a mean is the minimum sum possible. That is, the sum of squared deviations about a mean is less than the sum of squared deviations about any other number:

$$\overline{X}: \min \Sigma (X_i - \overline{X})^2 \qquad (4.3)$$

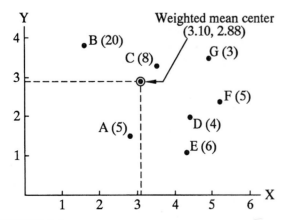

FIGURE 4.4
Graph of Point Locations, Frequencies (in Parentheses) and Weighted Mean Center

This important attribute is called the least-squares property of the mean.

The mean center is spatially analogous to the mean and has the same least-squares property as the mean. That is:

$$(\overline{X}_c \overline{Y}_c): \min \Sigma (X_i - \overline{X}_c)^2 + (Y_i - \overline{Y}_c)^2 \qquad (4.4)$$

In a Cartesian coordinate system based on location, deviations such as $(X_i - \overline{X}_c)$ and $(Y_i - \overline{Y}_c)$

TABLE 4.3

Worktable for Calculating Weighted Mean Center

Point	Locational coordinates*		Weight	Weighted coordinates	
	X_i	Y_i	f_i	$f_i X_i$	$f_i Y_i$
A	2.8	1.5	5	14.0	7.5
B	1.6	3.8	20	32.0	76.0
C	3.5	3.3	8	28.0	26.4
D	4.4	2.0	4	17.6	8.0
E	4.3	1.1	6	25.8	6.6
F	5.2	2.4	5	26.0	12.0
G	4.9	3.5	3	14.7	10.5

$n = 7 \qquad \Sigma f_i = 51 \qquad \Sigma f_i X_i = 158.1 \qquad \Sigma f_i Y_i = 147.0$

$$\overline{X}_{wc} = \frac{\Sigma f_i X_i}{\Sigma f_i} = \frac{158.1}{51} = 3.10$$

$$\overline{Y}_{wc} = \frac{\Sigma f_i Y_i}{\Sigma f_i} = \frac{147.0}{51} = 2.88$$

Weighted mean center coordinates: (3.10, 2.88)*

*See figure 4.4 for graph of point locations, frequencies, and weighted mean center.

are, in fact, *distances* between points. One standard procedure for measuring distances is based on straight-line or Euclidean distance. The **Euclidean distance** (d_i) separating point i (X_i, Y_i) from the mean center $(\overline{X}_c, \overline{Y}_c)$ is illustrated in figure 4.5 and defined by the Pythagorean theorem as follows:

$$d_i = \sqrt{(X_i - \overline{X}_c)^2 + (Y_i - \overline{Y}_c)^2} \qquad (4.5)$$

Thus, the mean center is the location that *minimizes the sum of squared distances* to all points. This characteristic makes the mean center an appropriate center of gravity for a two-dimensional point pattern, just as the mean is the center of gravity along a one-dimensional number line.

Euclidean Median

For many geographic applications, another measure of "center" is more useful than the mean center. Often, it is more practical to determine the central location that minimizes the sum of *unsquared*, rather than squared, distances. This location, which minimizes the sum of Euclidean distances from all other points in a spatial distribution to that central location, is called the **Euclidean median** (X_e, Y_e), or **median center.** Mathematically, this location minimizes the sum:

$$(X_e Y_e): \min \Sigma \sqrt{(X_i - X_e)^2 + (Y_i - Y_e)^2} \qquad (4.6)$$

Unfortunately, determining coordinates of the Euclidean median is methodologically complex. Computer-based iterative algorithms (step-by-step procedures) must be used to reach a solution. These algorithms evaluate a sequence of possible coordinates and gradually converge on the best location for the Euclidean median.

A weighted Euclidean median is a logical extension of the simple (unweighted) Euclidean median, and the same types of algorithmic procedures locate the weighted Euclidean median. The coordi-

nates of the weighted Euclidean median $(X_{we} Y_{we})$ will minimize the expression:

$$(X_{we} Y_{we}): \min \Sigma f_i \sqrt{(X_i - X_{we})^2 + (Y_i - Y_{we})^2} \qquad (4.7)$$

The weights or frequencies may represent population, sales volume, or any other feature appropriate to the spatial problem.

The location of the weighted Euclidean median is important to geographers for several practical reasons. For example, a classical problem in economic geography is the so-called Weber problem, which seeks to determine the "best" location for an industry. The optimal location minimizes the total cost of transporting the raw material to the factory and the finished product to the market. The weighted Euclidean median is the location that minimizes these transportation costs.

Perhaps the most extensively developed applications for the Euclidean median in geography are public and private facility location. Often an important goal in facility location is minimizing the average distance traveled per person to reach a designated or assigned facility. This efficiency-based objective is equivalent to minimizing the aggregate or total distance people must travel to use the service systemwide. The Euclidean median achieves this goal.

Consider, for example, the problem of locating the site for an urban fire station based on a predicted pattern of fires for the region. Using the past or present pattern of fires as a reasonable estimate of future fires, the optimal central location for the station could be defined as the site that minimizes the total (and hence, average) distance traveled by the fire equipment to reach fires. That location is determined by the Euclidean median.

In another application, suppose location analysts for an exclusive women's apparel chain wish to select an accessible site for a new store. Further, suppose that market analysis indicates that the demographic group most likely to shop in the store is women aged 45–65 who are members of households with incomes greater than $120,000. From the compilation of census tract information in the designated trade area, each tract could be weighted by the number of women having these age and income characteristics. The weighted Euclidean median will designate the location that minimizes the total (and average) distance traveled by these women to reach the potential store site.

Extending this procedure to the simultaneous location of multiple facilities within a spatial pattern of demand is known as the "location-allocation" problem or the "multiple facility location" problem. Suppose, for example, city health care planners wish to locate a set of neighborhood medical centers to

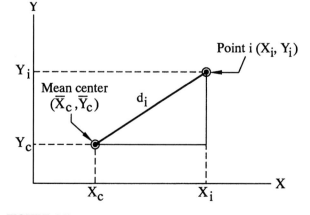

FIGURE 4.5
Calculation of Euclidean Distance (d_i) from Mean Center

provide selected types of remedial health care. Not only must a set of medical centers be located, but the potential clientele must be allocated to an appropriate facility, creating "catchment districts" or zones for each center. Problems such as these can be extremely complex and challenging, with both theoretical issues and practical applications receiving considerable attention from geographers.

In recent years, facility location strategies and modeling techniques have become much more sophisticated. One rapidly advancing research area concerns the application of Geographic Information Systems (GIS) to locational decision making, particularly as related to market analysis and the optimal siting of businesses. Several references at the end of this chapter are worth examining in this context.

4.2 SPATIAL MEASURES OF DISPERSION

Standard Distance

Just as the mean center serves as a locational analogue to the mean, **standard distance** is the spatial equivalent of standard deviation (table 4.1). Standard distance measures the amount of **absolute dispersion** in a point pattern. After the locational coordinates of the mean center have been determined, the standard distance statistic incorporates the straight-line or Euclidean distance of each point from the mean center. In its most basic form, standard distance (S_D) is written as follows:

$$S_D = \sqrt{\frac{\Sigma (X_i - \overline{X}_c)^2 + \Sigma (Y_i - \overline{Y}_c)^2}{n}} \qquad (4.8)$$

If equation 4.8 is modified algebraically, the number of required computations can be reduced considerably:

$$S_D = \sqrt{\left(\frac{\Sigma X_i^2}{n} - \overline{X}_c^2\right) + \left(\frac{\Sigma Y_i^2}{n} - \overline{Y}_c^2\right)} \qquad (4.9)$$

Using the same point pattern as in the earlier example (figure 4.1), standard distance is now calculated (table 4.4) and shown as the radius of a circle whose center is the mean center (figure 4.6).

Like standard deviation, standard distance is strongly influenced by extreme or peripheral locations. Because distances about the mean center are squared, "uncentered" or atypical points have a dominating impact on the magnitude of the standard distance.

Weighted standard distance is appropriate for those geographic applications requiring a weighted mean center. The definitional formula for weighted standard distance (S_{WD}) is

$$S_{WD} = \sqrt{\frac{\Sigma f_i (X_i - \overline{X}_{wc})^2 + \Sigma f_i (Y_i - \overline{Y}_{wc})^2}{\Sigma f_i}} \qquad (4.10)$$

TABLE 4.4

Worktable for Calculating Standard Distance

Point	Locational coordinates*			
	X_i	Y_i	X_i^2	Y_i^2
A	2.8	1.5	7.84	2.25
B	1.6	3.8	2.56	14.44
C	3.5	3.3	12.25	10.89
D	4.4	2.0	19.36	4.00
E	4.3	1.1	18.49	1.21
F	5.2	2.4	27.04	5.76
G	4.9	3.5	24.01	12.25

From earlier calculation of mean center:

$\overline{X}_c = 3.81 \quad \overline{Y}_c = 2.51 \quad \overline{X}_c^2 = 14.52 \quad \overline{Y}_c^2 = 6.30$

$n = 7 \qquad \Sigma X_i^2 = 111.50 \quad \Sigma Y_i^2 = 50.80$

$$S_D = \sqrt{\left(\frac{\Sigma X_i^2}{n} - \overline{X}_c^2\right)\left(\frac{\Sigma Y_i^2}{n} - \overline{Y}_c^2\right)}$$

$$= \sqrt{\left(\frac{111.55}{7} - 14.52\right)\left(\frac{50.80}{7} - 6.30\right)}$$

$$= 1.54$$

*See figure 4.6 for graph of locational coordinates, mean center, and standard distance.

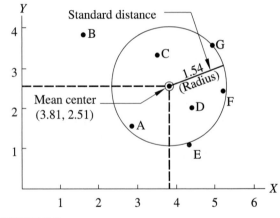

FIGURE 4.6
Graph of Point Locations, Mean Center, and Standard Distance

which may be rewritten in simpler form for computation as

$$S_{WD} = \sqrt{\left(\frac{\Sigma f_i (X_i^2)}{\Sigma f_i} - \overline{X}_{wc}^2\right) + \left(\frac{\Sigma f_i (Y_i^2)}{\Sigma f_i} - \overline{Y}_{wc}^2\right)}$$

$$(4.11)$$

A weighted standard distance can be computed using the same point pattern as before (table 4.5). A moderate disparity exists between the relative magnitudes of the unweighted and weighted standard

TABLE 4.5

Worktable for Calculating Weighted Standard Distance*

Point	f_i	X_i	X_i^2	$f_i(X_i)^2$	Y_i	Y_i^2	$f_i(Y_i)^2$
A	5	2.8	7.84	39.20	1.5	2.25	11.25
B	20	1.6	2.56	51.20	3.8	14.44	288.80
C	8	3.5	12.25	98.00	3.3	10.89	87.12
D	4	4.4	19.36	77.44	2.0	4.00	16.00
E	6	4.3	18.49	110.94	1.1	1.21	7.26
F	5	5.2	27.04	135.20	2.4	5.76	28.80
G	3	4.9	24.01	72.03	3.5	12.25	36.75

From earlier calculation of weighted mean center*:

$$\bar{X}_{wc} = 3.10 \quad \bar{Y}_{wc} = 2.88 \quad \bar{X}_{wc}^2 = 9.61 \quad \bar{Y}_{wc}^2 = 8.29$$

$$\Sigma f_i = 51 \quad \Sigma f_i(X_i)^2 = 584.01 \quad \Sigma f_i(Y_i)^2 = 475.98$$

$$S_{WD} = \sqrt{\left(\frac{\Sigma f_i(X_i)^2}{\Sigma f_i} - \bar{X}_{wc}^2\right) + \left(\frac{\Sigma f_i(Y_i)^2}{\Sigma f_i} - \bar{Y}_{wc}^2\right)}$$

$$= \sqrt{\left(\frac{584.01}{51} - 9.61\right) + \left(\frac{475.98}{51} - 8.29\right)}$$

$$= 1.70$$

*See figure 4.4 for graph of point locations and weighted mean center.

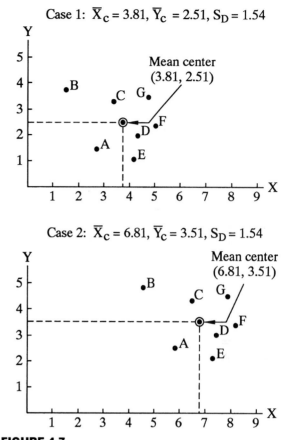

Case 1: $\bar{X}_C = 3.81$, $\bar{Y}_C = 2.51$, $S_D = 1.54$

Case 2: $\bar{X}_C = 6.81$, $\bar{Y}_C = 3.51$, $S_D = 1.54$

FIGURE 4.7
Arbitrary Placement of Coordinate Axes and Resultant Descriptive Spatial Statistics

distances (1.54 versus 1.70). This difference can be primarily explained by point B. Because this point is distant from the mean center and exerts a proportionally greater influence on the standard distance measure with its larger weight, point B causes the weighted standard distance to be larger than the unweighted standard distance.

Relative Distance

The coefficient of variation (standard deviation divided by the mean) is the nonspatial measure of **relative dispersion** (table 4.1). Unfortunately, a perfect spatial analogue to the coefficient of variation does not exist for measuring relative dispersion. Although it seems logical to divide the standard distance by the mean center to produce a relative dispersion index, this procedure will not provide meaningful results.

Consider a situation where the spatial statistics for the same point pattern are calculated twice using different positions for the coordinate system (figure 4.7). In case 1, the X coordinates of each point are three units lower and the Y coordinates one unit lower than in case 2. Notice, however, that the shift in the coordinate system does not affect standard distance. As a measure of absolute dispersion, standard

distance remains unchanged. Conversely, the coordinates of the mean center will change whenever the coordinate system is shifted. Because coordinate system location affects the mean center but not standard distance, a relative dispersion metric based on the ratio of these measures will be meaningless.

Despite these difficulties, some logical estimate of relative dispersion is necessary for spatial measurement. Consider the three point patterns in regions A, B, and C (figure 4.8, case 1). The distribution of points in each region has the same amount of absolute dispersion and the same standard distance. However, in small region A, the points have a high degree of relative dispersion, whereas they have a low relative dispersion in region C because the region is larger. The point patterns in regions D and E (figure 4.8, case 2) appear to have the same amount of relative dispersion. However, the point pattern in region D has a larger standard distance (absolute dispersion) than region E because of its larger size.

To derive a descriptive measure of relative spatial dispersion, the standard distance of a point pattern must be divided by some measure of regional magnitude. This divisor cannot be the mean center. One possible divisor is the radius (r_A) of a circle with the same area as the region being analyzed. A useful measure of relative dispersion, called **relative distance** (R_D), can now be defined:

$$R_D = \frac{S_D}{r_A} \qquad (4.12)$$

This relative distance measure allows direct comparison of the dispersion of different point patterns from different areas, even if the areas are of varying sizes.

To illustrate a simple application of relative distance, suppose now that the point pattern shown in figures 4.1 and 4.6 has two alternative locations in smaller region A or larger region B (figure 4.9). If region A were reshaped into a circle, it would have a radius of about 2, and region B would be a circle having a radius of about 5. Their respective relative distances would be quite different:

$$R_D \text{ (region A)} = \frac{S_D}{r_A} = \frac{1.54}{2} = 0.770 \qquad (4.13)$$

$$R_D \text{ (region B)} = \frac{S_D}{r_A} = \frac{1.54}{5} = 0.308 \qquad (4.14)$$

This measure of relative distance must be used with extreme caution. Geographers often have no control over the boundary and size of the study area. Study areas are often defined by political boundaries that have no logical relationship to the spatial pattern being investigated. A measure of relative dispersion based on the area of the study region clearly will be influenced by the location of the study area boundary.

Using a circle to represent the area of a region may not be particularly valid, especially if the shape of the region is highly irregular. An elongated region, for example, can have greater distances within its borders than a circular region of the same total area. Thus, the shape of a region can greatly affect the relative dispersion of a pattern.

4.3 LOCATIONAL ISSUES AND DESCRIPTIVE SPATIAL STATISTICS

Interpreting geostatistics must be done carefully. Because they are summary measures representing complex spatial patterns, results may sometimes be misleading or illogical. For example, the mean center or Euclidean median of high-income population in a

Case 1:
Same absolute dispersion; decreasing relative dispersion from region A to region C

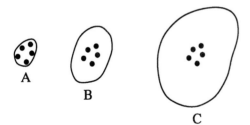

Case 2:
Same relative dispersion; decreasing absolute dispersion from region D to region E

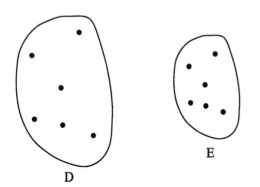

FIGURE 4.8
Comparisons of Absolute and Relative Point Pattern Dispersion

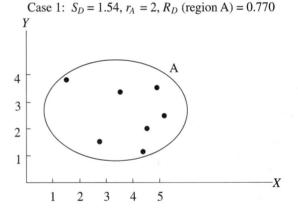

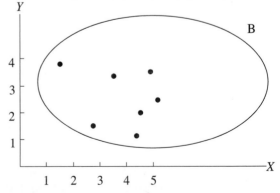

FIGURE 4.9
Effect of Study Area Size on Relative Distance

metropolitan area could be located in a low-income central city neighborhood that contains few high-income families.

The magnitude or location of each measure is influenced by all points in the distribution. A single anomalous point can severely modify the resultant descriptive spatial measures, making interpretation difficult. Geographers should view geostatistics as general indicators of location, rather than as precise measurement instruments to be used in isolation.

The descriptive analysis of point patterns could benefit from consideration of other possible pattern characteristics. In fact, a spatial or locational measure analogous to each nonspatial descriptive statistic can be created. Spatial statistics equivalent to skewness and kurtosis have potential application in geography. For example, knowing which patterns are more symmetric or skewed might offer spatial insights when comparing point patterns in geographic research. Geographers could also find it valuable to compare directly the degree of clustering and dispersal in different point patterns through measuring spatial kurtosis levels.

In earlier discussions about mean center, Euclidean median, and standard distance, reference was made only to straight-line or Euclidean distance based on the Pythagorean theorem. In certain spatial contexts, this may not be the most appropriate way to proceed. For example, in an urban geography study of various commercial activities (e.g., grocery stores, drug stores, etc.), neither the mean center nor the Euclidean median may be the most suitable measure of "center." In those situations where travel is confined to a grid or rectilinear street pattern, a straight-line or Euclidean **distance metric** underrepresents the actual travel distances and could result in a misplaced center and inaccurately low measure of absolute dispersion.

A general distance formula is available to handle various geographic situations where "friction of distance" or other influences preclude simple straight-line measure. Recall that the formula measuring the Euclidean distance (d_{ij}) of point i (X_i, Y_i) from point j (X_j, Y_j) is

$$d_{ij} = \sqrt{(X_i - X_j)^2 + (Y_i - Y_j)^2} \qquad (4.15)$$

If this Euclidean distance metric is now generalized to allow non-Euclidean distance measurement, the result is

$$d_{ij} = ((X_i - X_j)^k + (Y_i - Y_j)^k)^{1/k} \qquad (4.16)$$

Consider k the general distance metric, and when k equals two, the formula is conventional Euclidean distance. When k equals one, however, the formula measures distances when movement is restricted to

a rectangular or grid system. The term **Manhattan distance** describes the restrictive movement typical of travel in the New York City borough of Manhattan. Measuring the distance between points i and j in Manhattan space, where k equals one, gives

$$d_{ij} = |X_i - X_j| + |Y_i - Y_j| \qquad (4.17)$$

The "center" point in Manhattan space is the **Manhattan median**, which minimizes the sum of absolute deviations between itself and all other points in the pattern. Absolute values are used to ensure that all distances are positive. Formally, the Manhattan median (X_m, Y_m) is that location that minimizes the sum:

$$(X_m Y_m): \min \Sigma |X_i - X_m| + |Y_i - Y_m| \qquad (4.18)$$

Unfortunately, a single unique location for the Manhattan median cannot be found in a spatial pattern having an even number of points. That is, no single "middle" point exists when the number of points is even. However, a unique set of coordinates can always be determined for the Manhattan median with an odd number of points. This situation parallels the calculation of the "nonspatial" median (section 3.1), since both statistical procedures differ in problems having an even or odd number of points. A further difficulty with the Manhattan median is that its location will change if the coordinate axes are shifted.

Depending on the geographic context, the distance metric k could assume many values (figure 4.10). In case 1, a straight line is traveled, and the Euclidean distance metric ($k = 2$) is used. Case 2 illustrates Manhattan space ($k = 1$), where distances are measured along a gridlike street pattern. In other geographic situations, distances might best be measured with a metric value somewhere between Euclidean and Manhattan. For example, in case 3, an intermediate-value distance metric ($k = 1.5$) best estimates the distance separating points i and j. In case 4, travel from point i to j must be around the intervening lake. Since a circuitous route must be traveled, a k value less than one is needed ($k = 0.6$).

KEY TERMS

absolute dispersion, 56
center of gravity, 53
distance metric, 59
Euclidean distance, 55
Euclidean median (median center), 55
geographic "center of population", 53
geostatistics, 52
Manhattan distance and median, 59
mean center (unweighted and weighted), 53

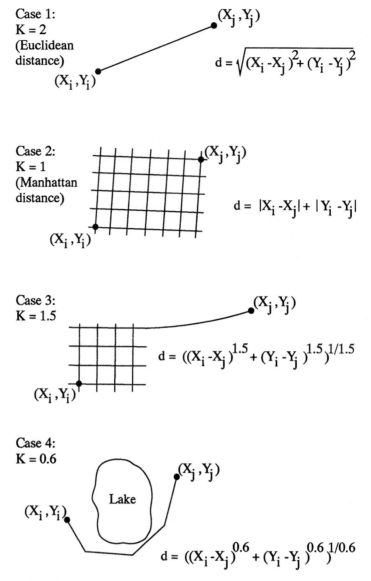

FIGURE 4.10
Distance Metrics under Various Spatial Conditions

MAJOR GOALS AND OBJECTIVES

If you have mastered the material in this chapter, you should now be able to do the following.

1. Understand the concept of central tendency as applied in a spatial context and explain the distinctive characteristics of the mean center and Euclidean median.

2. Define the spatial measures of dispersion (standard distance and relative distance) and recognize their potential applications in geographic problem solving.

3. Recognize when a point pattern in a study region has a large or small absolute dispersion and a large or small relative dispersion.

4. Identify the potential limitations and locational issues associated with the application of descriptive spatial statistics.

5. Understand the concept of distance metric and distinguish between Euclidean distance, Manhattan distance, and other distance metrics under various spatial conditions.

REFERENCES AND ADDITIONAL READING

Abler, R. F., J. S. Adams, and P. R. Gould. 1971. *Spatial Organization: The Geographer's View of the World.* Englewood Cliffs, NJ: Prentice-Hall.

Birkin, M. G., G. Clarke, M. Clarke, and A. Wilson. 1996. *Intelligent GIS, Location Decisions and Strategies Planning.* New York: John Wiley and Sons.

Davies, R. and D. Rogers. 1984. *Store Location and Store Assessment Research.* New York: John Wiley and Sons.

Drezner, Z. (editor). 1996. *Facility Location: A Survey of Applications and Methods.* Heidelberg: Springer-Verlag.

Ebdon, D. 1985. *Statistics in Geography: A Practical Approach* (2nd edition). Oxford: Basil Blackwell.

Kuhn, H. W. and R. E. Kuenne. 1962. "An Efficient Algorithm for the Numerical Solution of the Generalized Weber Problem in Spatial Economics." *Journal of Regional Science* 4:21–33.

Menecke, E. 1997. "Understanding the Role of Geographic Information Technologies in Business: Application and Research Directions." *Journal of Geographic Information and Decision Analysis* 1: 44–68.

Neft, D. S. 1966. *Statistical Analysis for Areal Distributions* (Monograph Series No. 2). Philadelphia: Regional Science Research Institute.

Scott, A. J. 1971. *Introduction to Spatial Allocation Analysis* (Resource Paper No. 9). Washington, DC: Assoc. of Amer. Geographers.

Taylor, P. J. 1977. *Quantitative Methods in Geography: An Introduction to Spatial Analysis.* Boston: Houghton, Mifflin.

THE TRANSITION TO INFERENTIAL PROBLEM SOLVING

Probability

Studying spatial patterns found on the physical and cultural landscape is a central concern of geographers. They seek to develop descriptions and explanations of existing patterns and to understand the processes that create these distributions. In some cases, they attempt to predict future occurrences of geographic patterns. In short, the core of geographic problem solving is the description, explanation, and prediction of geographic patterns and processes.

In earlier chapters, the focus was on ways to describe or summarize spatial data. Much of the remainder of the book involves methods used for exploring relationships between spatial patterns and for understanding the nature or characteristics of processes that create these patterns. The concept of probability occupies a central position in the chapters that follow.

Chapter 5 provides an introduction to probability and the important role it plays in geographic problem solving. The first section discusses the nature of geographic patterns and the processes that produce them. Included here are examples of deterministic and probabilistic processes in geography. Section 5.2 introduces basic terms and concepts of probability, including definitions and rules for using probability in geographic analysis. Sections 5.3 through 5.5 cover specific probability distributions important in geographic problem solving. The binomial distribution, discussed in section 5.3, concerns probability of events having only two possible outcomes and is particularly useful for studying multiple events or trials. Section 5.4 discusses the Poisson distribution, which is used to analyze patterns that are random over time or space. The most important probability distribution to geographers is the normal distribution, introduced in section 5.5. The concluding section of chapter 5 presents a probability mapping technique, which extends the procedure of normal probabilities into a spatial context.

5.1 DETERMINISTIC AND PROBABILISTIC PROCESSES IN GEOGRAPHY

Real-world processes that produce physical or cultural patterns on the landscape are often complex and are not usually totally identifiable. Two general categories can be used to describe the nature of these processes—deterministic and probabilistic. **Deterministic processes** create patterns with total certainty, since the outcome can be exactly specified with 100 percent likelihood. Because of uncertainty in human behavior and decision making, virtually no cultural processes are completely deterministic, but some physical processes established through scientific principles fall into this category. For example, the length of time insolation (solar radiation) strikes a point on the earth's surface is determined by both latitude and day of the year. Given these two components, geographers can determine the exact hours of daylight and darkness at any location.

The second category, **probabilistic processes,** concerns all situations that cannot be determined with complete certainty. Given the complex character of physical and cultural patterns and processes, most geographic situations fall into this category. For example, the number of hours of sunlight is considered deterministic, although the amount of sunlight reaching the ground is probabilistic, since cloud cover and particulate matter absorb and reflect solar energy as it passes through the atmosphere.

Probabilistic processes can be subdivided into two useful categories. With **random processes,** the same probability must be assigned to all outcomes, because each outcome has an equal chance of occurring. Drawing a card from a deck, rolling a die, or tossing a coin are all examples of random processes. Such situations may be considered as examples of maximum uncertainty. With **stochastic processes,** the likelihood or probability of any particular outcome can be specified, and not all outcomes are equal. For example, the probability of a hurricane hitting the coast of Florida is higher during September than it is during March.

Forecasting future use for a currently undeveloped parcel of land is one way to illustrate the nature of stochastic processes in geography. Land-use selection entails numerous complex factors, such as the monetary value of the parcel, its accessibility to regional activities, government restrictions (like zoning), and the physical characteristics of the land (e.g., slope, drainage, and soil type). Knowledge of these factors can be used to estimate the probability that a land parcel will have a given use. For example, a particular parcel of land could be developed as an office park, an industrial park, a shopping center, or a residential subdivision. On the other hand, it could simply be left undeveloped. Because the land development process is complex, future land use cannot be specified with certainty. The most that can be done in this situation is to specify the likelihood or probability of each potential land use.

Some spatial patterns are totally unpredictable. For example, the location of tornado touchdown points within a region is the result of random meteorological processes. Sometimes the size of the study area affects the degree of randomness. For example, the number of tornadoes next year within Kansas can be estimated as a stochastic probability. However, if the focus is narrowed to a small area within Kansas, the exact location of a particular tornado touchdown point is random.

In summary, most geographic patterns result from processes that are stochastic in nature. Few patterns studied by geographers occur as a result of either totally deterministic or totally random processes. In addition to examining spatial distributions, geographers also study temporal patterns that can be described as occurring from either deterministic or probabilistic processes. In such investigations, use of probability again plays an integral role, since most temporal patterns are neither completely deterministic nor totally random.

5.2 BASIC PROBABILITY TERMS AND CONCEPTS

Since most spatial and temporal patterns are produced by processes that have some degree of uncertainty, geographers need to understand and use probability for solving problems. For example, every location on the earth's surface receives a variable amount of precipitation. These data can be recorded over time and across space, and precipitation patterns can be summarized using such calculations as the mean and standard deviation. However, since precipitation results from complex atmospheric processes, its prediction can only be stated in terms of probabilities, not exact certainty. Therefore, geographers make statements such as "50 percent of the time snowfall in January exceeds 20 inches" or "9 years out of 10, at least 5 inches of rain fall in June." Although probabilities can be stated, the exact amount of snowfall next January or the exact amount of rainfall next June cannot be determined.

The study of probability focuses on the occurrence of an **event,** which can usually result in one of several possible **outcomes.** Once all possible outcomes have been considered for an event, probability represents the likelihood of a given result or the chance that any outcome actually takes place. Suppose a simple **random experiment** is conducted by rolling a single standard die having six faces—1, 2, 3, 4, 5, and 6. What is the likelihood of rolling a 6 with a single toss of the die? In this experiment, the event is the roll of the die, and the possible outcomes are the 6 sides of the die. In effect, rolling the die is taking a sample observation from the infinitely large population of all rolls of that die. The die could be rolled 20 times to collect a sample of size 20. If the die is not imbalanced or "loaded," each outcome or face of the die has an equal chance of occurring, and the likelihood of rolling a 6 is one out of six, or .167. The same probability also exists that a 1, 2, 3, 4, or 5 will be thrown.

Collecting sample data of outcomes from such experiments (when the sample is only a portion of the population) and then calculating the probabilities of different outcomes is the basis for statistical inference. An understanding of probability concepts is therefore essential if inferences are to be made about statistical populations. In fact, much of the remainder of this book provides a transition to inferential problem solving (Part III) or deals directly with inferential problem solving in geography (Parts IV and V).

Probability can be thought of as **relative frequency**—the ratio between the absolute frequency for a particular outcome and the frequency of all outcomes:

$$P(A) = F(A)/F(E) \qquad (5.1)$$

where $P(A)$ = probability of outcome A occurring
$F(A)$ = absolute frequency of outcome A
$F(E)$ = absolute frequency of all outcomes for event E

Probabilities can also be interpreted as percentages when the denominator of equation 5.1 is converted to 100. In the example of the die roll, the probability of a 6 (.167) can also be described as an outcome that has a 16.7 percent likelihood of occurring.

Some of the basic terms and concepts of probability can be observed through a simple geographic example. Suppose a day is classified as wet (W) if measurable precipitation (defined as at least 0.01 of an inch) falls during a 24-hour period, while the day is termed dry (D) if measurable precipitation does not occur. This is a valid classification, with clearly defined operational definitions of wet and dry. The two categories are **mutually exclusive,** since it is not possible to assign a particular day to more than one category because the categories of wet and dry do not overlap. For example, by keeping a record of wet and dry days over a 100-day period, absolute frequencies of precipitation can be determined and relative probabilities calculated from the data (figure 5.1). In this example, 62 days are categorized as dry and 38 as wet. The probability of a wet day occurring $P(W)$ is

$$P(W) = \frac{\text{number of wet days}}{\text{total days}} = \frac{38}{100} = .38 \qquad (5.2)$$

Similarly, the probability of a dry day occurring $P(D)$ is

$$P(D) = \frac{\text{number of dry days}}{\text{total days}} = \frac{62}{100} = .62 \qquad (5.3)$$

Thus, a 38 percent chance exists that a day will have measurable precipitation, and a 62 percent chance exists that a day will be dry.

Several **rules of probability** guide the decision maker. The maximum probability for any outcome is 1.0, indicating total certainty or perfect likelihood of a particular occurrence. The lowest probability for any outcome is 0.0, suggesting no chance of occurrence. Most outcomes have probabilities falling be-

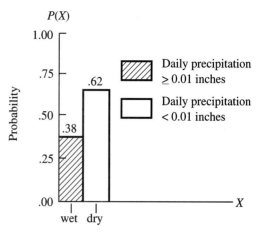

FIGURE 5.1
Relative Frequency of Wet and Dry Days from a 100-Day Period

tween these maximum and minimum values. Thus, for any outcome (A),

$$0.0 \le P(A) \le 1.0 \qquad (5.4)$$

Another straightforward definition concerns the **complement** of an outcome. If (A) contains all those events in which a particular outcome occurs, then the complement of (A), (noted $\overline{A}$), includes all those events where that particular outcome does not occur. The sum of the probabilities of an outcome and its complement is 1.0:

$$P(A) + P(\overline{A}) = 1.0 \qquad (5.5)$$

In the precipitation example, the probability of a wet day is .38, and the probability of a dry day is .62. Since these are complementary outcomes and they are mutually exclusive, their probabilities must total 1.0:

$$P(\overline{W}) = P(D) \qquad (5.6)$$

so $P(W) + P(\overline{W}) = P(W) + P(D) = 1.0 \qquad (5.7)$

Return to the example of the die for a moment. With a single toss of the die, it is known that the probability of rolling a 6 is 1/6, or .167. What is the probability of tossing the die twice in succession and getting a 6 both times? These two events, the first roll of the die and the second roll of the die, are **statistically independent** events. Statistical independence of events exists when the probability of one event occurring is not influenced or affected by whether another event has occurred. The chance of rolling a 6 on the second roll is not at all influenced by what happened on the first roll. Many of the inferential statistics discussed later in the book are based on this property of statistical independence. This means that many of the sampling methods developed in the following chapters must be designed to incorporate this assumption of independence.

If events are independent, the simple *multiplication rule* of probability can be applied. For independent events A and B, this rule states

$$P(A \text{ and } B) = P(A) \cdot P(B) \qquad (5.8)$$

Therefore, if event A is the first roll of the die, and event B is the second roll of the die,

$$P(A \text{ and } B) = P(A) \cdot P(B) = 1/6 \cdot 1/6 = 1/36 \quad (5.9)$$

That is,

$$P(A \text{ and } B) = P(A) \cdot P(B) = .167 \cdot .167 = .02778 \quad (5.10)$$

There is 1 chance in 36 (somewhat less than a 3 percent chance) of tossing a die twice in succession and getting a 6 with both tosses.

The multiplication rule of probability can be extended using similar logic. For example, if a die is rolled three times, the likelihood of rolling a 6 on the first roll (event A), another 6 on the second roll (event B), and yet another 6 on the third roll (event C) is

$$P(A \text{ and } B \text{ and } C) = P(A) \cdot P(B) \cdot P(C)$$
$$= 1/6 \cdot 1/6 \cdot 1/6 = 1/216 \qquad (5.11)$$

That is,

$$P(A \text{ and } B \text{ and } C) = P(A) \cdot P(B) \cdot P(C)$$
$$= .167 \cdot .167 \cdot .167 = .00463 \qquad (5.12)$$

The chance of tossing three 6s in succession on a die is very low (less than one-half of one percent.)

Consider how these basic probability rules can be applied to 1992 U.S. immigration data (table 5.1). These data show the total number of immigrants by census region destination, as well as the number of immigrants from selected countries of birth settling in each census region. Some categories in the table are mutually exclusive. For example, some immigrants initially settled in the Northeast, whereas others first settled in the South. These are mutually exclusive (nonoverlapping) categories because no single immigrant can initially settle in more than one census region. Similarly, no single immigrant can have more than one country of birth. Within certain portions of the table, however, the categories are not mutually exclusive. For example, a single immigrant could certainly be from Mexico *and* move to the Northeast and thus be included in both of these (nonmutually exclusive) categories. In fact, 2,878 such individuals existed.

TABLE 5.1

Number of Immigrants from Selected Countries to the United States Census Regions, 1992

Census region destination	Country of birth (origin)			
	Total	Mexico	Vietnam	Philippines
Northeast	252,187	2,878	9,254	11,088
Midwest	94,144	12,650	7,387	4,704
South	212,106	45,662	19,523	8,263
West	404,205	152,191	41,499	34,871
U.S. Total*	973,977	213,802	77,735	61,022

*Includes additional destinations: Guam, Puerto Rico, Northern Mariana Island, Virgin Islands, and other or unknown areas not shown separately.
Source: U.S. Immigration and Naturalization Service, *Statistical Yearbook,* annual.

Suppose the task is to calculate the probability of selecting at random from a list of all 973,977 immigrants an immigrant from Vietnam *or* an immigrant from the Philippines. The concept of randomness suggests that each individual on this very lengthy list has an equal chance of being selected, and the selection is not biased in any way. Since these occurrences do not overlap, the *addition rule for mutually exclusive events* can be applied. In general form, this rule is as follows:

$$P(A \text{ or } B) = P(A) + P(B) \qquad (5.13)$$

where $P(A \text{ or } B)$ = probability of event A or event B
$P(A)$ = probability of event A
$P(B)$ = probability of event B

If event A is the selection of an immigrant from Vietnam and event B is the selection of an immigrant from the Philippines, then the probability of selecting an immigrant from one or the other of these countries is the following:

If $\qquad P(A) = 77{,}735/973{,}977 = .0798 \qquad (5.14)$

and $\qquad P(B) = 61{,}022/973{,}977 = .0627 \qquad (5.15)$

then $\qquad P(A \text{ or } B) = .0798 + .0627 = .1425 \qquad (5.16)$

Using similar logic for more than two events, the addition rule for mutually exclusive events can be generalized. For example, with three mutually exclusive events—A, B, and C—the addition rule becomes

$$P(A \text{ or } B \text{ or } C) = P(A) + P(B) + P(C) \quad (5.17)$$

Identify event A as the selection of an immigrant from Vietnam, event B as the selection of an immigrant from the Philippines, and event C as the selection of an immigrant from Mexico. The probability of selecting an immigrant from any one of these three countries is their sum, according to the addition rule:

If
$$P(A) = 77,735/973,977 = .0798 \quad (5.18)$$

and
$$P(B) = 61,022/973,977 = .0627 \quad (5.19)$$

and
$$P(C) = 213,802/973,977 = .2195 \quad (5.20)$$

then

$$P(A \text{ or } B \text{ or } C) = P(A) + P(B) + P(C) \quad (5.21)$$
$$= .0798 + .0627 + .2195 = .3620$$

When randomly selecting a single immigrant from all immigrants in 1992, the probability of choosing a person from Vietnam, the Philippines, or Mexico is .3620 (i.e., a 36.20 percent chance).

Suppose the task is to determine the probability of selecting an immigrant from Vietnam *or* an immigrant settling in the West, *or both*. These are not mutually exclusive events, as many Vietnamese immigrants certainly settled in the West in 1992. In fact, according to table 5.1, that number is known to be 41,499. For this situation, an *addition rule for nonmutually exclusive events* must be applied:

$$P(A \text{ or } B) = P(A) + (B) - P(A \text{ and } B) \quad (5.22)$$

If event A is an immigrant from Vietnam, and event B is an immigrant settling in the West, then the following is known:

If
$$P(A) = 77,735/973,977 = .0798 \quad (5.23)$$

and
$$P(B) = 404,205/973,977 = .4150 \quad (5.24)$$

and
$$P(A \text{ and } B) = 41,499/973,977 = .0426 \quad (5.25)$$

then
$$P(A \text{ or } B) = P(A) + P(B) - P(A \text{ and } B) \quad (5.26)$$
$$= .0798 + .4150 - .0426 = .4522$$

Note that the 41,499 Vietnamese settling in the West have been counted twice: those immigrants are included in the numerator of both $P(A)$ and $P(B)$. Therefore, this outcome, $P(A \text{ and } B)$, needs to be subtracted to avoid double counting or duplication. Thus, if a single immigrant is randomly selected from a list of all 973,977 immigrants to the United States in 1992, there is a .4522 probability (45.22 percent chance) of selecting an immigrant from Vietnam, or an immigrant settling in the West, or both.

Many other rules of probability exist for more complex applications. However, because this discussion of probability is only introductory, readers should examine the sources listed at the end of the chapter for more thorough treatments of the subject.

Probabilities from Known Mathematical Distributions

The probability of outcomes in certain problems follows consistent or typical patterns. Such patterns, called **probability distributions,** relate closely to frequency distributions discussed in section 2.5. In a frequency distribution, the frequency of occurrences appears on the vertical axis, and in a probability distribution, the probability of occurrence is displayed on the vertical axis. In both cases, the horizontal axis shows the actual outcomes, occurrences, or values of the variable being studied.

Recall that variables are termed discrete or continuous depending on whether the values occur as distinct whole numbers (discrete) or decimal values (continuous). Probability distributions for discrete outcomes are termed **discrete probability distributions,** whereas those for outcomes that can occur at an infinite number of points are termed **continuous probability distributions.** Although numerous examples of both types of distributions are used in geography, this discussion focuses on three commonly used distributions: binomial, Poisson, and normal.

5.3 THE BINOMIAL DISTRIBUTION

The **binomial** is a discrete probability distribution associated with events having only two possible outcomes. Binary outcomes are often described as zero–one, yes–no, or presence–absence problems, and many geographic situations fit a binary framework. Consider these examples: A location either has measurable precipitation over a 24-hour period or it does not; a person is either employed or unemployed; a river is either above or below flood stage; a respondent to an opinion poll either favors or opposes an issue. In each of these situations, only one of two outcomes is possible, assuming that an undecided or uncertain result is not possible.

The binomial distribution is especially useful in examining probabilities from multiple events or trials. Geographers are often interested in evaluating temporal sequences, phenomena that either occur or don't occur at some location through time. In many of these situations, the probability of a single event can usually be determined quite easily, perhaps using historical data. For example, the probability of a river in Bangladesh reaching flood stage during a given year may be .40. Thus, on average, flooding occurs four years out of ten. With this information, the binomial distribution may be used to determine the probability of flooding over other time periods (perhaps 15 years out of 30). In this example, the assumption is that a river reaching flood stage during a given year is an event independent of whether the river reached flood stage in other years.

When using binomial probabilities, the focus is on one of the two possible outcomes, termed the *given* outcome (X). The binomial distribution is shown mathematically as follows:

$$P(X) = \frac{n! \, p^X \, q^{n-X}}{X!(n-X)!} \qquad (5.27)$$

where n = number of events or trials

p = probability of the *given* outcome in a single trial

q = $\bar{p}$ = $1 - p$ or the probability of the *other* outcome in a single trial

X = number of times the *given* outcome occurs within the n trials

$n!$ = n **factorial**:

If $n > 0$, $n! = [n(n-1)(n-2) \ldots (2)(1)]$

If $n = 0$, $n! = 1$

For example, $5! = (5)(4)(3)(2)(1) = 120$

The best way to illustrate the characteristics and practical uses of the binomial probability is with a geographic example. Suppose a vegetable grower is seeking a new location to start a business. One of the key variables in site selection is the probability of adequate precipitation that will reduce the necessity of expensive irrigation. The grower has determined that at least three inches of precipitation are needed during the growing season to avoid irrigation and that irrigation can be afforded only one year in five to make a profit. Precipitation data collected for a potential site show that in 21 of the last 25 years rainfall exceeded three inches during the growing season. The historical record therefore suggests that the probability of a given year requiring supplemental irrigation is 4/25, or 0.16. What is the probability that the grower can meet the requirement of having to irrigate only one year in five?

Over the specified five-year period, the farmer faces six possible outcomes: from none of the five years requiring irrigation up to all five years requiring irrigation. However, only two of the six outcomes—no years and one year—would result in a profitable situation for the grower; each of the other four outcomes would be too costly. Thus, the probability that the grower will meet the profit requirement is found by summing the probabilities for the two suitable outcomes. The critical values and calculations of the binomial probabilities for all possible outcomes are shown in table 5.2. The binomial probabilities for the problem are listed in table 5.3 and plotted graphically in figure 5.2.

TABLE 5.2

Critical Values and Binomial Probabilities of Suitable Outcomes for Vegetable Grower

$$P(X) = \frac{n! \, p^x q^{n-x}}{X!(n-X)!}$$

Critical values:

X = 0, 1, 2, 3, 4, 5
n = 5 years
p = 4/25 = .16
q = $(1-p)$ = .84
$n!$ = (5) (4) (3) (2) (1) = 120

When $X = 0$: $P(0) = \dfrac{5! \, (.16)^0 (.84)^5}{0! \, (5)!} = \dfrac{120(1)(.418)}{1(120)} = .418$

When $X = 1$: $P(1) = \dfrac{5! \, (.16)^1 (.84)^4}{1! \, (4)!} = \dfrac{120(.16)(.498)}{1(24)} = .398$

When $X = 2$: $P(2) = \dfrac{5! \, (.16)^2 (.84)^3}{2! \, (3)!} = \dfrac{120(.026)(.593)}{2(6)} = .152$

When $X = 3$: $P(3) = \dfrac{5! \, (.16)^3 (.84)^2}{3! \, (2)!} = \dfrac{120(.004)(.706)}{6(2)} = .029$

When $X = 4$: $P(4) = \dfrac{5! \, (.16)^4 (.84)^1}{4! \, (1)!} = \dfrac{120(.001)(.84)}{24(1)} = .003$

When $X = 5$: $P(5) = \dfrac{5! \, (.16)^5 (.84)^0}{5! \, (0)!} = \dfrac{120(.000)(1)}{120(1)} = .000$

TABLE 5.3

Binomial Probabilities of Vegetable Grower Needing Irrigation Over a Five-Year Period

Number of years out of five	Binomial probability	Type of outcome
0	.418	Suitable
1	.398	Suitable
2	.152	Unsuitable
3	.029	Unsuitable
4	.003	Unsuitable
5	.000	Unsuitable
Total	1.000	

Total probability of suitable outcomes = .816
Total probability of unsuitable outcomes = .184

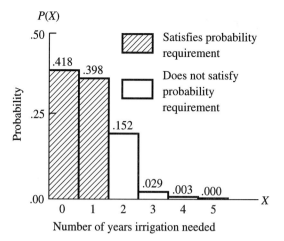

FIGURE 5.2
Probability of Vegetable Grower Needing Irrigation for 0 to 5
Years

The vegetable grower would have a .418 probability of needing no irrigation and a .398 chance of needing one year of irrigation during the next five years. Adding the binomial probabilities for no years and one year of irrigation (.418 + .398) shows that the grower has an 81.6 percent chance of meeting the profitability requirement at this potential site during the five-year interval. This level of risk may or may not be acceptable to the grower.

To summarize, the binomial distribution is applicable for those geographic problems in which the following conditions are met:

- The objective is to determine the probability of multiple (*n*) independent events or trials.
- Each event or trial has two possible outcomes, one termed the *given* outcome (*X*), with associated probability *p*, and a complementary or *other* outcome having probability $q = \bar{p} = 1 - p$.
- The probabilities *p* and *q* must remain stable or consistent over the duration of the study period or over successive trials through time.

5.4 THE POISSON DISTRIBUTION

Some probability problems in geography involve the study of events that occur repeatedly and randomly over either time or space. For example, the placement of calls to an emergency response dispatcher might be considered random over a short period of time. At certain spatial levels, multiple occurrences of weather-related phenomena, such as thunderstorms, tornadoes, and hurricanes, may occur with little spatial predictability. Geographers also study various cultural entities, some of whose patterns may be the result of random processes. In instances where events occur repeatedly and at random (i.e.,

independent of past or future occurrences), the **Poisson probability distribution** can be used to analyze how frequently an outcome occurs during a certain time period or across a particular area. Other geographic applications of Poisson involve the analysis of existing frequency count data to determine if a random distribution exists.

Suppose, for example, farmers in the prairie provinces of Canada are concerned about the frequency of the devastating effects of hailstorms on their summer wheat yields. Because of the uncertain nature of such weather events, hailstorm occurrence varies greatly from year to year and from province to province. Farmers may want to know the probability of experiencing two hailstorms per year. Such information can be useful in deciding whether to invest in insurance against hail damage. The current occurrence of hailstorms is independent of past or future occurrences and can be considered random. Therefore, Poisson is the valid probability model for describing hailstorm frequency over time across the prairie provinces of Canada.

The use of Poisson in a temporal situation can be illustrated with a small set of hypothetical data showing hailstorm frequency over a 35-year period in a small portion of Manitoba (table 5.4). During these 35 years, 10 years were completely free of hailstorms, and four hailstorms occurred in one year. The most common frequency (mode) was one hailstorm per year, which occurred 13 of the 35 years. Since a total of 42 hailstorms occurred during the period, the mean number of hailstorms per year was 1.2.

The hailstorm probability can be calculated for any frequency of occurrence. For example, since the data cover 35 years and 10 of these years did not have a hailstorm, the resultant probability of no hailstorm occurring in a year is 10/35, or .286. Thus, given this set of data, this portion of Manitoba has a

---

TABLE 5.4

Observed Hailstorm Occurrence over a 35-Year Period for a Region in Manitoba, Canada

Number of hailstorms per year	Observed frequency of years	Total hailstorms	Observed probability of occurrence
0	10	0	.286
1	13	13	.371
2	8	16	.229
3	3	9	.086
4	1	4	.029
5+	0	0	.000
Total	35	42	1.001*

*Observed probabilities do not total to 1.0 due to rounding.

28.6 percent chance of avoiding a hailstorm in any particular year.

If the process producing the pattern is truly random, the probabilities of occurrence will follow the Poisson distribution. The mean frequency is the average number of hailstorms per time period or geographic area. Knowing only this mean value is sufficient to allow all Poisson probabilities to be calculated:

$$P(X) = \frac{e^{-\lambda}(\lambda^X)}{X!} = \frac{\lambda^X}{e^\lambda(X!)} \qquad (5.28)$$

where X = frequency of occurrence
λ = mean frequency
e = mathematical (exponential) constant (approximately 2.72)
$X!$ = X factorial

Like the binomial distribution, Poisson requires discrete or integer outcomes (X) to represent the number or frequency of occurrences. Unlike the binomial, however, the Poisson distribution is not binary. In the hailstorm example, discrete outcomes represent the different number (frequency counts) of hailstorms occurring annually. The mean outcome or average number of hailstorms per year (λ) is 1.2. Since both e and λ are constants, the expression e^λ in equation 5.28 is also a fixed value ($e^\lambda = 2.72^{1.2} = 3.32$). The exponential function on a calculator or computer is needed for this calculation. Using this constant e^λ value, Poisson probabilities are calculated for a series of hailstorm frequencies ($X = 0, 1, 2, 3, 4$) and presented in table 5.5.

If hailstorms occur randomly with a mean of 1.2 per year, the likelihood of no storm during a given year is .301 (table 5.5). In other words, extending this logic to a period of 100 years, about 30 of those years should be free of hailstorms. The most likely occurrence is one hailstorm per year, which should occur at a probability of .361, or about 36 years out of 100 (figure 5.3).

Geographers often want to compare the frequencies obtained from a set of observed data to frequencies generated by the Poisson or random distribution. The expected frequencies are determined by multiplying the Poisson probabilities by the total frequency (N), which is 35 years in the current example. If the difference between the observed and expected frequencies is small, it is likely that a random process generated the pattern. Conversely, if the differences are large, it is less likely that the observed pattern is random. Table 5.6 shows both the actual and expected Poisson frequencies of hailstorm occurrence. By noting the similarities in the observed and expected frequencies, it appears that the occurrence of hailstorms is random.

TABLE 5.5

Critical Values and Poisson Probabilities for Expected Hailstorm Frequency per Year

$$P(X) = \frac{\lambda^X}{e^\lambda(X!)}$$

Critical values:

X = 0, 1, 2, 3, 4
N = 35 years
f = total frequency = 42 hailstorms
λ = mean frequency per year = f/N = 42/35 = 1.2
e = exponential (approximately 2.72)
e^λ = $2.72^{1.2}$ = 3.32

$$P(0) = \frac{1.2^0}{3.32(0!)} = \frac{1}{3.32} = .301$$

$$P(1) = \frac{1.2^1}{3.32(1!)} = \frac{1.2}{3.32} = .361$$

$$P(2) = \frac{1.2^2}{3.32(2!)} = \frac{1.44}{6.64} = .217$$

$$P(3) = \frac{1.2^3}{3.32(3!)} = \frac{1.728}{19.92} = .087$$

$$P(4) = \frac{1.2^4}{3.32(4!)} = \frac{2.074}{79.68} = .026$$

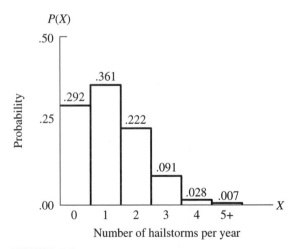

FIGURE 5.3
Poisson (Expected) Probabilities for Number of Hailstorms per Year

TABLE 5.6

Observed and Expected (Poisson) Frequencies for Hailstorm Occurrence per Year

Number of hailstorms per year	Observed frequencies (years)	Poisson probabilities	Expected frequencies (years)
0	10	.301	10.5
1	13	.361	12.6
2	8	.217	7.6
3	3	.087	3.1
4	1	.026	0.9
5+	0	.008	0.3
Total	35	1.000	35.0

Since geographers primarily study spatial patterns rather than temporal sequences, the Poisson probability distribution is often used to investigate the degree of randomness in point patterns. In the tempo-ral Poisson example, the period of study was divided into equal time intervals, as years. In the spatial equivalent of Poisson, the geographic region under study is divided into spatial areas, usually a series of regular-sized square cells known as *quadrats*. The number of occurrences of an item or phenomenon being studied is recorded for each of the quadrats covering the study area. Given the mean number of occurrences per quadrat, or cell, across the region, the Poisson distribution shows the probability of a quadrat containing a certain frequency of occurrences.

The use of Poisson probabilities to examine spatial point patterns for randomness is demonstrated with an example of tornado touchdown sites in Illinois (figure 5.4). Because tornadoes are produced by complex processes, their locational pattern is commonly thought to be random across small geographic regions. To calculate the expected or Poisson pattern of tornado occurrence that would be found under an assumption of randomness, a set of quadrats must be placed over the study area (figure 5.5).

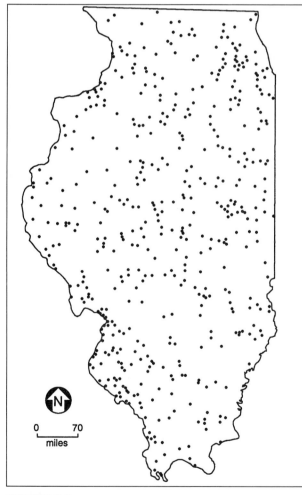

FIGURE 5.4
Spatial Pattern of Tornado "Touchdowns" in Illinois, 1916–1969 (*Source:* Modified from Wilson, John W. and Stanley A. Changnon, Jr. 1971, *Illinois Tornadoes*. Illinois State Water Survey Circular 103, pp. 10, 24.)

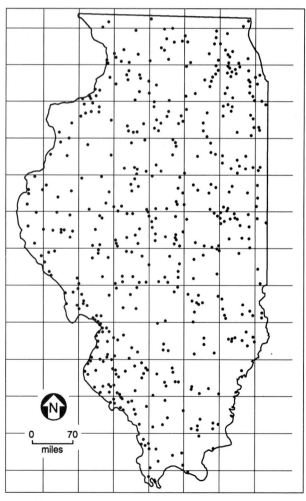

FIGURE 5.5
Illinois Tornado Pattern with Quadrats Superimposed

Because the boundary of Illinois is irregular, some of the quadrats near the border lie only partially inside the study area. In fact, of the 85 cells covering the state, only 47 lie totally inside the state boundary. Of the remaining 38 cells, 22 have less than half of their area inside the study region. Which cells should be included in the analysis? A procedural method must be selected that is as objective, consistent, and unbiased as possible. One logical way to proceed would be to include only those cells with more than half of their area in Illinois. If Poisson probabilities for the tornado pattern are determined in this fashion, 63 quadrats (containing 450 of the 480 tornado touchdowns on the map) are analyzed, representing almost 94 percent of the tornadoes that occurred in Illinois over the 54-year period.

The frequency of tornadoes is determined for each cell by counting the number of points inside each quadrat. The observed frequencies range from a low of 1 tornado to a high of 18 tornadoes (table 5.7). The most frequent occurrence or mode is five points per quadrat, which is found in ten cells. The average cell frequency (λ)—total number of tornadoes (450) divided by the number of cells (63)—is 450/63, or 7.14. Thus, for this set of quadrats, an average cell contains 7.14 points (tornadoes).

The probability of tornado occurrence under an assumption of randomness is determined from the Poisson equation (equation 5.28) using the mean cell frequency. The calculations for zero, one, and two tornadoes per cell are shown in table 5.8, and figure 5.6 shows the Poisson probabilities graphically. To compute the expected cell frequencies, each Poisson probability is multiplied by the total number of cells. For example, since the Poisson probability of a cell having four tornadoes is .086, 8.6 percent of the 63 cells, or 5.42 cells, should have four tornadoes (table 5.9). The largest frequency expected for a random pattern of tornadoes is 7 points per cell, which should occur in 9.39 cells of the study area. This maximum expected value is consistent with the mean cell frequency of 7.14 points per quadrat.

Although the observed frequencies appear to match the calculated Poisson frequencies somewhat, notable discrepancies exist between the two frequencies. Inferential statistics can be used to determine whether the expected and observed frequencies are significantly different. This procedure is discussed in section 12.1, where the focus is inferential statistics for point patterns.

The geographer must make several methodological decisions when using quadrats to examine spatial point patterns. In addition to deciding how to handle quadrats partially outside the study area, researchers must consider the important issue of quadrat size. How would the Poisson probabilities have differed if more quadrats of smaller size had been placed over the pattern of tornadoes? How

---

TABLE 5.7

Observed Frequency of Tornado Occurrence per Cell for Illinois

Number of tornadoes per cell	Observed frequencies of cells	Total tornadoes	Observed probability of occurrence
0	0	0	.000
1	1	1	.016
2	2	4	.333
3	7	21	.111
4	4	16	.063
5	10	50	.159
6	5	30	.079
7	8	56	.127
8	6	48	.095
9	8	72	.127
10	3	30	.048
11	3	33	.048
12	0	0	.000
13	0	0	.000
14	4	56	.063
15	1	15	.016
16	0	0	.000
17	0	0	.000
18	1	18	.016
19+	0	0	.000
Total	63	450	1.000

TABLE 5.8

Critical Values and Poisson Probabilities of Expected Tornado Frequency per Cell for Three Outcomes

$$P(X) = \frac{\lambda^X}{e^\lambda(X!)}$$

Critical values:

X = 0, 1, 2, 3, 4, . . . , 17, 18
N = 63 cells
f = total frequency = 450 tornadoes
λ = mean frequency per cell = f/N = 450/63 = 7.14
e = exponential (approximately 2.72)
$e^\lambda = 2.72^{7.14} = 1261.43$

$$P(0) = \frac{7.14^0}{1261.43(0!)} = \frac{1}{1261.43} = .001$$

$$P(1) = \frac{7.14^1}{1261.43(1!)} = \frac{7.14}{1261.43} = .006$$

$$P(2) = \frac{7.14^2}{1261.43(2!)} = \frac{50.98}{2522.86} = .020$$

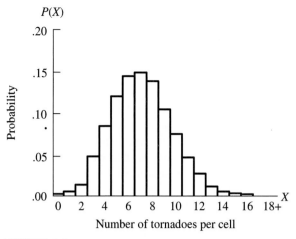

FIGURE 5.6

Poisson (Expected) Probabilities for Number of Tornadoes per Cell

TABLE 5.9

Observed and Expected (Poisson) Frequencies for Tornado Occurrence per Cell

Number of tornadoes per cell	Observed frequencies (cells)	Poisson probabilities	Expected frequencies (cells)
0	0	.001	.06
1	1	.006	.38
2	2	.020	1.26
3	7	.048	3.02
4	4	.086	5.42
5	10	.123	7.75
6	5	.146	9.20
7	8	.149	9.39
8	6	.133	8.38
9	8	.105	6.62
10	3	.075	4.73
11	3	.049	3.09
12	0	.029	1.83
13	0	.016	1.01
14	4	.008	0.50
15	1	.004	.25
16	0	.002	.13
17	0	.001	.06
18	1	.000	.00
19+	0	.000	.00
Total	63	1.001*	62.98*

*Poisson probabilities do not total to 1.0, and expected frequencies do not total to 63 due to rounding.

does a decision to include quadrats only partially inside the study area affect results? These questions will be examined in section 12.1.

In summary, the Poisson distribution is appropriate for geographic problems that meet the following criteria:

- The objective is to determine how frequently an outcome occurs where events occur repeatedly and at random over time or across space. In other

instances, the objective is to analyze a set of existing frequency count data to determine if the data are randomly distributed through time or to analyze a point pattern to determine if the points are randomly distributed across space.

- The time period being analyzed is divided into discrete units (e.g., years) or the study area being analyzed is divided into discrete areal subdivisions (e.g., quadrats).
- The frequency distribution of the occurrence of multiple events by discrete time unit or the frequency distribution of a point pattern by quadrat is estimated under the assumption that a random process is operating. Alternatively, the existing frequency distribution is analyzed to determine if it has been generated by a process operating randomly through time or across space.

5.5 THE NORMAL DISTRIBUTION

The most generally applied probability distribution for geographic problems is the **normal distribution.** When a set of geographic data is normally distributed, many useful conclusions can be drawn, and various properties of the data can be assumed. The normal distribution provides the basis for sampling theory and statistical inference, both of which are discussed in later chapters. The discussion in this section shows how probability statements are made from data sets that are normally distributed.

Although a normal distribution is fully described with a rather complex mathematical formula, it can be generally understood by a simple graph that shows the frequency of occurrence on the vertical scale for the range of values displayed on the horizontal axis (figure 5.7). Since the values on the horizontal axis are not restricted to integers, the normal curve is an example of a continuous distribution. The most striking feature of the normal curve is its symmetry; the lower (left-hand) and upper (right-hand) ends of the frequency distribution are balanced. This symmetric pattern of values in a normal distribution means that no skewness exists in the data. The central value of the data represents the peak or most frequently occurring value. In a normally

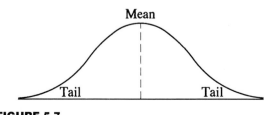

FIGURE 5.7

The Normal Distribution

distributed set of data, this position corresponds to all three measures of central tendency—the mean, median, and mode.

A normal curve is sometimes referred to as bell-shaped in appearance. This characteristic shape represents a frequency distribution with an intermediate amount of kurtosis. The tails of a normal distribution are the portions of the curve with the lowest frequencies located farthest from the center. Note that the frequency of values in a set of normally distributed data declines in a gradual manner in both directions away from the mean.

Because of its particular shape and mathematical definition, the normal distribution is very useful in making probability statements about actual outcomes in geographic problems. For example, if the amount of precipitation a location receives is normally distributed over a series of years, the probability of this site receiving a given amount of precipitation in a year can be calculated mathematically. The normal distribution provides the theoretical basis for sampling and statistical inference, a primary focus of the rest of the book.

The way in which areas are distributed under the normal curve provides the basis for making probability estimates. The total area under the normal curve represents 100 percent of the outcomes. The percentage of values within any portion of the curve along the horizontal axis can be determined. For example, because of the symmetry of the normal distribution, 50 percent of the values must lie under the curve and to the right of the central, or mean, value. Since the normal curve is also a probability distribution, a value taken from a normal distribution has a .50 probability of falling above the mean.

Given the symmetric form of the normal curve, it is clear that 50 percent of all values are greater than the mean. However, a methodology is needed to determine percentages for other intervals under the normal curve. Since integral calculus is used to calculate areas under a mathematical distribution, it could be applied here to determine areas under the normal curve. A simple alternative is to use a table of normal values, showing the proportion of total area under any part of the normal curve. This table is derived mathematically from a theoretical normal distribution.

To use the table of normal values, data must be standardized. On a standardized scale, each observation is assigned a **standard score** (also called a **normal deviate**), which indicates how many standard deviations separate a particular value from the mean of the distribution. Standard scores can be either positive or negative. For units of data greater than the mean, the corresponding standard scores are positive; for values less than the mean, standard scores

are negative. A standard score of 1 represents a value that is one standard deviation *above* the mean, whereas a score of −1 is one standard deviation *below* the mean. The mean corresponds to a standard score or normal deviate of 0. The larger the standard score (in either the positive or negative direction), the farther the value lies above or below the mean.

For any standard score, the table of normal values provides a probability that can be interpreted in two ways. First, it gives the probability of a value falling between the mean and that standard score location for a set of data that is normally distributed. Second, when the probability is multiplied by 100, it shows the percentage of all values in the normal distribution that lie between the mean and this location. Thus, the table of normal values provides information to determine the area under the normal curve for any interval.

Consider these examples using the table of normal values (appendix, table A). The probability value associated with a standard score of 1.0 is .3413 (or 34.13 percent) (figure 5.8). In a normally distributed set of data, approximately 34 percent of the values lie between the mean and one standard deviation above the mean. Since the normal curve is symmetric, 34.13 percent of the values also lie between the mean and one standard deviation *below* the mean. Therefore, by combining these two areas, approximately 68 percent of all values in a normal distribution lie within one standard deviation on either side of the mean.

Similarly from the normal table, the probability of values lying between the mean and a standard score of 2 is .4772. Thus, almost 48 percent of values in a normally distributed set of data lie between the mean and two standard deviations above the mean, and about 95 percent of the values are within two standard deviations on either side of the mean (figure 5.8). The remaining 5 percent of the values are in the two tails of the distribution, where the standard score is either greater than 2.0 or less than −2.0.

Although the table of normal values provides probabilistic information on a standardized scale,

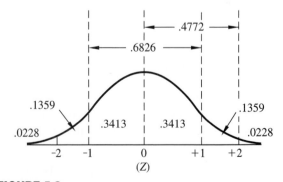

FIGURE 5.8
Selected Areas Under the Normal Curve

geographic data can be measured on various scales. To apply the normal probabilities to specific sets of data, values must first be converted from their original units of measurement (X) to the unitless standardized (Z) scale. In this process, data values are represented by their relative position in comparison to the mean:

$$Z_i = \frac{X_i - \overline{X}}{s} \qquad (5.29)$$

where Z_i = Z-score or standard score for the *i*th value
X_i = observation *i*
$\overline{X}$ = mean of the data
s = standard deviation of the data

The numerator of equation 5.29 shows the deviation of observation *i* from the mean of the data set. This deviation is then divided by the standard deviation of the distribution. The resulting Z-value (or standard score) can be interpreted as the number of standard deviations an observation lies above or below the mean. If the value under consideration is greater than the mean, the deviation is positive, and Z will be greater than zero. If the observation is less than the mean, the deviation and resulting Z-score will be negative. If X equals the mean, the value does not deviate from the mean, and the Z-score will be zero.

Any set of data can be converted from its original units of measurement into the corresponding set of standardized values. However, to estimate probabilities using the table of normal values, the original data must be normally distributed. If the data is not normal, it is invalid to use the table of normal values to estimate probabilities. The statistical test for normality in a data set is discussed in section 11.1.

Annual precipitation data for Washington, D.C., (table 2.6) are now used to make probability estimates. These data are normally distributed over a 40-year period, with a mean of 39.95 inches and a standard deviation of 7.5 inches. What is the probability of annual precipitation in Washington, D.C., exceeding 48 inches? A three-step process can be used to estimate probabilities of precipitation from a normal distribution. (1) Calculate the standard score corresponding to the 48-inch precipitation level. (2) Using the table of normal values, determine the probability associated with this standard score. (3) Evaluate this probability value to answer the specific research question.

Step 1: Calculate the standard score:
The standard score corresponding to 48 inches is calculated, where $\overline{X} = 39.95$ and $s = 7.5$.

$$Z_i = \frac{X_i - \overline{X}}{s} = \frac{48.0 - 39.95}{7.5} = \frac{8.05}{7.5} = 1.07 \quad (5.30)$$

Thus, a precipitation level of 48 inches is 1.07 standard deviations *above* the mean precipitation of 39.95 inches (see figure 5.9).

Step 2: Determine the probability from the normal table:
Using the table of normal values, the Z-score of 1.07 corresponds to a probability level of .3577. Thus, almost 36 percent of the values under the normal curve lie between the mean and 1.07 standard deviations. In other words, in about 36 years out of 100, precipitation in Washington, D.C., should fall between 39.95 and 48 inches.

Step 3: Evaluate the probability value:
Although the table of normal values always determines probabilities for areas under the curve in relation to the mean, the actual probability being sought may represent a different part of the curve. For this problem, the answer lies in the shaded portion of the curve above (to the right of) the Z-score for 1.07. Because the proportion of the total area under the normal curve above the mean is .5000, the correct answer is found by subtracting the probability in step 2 from .5000 (.5000 − .3577 = .1423). Therefore, in 14 years out of 100, annual precipitation in Washington, D.C., should exceed 48 inches.

Probability questions from a normally distributed data set can be stated in another way. Still using the Washington, D.C., precipitation data: What amount of precipitation is likely to be exceeded with a probability of .90? That is, what amount of precipitation will be exceeded nine years out of ten? To answer this question, the three-step methodology is altered.

Step 1: Determine the probability from the normal table:
As shown in figure 5.10, precipitation in Washington, D.C., will be exceeded 90 percent of the time at the position indicated by the shading. The total

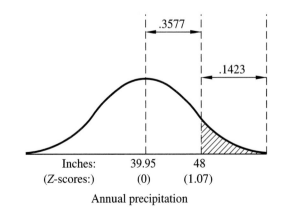

FIGURE 5.9
Determining the Probability of Annual Precipitation Exceeding 48 Inches in Washington, D.C.

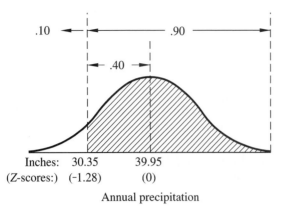

FIGURE 5.10
Determining the Amount of Precipitation Exceeded Nine Years Out of Ten in Washington, D.C.

shaded area represents .90 of the total area under the curve, while the portion to the left of the mean is .40. Therefore, the Z-score corresponding to a probability of .40 must be determined.

Step 2: Calculate the standard score:

According to the table of normal values, a probability of .40 lies between a Z-score of 1.28 (when $Z = 1.28$, the probability is .3997) and a Z-score of 1.29 (when $Z = 1.29$, the probability is .4015), but is considerably closer to $Z = 1.28$. Since the location is below (or less than) the mean, the value is *minus* 1.28. Thus, the precipitation that occurs at least 90 percent of the time (nine years out of ten) is at a value 1.28 standard deviations below the mean.

Step 3: Calculate the precipitation value (X_i):

Using equation 5.29, the precipitation value corresponding to a Z-score of -1.28 is determined:

$$-1.28 = \frac{X_i - 39.95}{7.50} \qquad (5.31)$$

and $X_i = -1.28(7.50) + 39.95 = 30.35$ inches (5.32)

Therefore, nine years out of ten, precipitation in Washington, D.C., will likely exceed 30.35 inches, and one year in ten precipitation should be less than 30.35 inches.

In summary, the normal curve is a very useful probability distribution in geography. For normally distributed data, the proportions of total area under the normal curve are fixed, allowing a variety of probability statements to be made. The normal curve serves as the basis for sample estimation and inferential statistics, discussed in upcoming chapters.

5.6 PROBABILITY MAPPING

The previous section discussed how to make probabilistic statements about an event at a single location. This technique can be extended so that probability es-

timates can be calculated for multiple locations distributed across a region. If probability data are plotted on a map, the resulting information represents a **probability map,** or "probability surface," showing spatial variation in the variable under consideration.

Suppose annual precipitation data are collected for a set of cities in the United States, and the mean and standard deviation of each distribution are calculated. Assuming all data sets are normal, the level of precipitation exceeded nine years out of ten could be determined for each location, using the technique discussed in the previous section. A probability map can be produced by assigning each probability value to its map location and by connecting equal probability values with isolines. Just as contours show elevation patterns in an area, the probability surface would show the spatial pattern of precipitation probability.

What additional information is provided by a map of precipitation probability that is not provided by a simple annual precipitation map? Probability maps consider both the central tendency and variability of the data *at each location*. In fact, the key advantage of probability maps over maps of central tendency (e.g., a map of average annual precipitation) is their consideration of the variability of values at each location.

The following example illustrates the importance of variability in the construction of probability maps. Consider two cities, A and B. They have equal average annual precipitation (50 inches), but very different levels of variability around those averages (figure 5.11). Note the contrast between the minimum annual precipitation expected nine years out of ten in city A versus city B. City A has little precipitation variability from year to year, and the minimum expected precipitation is only slightly lower than the mean (43.6 vs. 50). By contrast, city B has great variability in precipitation over time, making the minimum expected precipitation considerably lower than the mean (24.4 vs. 50). When mapping the minimum annual precipitation expected nine years out of ten for many different cities, the resulting probability map has a spatial pattern that can differ greatly from that of average annual precipitation. In fact, a map of average annual precipitation is comparable to a probability map of minimum precipitation expected five years out of ten, or one-half of the time. Since the data analyzed are assumed to be normally distributed, the mean precipitation value at each location corresponds to a .50 probability.

Cooling degree day information for a set of cities across the United States can illustrate how probability maps are constructed and interpreted. A *cooling degree day* is a surrogate measure for the amount of cooling energy needed to produce a "comfortable" indoor climate. It is a critical index for meas-

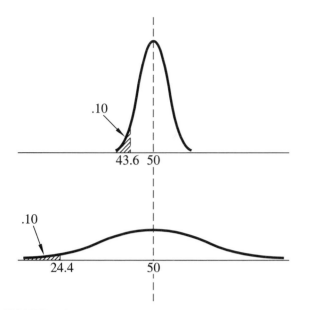

FIGURE 5.11

Comparison of Expected Precipitation in Two Cities with the Same Average Annual Precipitation, but Very Different Precipitation Variability

uring yearly energy demand in warm climates and in cooler climates experiencing hot summers. A cooling degree day is the opposite of a heating degree day, which estimates the amount of energy needed to produce indoor warmth in cold or cooler climates.

The cooling degree day statistic is calculated from the average daily temperature at a site, which is the mean of the high and low temperatures for that particular day. Sixty-five is then subtracted from the average daily temperature in degrees Fahrenheit. (If temperature is measured in degrees Celsius, the constant 18.3 is used, rather than 65.) If the cooling degree day statistic is negative, no cooling degree days are recorded, and the assumption is that air-conditioning will not be needed on that day. These daily values totaled over a year define the annual number of cooling degree days at that location.

One-hundred-three large cities in the conterminous United States, excluding Alaska and Hawaii, have been selected as locations for creating a probability map of cooling degree days. This map shows the number of cooling degree days that should be exceeded nine years out of ten (.90 probability). A similar procedure could be followed for any other probability level. For example, how many cooling degree days should occur 95 years out of 100 (.95 probability) or eight years out of ten (.8 probability)?

Using the means and standard deviations for the cooling degree day data of the 103 cities, probability values are calculated according to the standard score formula (equation 5.29). The calculation for Birmingham, Alabama, indicates that a cooling degree day level of 1,632.94 will be exceeded 90 percent

of the time (figure 5.12). Following this procedure, values for all cities are placed on a map of the United States, and isolines are drawn connecting locations having the same estimated number of cooling degree days (figure 5.13). Note that isoline mapping always requires interpolation to proceed from a finite set of point data to a continuous map surface.

The probability map of cooling degree days is roughly analogous to the spatial pattern of solar energy or heat received across the United States. Since the map portrays a situation expected to occur nine years out of ten, it provides a reasonable spatial estimate of the amount of energy needed for air-conditioning.

The general east to west trend of isolines on this probability map reflects the important influence of latitude on the distribution of heat. This situation is demonstrated by the regular north to south increase in cooling degree days and air-conditioning need—for example, from Minnesota to Louisiana. When isolines dip southward, the number of cooling degree days decreases, because higher elevations have cooler summer temperatures. This pattern is

Step 1: Calculate X and s: $\overline{X} = 1796.43$

$s = 127.73$

Step 2: Use Z table to determine standard score when $P = .90$

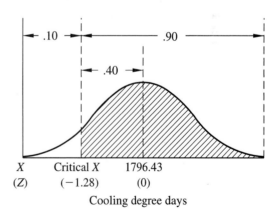

Step 3: Calculate the critical cooling degree day value (X):

$$Z = \frac{X - \overline{X}}{s}$$

then

$$X = \overline{X} + (Z)(s)$$
$$= 1796.43 + (-1.28)(127.73)$$
$$= 1632.94$$

FIGURE 5.12

Calculation of Cooling Degree Days Exceeded Nine Years Out of Ten for Birmingham, Alabama

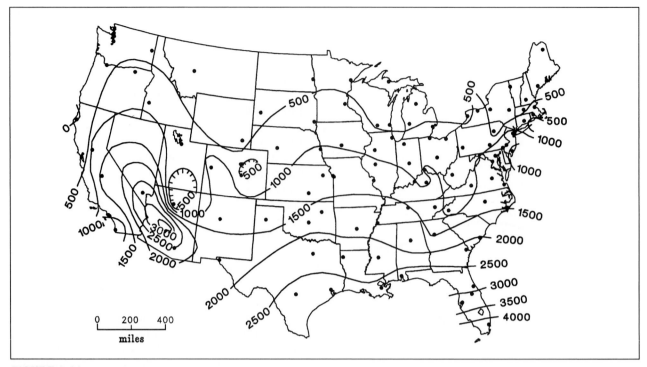

FIGURE 5.13
Number of Cooling Degree Days Expected 9 Years Out of 10: An Estimate of Air-Conditioning Need (*Source:* Calculated from data from National Climatic Data Center, U.S. Dept. of Commerce.)

especially evident in the Appalachian Mountains of West Virginia, western Virginia and North Carolina, and northern Georgia.

Isolines that curve to the north suggest warmer temperatures are found in these areas during the summer. This pattern is seen immediately west of the Rocky Mountains—for example, across Nevada and eastern Oregon. The moderating influence of the Pacific Ocean is seen along the California coast, keeping the number of cooling degree days lower than the corresponding latitudes several hundred miles inland.

A peak of cooling degree days occurs near Phoenix, Arizona; so this region will need more air-conditioning than adjacent areas. The significant decline in elevation from Flagstaff, Arizona, to Phoenix results in a complex regional isoline pattern. Flagstaff has an elevation of nearly 7,000 feet and requires virtually no air-conditioning. Note the cooling degree day value of Flagstaff is under 500. On the other hand, Phoenix, in the Sonoran Desert, has a lower elevation of about 1,100 feet, and needs a great deal of air-conditioning.

Interpretation of the cooling degree day pattern in Arizona illustrates a common problem associated with isoline mapping. The isoline pattern is influenced by the density and placement of control points from which the isolines are drawn. If data from additional weather stations were used, the isolines that

make up the probability surface could be located more precisely.

Probability maps can be constructed for many spatial variables. The technique seems most directly applicable for the analysis of natural phenomena in climatology, meteorology, and environmental studies. For example, geographers could construct probability maps of atmospheric particulates, ozone levels, major winter storms, timing of killing frosts, or the pH levels of acid deposition. In some instances, probability mapping could be extended to selected topics in human geography, such as disease, unemployment, or poverty. Extreme care must be taken, however, because virtually all variables in human geography are not distributed continuously over space.

KEY TERMS

binomial distribution, 69
complement, 67
deterministic processes, 65
discrete and continuous probability
  distributions, 69
events and outcomes, 66
factorial, 70
mutually exclusive, 67
normal distribution, 75
Poisson probability distribution, 71
probabilistic processes, 66

MAJOR GOALS AND OBJECTIVES

If you have mastered the material in this chapter, you should now be able to do the following.

1. Understand the general nature of deterministic and probabilistic processes in geography.
2. Recognize the distinction between stochastic and random probabilistic processes.
3. Explain the concept of randomness and recognize the characteristics of a random experiment.
4. Explain the probability concept of relative frequency and recognize applications in geography.
5. Understand the basic rules of probability.
6. Understand the characteristics of the binomial probability distribution and identify potential applications in geography.
7. Explain the characteristics of the Poisson probability distribution and identify potential applications in geography.
8. Explain the nature of the normal distribution, know how to use the table of normal values, and identify potential applications in geography.
9. Understand the calculation and interpretation of standard (Z) scores.
10. Describe the technique of probability mapping and recognize the types of spatial variables for which probability maps can be constructed.

REFERENCES AND ADDITIONAL READING

Bulmer, M. G. 1967. *Principles of Statistics.* Edinburgh: Oliver and Boyd.

Gregory, S. 1978. *Statistical Methods and the Geographer.* London: Longman.

Matthews, J. 1981. *Quantitative and Statistical Approaches to Geography: A Practical Manual.* Oxford: Pergamon.

Silk, J. 1990. *Statistical Concepts in Geography.* London: Allen and Unwin.

Taylor, P. J. 1977. *Quantitative Methods in Geography: An Introduction to Spatial Analysis.* Boston: Houghton, Mifflin.

Unwin, D. 1981. *Introductory Spatial Analysis.* London: Methuen.

Winkler, R. L. and W. L. Hays. 1975. *Statistics: Probability, Inference, and Decision.* New York: Holt, Rinehart & Winston.

Basic Elements of Sampling

6.1 **Basic Concepts in Sampling**
6.2 **Types of Probability Sampling**
6.3 **Spatial Sampling**

Sampling was discussed briefly in chapter 1 as an essential component of the scientific research process. In many aspects of geographic research and problem solving, statistical techniques are incorporated into that research procedure. In fact, a basic knowledge of sampling procedures and methodology is valuable, whatever the area of geographic study.

How does sampling fit into the overall scheme of statistical analysis and scientific research? Earlier chapters include some brief comments about the nature of these relationships. Recall the fundamental distinction between descriptive and inferential statistics made in chapter 1. The ultimate aim of inferential statistics is to generalize about certain characteristics of a large group, based on information obtained from a portion (often a very small portion) of that large group. That is, characteristics regarding a statistical **population** (the universe of all individuals) are *inferred* from information obtained from a **sample** (a subset or portion of individuals selected from the population for detailed analysis). These concepts are further clarified in the chapter on probability, where the importance of randomness and random processes is discussed in the context of making generalized probability statements about a large population (the conceptual notion of an infinite number of die rolls) using only sample observations (such as a few rolls of a die).

This chapter introduces some basic sampling concepts and discusses the important roles of sampling in geographic problem solving and research. Without a properly drawn sample, valid generalizations or statistical inferences about the population may not be possible. Section 6.1 presents an over-view of the advantages of sampling and discusses the sources of sampling error and the steps involved in a well-designed sampling procedure. The various types of probability sampling are reviewed in section 6.2. All probability samples contain an element of randomization, but simple random sampling is often not the best choice of sample design. Systematic, stratified, cluster, and hybrid sample designs are often useful and are examined here. Section 6.3 discusses the circumstances under which spatial sampling is necessary and reviews different types of spatial sampling. Special attention is given to point sampling designs.

6.1 BASIC CONCEPTS IN SAMPLING

Sampling is an essential skill for virtually all areas of geographic study. A biogeographer interested in the spatial pattern of environmental change associated with high-intensity recreation use in a national park cannot examine conditions everywhere in the park. A representative sample of study sites needs to be selected for detailed analysis of these human-environment relations. An urban geographer wishing to examine the locational variation of housing quality in a metropolitan area must select a sample of homes for detailed study. A behavioral geographer conducting research on natural hazards may distribute questionnaires to a representative sample to learn public attitudes toward alternative flood-management policies along the floodplain of a river. A medical geographer concerned with neighborhood variations in the use of hospital emergency rooms as primary care centers would use sampling to select the neighborhoods and the hospitals to include in the study.

These examples illustrate the geographer's use of sampling in both spatial and nonspatial contexts. Spatial sampling takes place when the biogeographer selects *locations* to examine environmental change in the national park. Likewise, the neighborhoods the medical geographer chooses to examine patterns of hospital use also constitute a spatial sample. On the other hand, the geographer conducting the study on attitudes toward natural hazards along the river floodplain may have selected individuals from a nonspatial list of households in the area. The urban geographer conducting the study of housing quality could have taken a sample of homes from either a nonspatial list (e.g., tax rolls) or a spatial source (e.g., a map).

Advantages of Sampling

A variety of both practical and theoretical reasons make sampling preferable to complete enumeration or census of an entire population. The advantages of sampling may be summarized as follows:

- *Sampling is a necessity in many geographic research problems.* If the population being studied is extremely large (or even theoretically infinite), completing a total enumeration is not possible. When studying the reasons why families in the United States change residence, sampling is the only choice—all those who move cannot possibly be contacted and surveyed individually. A geographer analyzing global changes in the nature and spatial extent of tropical rain forests will find it impossible to have total spatial coverage, since the number of locations in a rain forest is infinite.
- *Sampling is an efficient and cost-effective method of collecting information.* An appropriate amount of data concerning the population can be obtained and analyzed quickly with a sample. Not only is sample information collected in less time, but sampling also keeps expenditures lower and logistical problems to a minimum. The overall scale of effort (time, cost, personnel, logistics, etc.) is made practicable with sampling as opposed to conducting an examination of the entire population.
- *Sampling can provide highly detailed information.* In geographic problems where in-depth analysis is necessary, only a small number of individuals or locations can be included in the study. These few elements in the sample could then be closely scrutinized with the collection of a comprehensive set of information. A study of shopping behavior patterns, for example, might require numerous questions about the number of shopping trips, loca-tions visited while shopping, attitudes about alternative stores, as well as demographic or socio-economic information concerning household members. Such detailed analysis can be obtained only through sampling.
- *Sampling allows repeated collection of information quickly and inexpensively.* Many geographic research problems require detailed information collected over a specific period of time or focus on spatial changes that occur rather quickly. For example, a geographer studying the attitudes of citizens living in a barrier island community toward alternative coastal zone management strategies may want to follow a sample of individuals through time as pertinent legislation moves through the political process. Their attitudes may change over time, especially if a hurricane hits the island during the study period. An urban geographer analyzing the spatial pattern of growth may wish to focus on the views of residents in a neighborhood before, during, and after the construction of a nearby shopping mall. With dynamic situations such as these, sampling is required—information from all individuals in the population could not possibly be collected without unreasonable effort and cost.
- *Sampling can provide a high degree of accuracy.* With sampling, an acceptable level of quality control can be assured. Complete and accurate questionnaire returns can be obtained if a small number of well-trained personnel conduct all of the interviews. This procedure may provide more accurate results than a complete census requiring a larger number of personnel, some of whom may not be fully trained. The 1990 U.S. Census of Population illustrated the difficulties of acquiring accurate information from everyone. Considerable controversy developed because many Americans were literally not counted. The Census Bureau has used statistical modeling procedures to estimate actual population counts from samples. In preparation for the 2000 census, considerable discussion has centered on the issue of sampling; some professionals advocated that *all census data* be obtained through carefully designed samples. In 1990, some census variables came from total enumeration of the population, whereas other variables were obtained from samples.

Sampling and the Census 2000 Debate

Despite the many clear and obvious advantages of statistical sampling, the decision to use sampling (or how to use sampling) in a specific practical situation can often be controversial, with multifaceted implications. Real-world difficulties and political policy issues can arise concerning the implementation of sampling. In addition, many citizens are not aware of the advantages of sampling over total enumeration and will not be strong advocates of sampling.

Recent political and legal developments in the United States highlight some of these controversies and the practical implications of sampling. In late 1998, the Supreme Court heard a case regarding the constitutionality of statistical sampling in the 2000 census. The Constitution specifies a decennial enumeration of the population. The basic constitutional issue is whether statistical sampling meets the definition of "enumeration." At issue is the allocation of billions of dollars in federal funding and the redrawing of congressional district boundary lines, both of which are based on population estimates. Must these population estimates be based on a complete counting (total enumeration) of all Americans, or can statistical sampling be used in some way to estimate population?

Advocates of sampling argue that sampling procedures are necessary for significantly improving the accuracy and cost efficiency of the census. They note that some 4 million people were missed in the 1990 census, but this undercount was later corrected using follow-up sampling methods. The argument is that minorities and the poor were underrepresented in the 1990 count and that certain regions of the country, especially urban areas, received insufficient funding from the federal government as a result. Opponents counter that sampling is not necessarily a more accurate method of counting, especially when performed under severe time constraints. They note that because sampling relies on the use of estimates, and many different methods of estimation are available, the one chosen will likely have profound political and economic consequences. There may be a real temptation to choose an estimate that favors the political goals of those in power. Both sides of the debate are vulnerable to charges that their positions on the issue are politically motivated.

Regardless of the Supreme Court vote on the constitutionality of sampling in the U.S. census, general issues related to the suitability of sampling versus total enumeration will continue to be discussed in a variety of contexts. Many people remain skeptical of the advantages of sampling over total enumeration, even though strong theoretical evidence supports statistical sampling as the better alternative.

Sources of Sampling Error

A central goal of sampling is to derive a truly *representative* set of values from a population. A representative sample will accurately reflect the actual characteristics of the population without bias. To ensure an unbiased representative sample, an element of **randomness** must be incorporated into the sample design procedure. However, just having randomness built into a sampling plan does not guarantee an unbiased sample, for many other **sources of sampling error** are possible.

Unfortunately, it is not possible to know with absolute certainty whether a sample is totally representative. The very nature of sampling means that everything cannot be known about the population containing the sample. Because only sample data are available, some uncertainty will always be associated with sample estimates, and some sampling error will occur. The geographer must try to minimize sampling error given various practical constraints such as cost or time.

The measurement concepts of precision and accuracy help categorize the many sources of sampling error. The results from a small sample may not be very exact or precise. Increasing the sample size, which permits a more exact estimate (figure 6.1, line A), can reduce imprecision. Larger samples, however, are invariably more costly and time-consuming to obtain (figure 6.1, line B). Satisfactory resolution of this difficult trade-off between sample precision and the effort required to sample is important in most real-world problems involving sampling.

Sample inaccuracy is a more complex issue. Systematic bias can enter sampling in many different ways. Some inaccuracies are the result of problems with the sampling procedure itself. For example, the elements or individuals in a population could be "mismatched" with the set of elements or individuals from which a sample is taken. An urban geographer studying new home construction patterns might use a list of building permits as a source of data. If this list contains home renovations as well as new home construction, some of the renovations might be erroneously included in the new home construction sample.

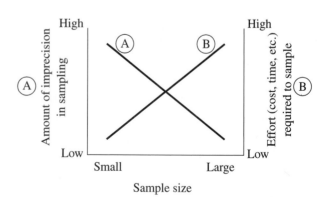

FIGURE 6.1

The Influence of Sample Size on Amount of Imprecision and Effort Required to Sample

Selecting an improper sampling design is another potential source of inaccuracy. The complexity of a problem might call for a more sophisticated approach, rather than a simple random design. In other sampling problems, the method of data collection used may be inappropriate. For example, a geographer studying population mobility might decide to use a mail questionnaire to keep costs down when telephone interviews would provide more accurate results.

Other types of inaccuracy may result from some operational, logistic, or personnel problem not directly connected to the actual sampling procedure. Inconsistencies in collecting field data or interviewing could adversely affect the sample results. Errors could be made in the editing, recording, or tabulating of information. Even forces of nature beyond the researcher's control could bias results. For example, a geographer examining land-use patterns along the floodplain of a large river could find the entire project in jeopardy if severe flooding occurs during the study period.

Simply stated, the quality of statistical problem solving in geography depends heavily on samples that have been properly designed and collected. Procedures that reduce sampling error to a tolerable or acceptable level must be carefully developed. Most sources of error can be reduced substantially (and perhaps even avoided or eliminated) if all steps of the sampling procedure are carefully planned and evaluated *before* the full set of sample data is collected and analyzed.

Steps in the Sampling Procedure

If faced with a geographic research problem in which the collection of sample information is necessary, a number of steps need to be followed (table 6.1 and figure 6.2). Collecting sample data is *not* the first action taken. The researcher must first anticipate and resolve various problems that might cause sampling error or other difficulties. Numerous safeguards or checks should be incorporated into the sampling procedure whenever necessary. The characteristics and relevant issues at each step of a well-designed sampling procedure are now summarized.

Step 1: As hypotheses are formulated in scientific research, variables in the problem must be defined both conceptually and operationally. In

TABLE 6.1

Steps in the Sampling Procedure

Step 1: Conceptually Define Target Population and Target Area

- **Target population:** the complete set of individuals from which information is to be collected.
- **Target area:** the entire region or set of locations from which information is to be collected.

Step 2: Designate Sampled Population and Sampled Area from Sampling Frame

- **Sampling frame:** the practical or operational structure that contains the entire set of elements from which the sample will actually be drawn.
- **Sampled population:** the set of all individuals contained in the sampling frame, from which the sample is actually drawn.
- **Sampled area:** the set of all locations within the study area boundary line that delimits the spatial sampling frame, from which the sample is actually drawn.

Step 3: Select Sampling Design

- Probability sampling preferred over nonprobability sampling.
- Types of probability samples include random, systematic, stratified, cluster, and hybrid designs.
- Spatial and nonspatial variations in sampling design exist.

Step 4: Design Research Instrument and Operational Plan

- Methods of data collection include direct observation, field measurement, mail questionnaire, personal interview, telephone interview.
- Establish protocols for handling all problems or situations that can be anticipated in the sampling procedure.
- Complete miscellaneous logistic and procedural tasks in the preparation of sample taking.

Step 5: Conduct Pretest

- Complete trial run or pilot survey of sample data collection method.
- Correct all discovered problems that could lead to sampling error.
- Pretest results may be used to determine sample size.

Step 6: Collect Sample Data

- Consistency in collection methods and procedures is essential.
- Ensure overall high level of quality control.

Step 1: Conceptually define target population and target.

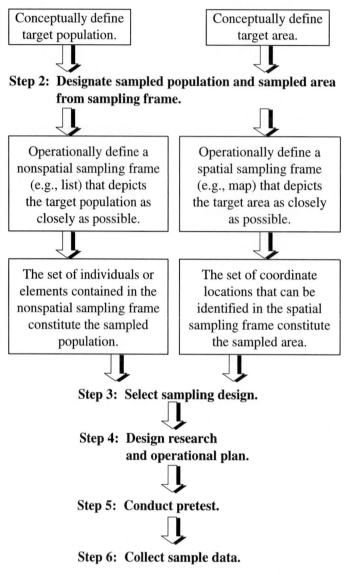

FIGURE 6.2
Steps in the Sampling Procedure

sampling, the population and area "targeted" for study must be defined. The **target population** is the complete set of individuals from which information is collected, whereas the **target area** is the entire region or set of locations from which information is gathered. Precise delineation of the target population and target area is not always a simple task. Suppose, for example, an urban geographer wishes to study the spatial variation of public attitudes regarding several proposed revitalization projects in the central business district (CBD) of a large metropolitan area. What set of individuals should make up the target population? Where should the boundaries of the target area be placed? Arguments could be made in support of any of the following alternatives:

- all city residents active in local civic groups
- all city landowners
- all city residents
- all city and suburban residents
- all area residents plus nonresident visitors

Step 2: Once the target population and target area have been defined conceptually, an operational sampling frame(s) must be created. A **sampling frame** is defined as the practical or operational structure that contains the entire set of elements from which the sample will actually be drawn. This structure is sometimes a comprehensive list of individuals (nonspatial) and sometimes the boundary line delimiting the extent of the study area (spatial).

Sometimes both spatial and nonspatial sampling frames are components of the same problem. A **sampled population** is the set of all individuals contained in the sampling frame, from which the sample is actually drawn. A **sampled area** is the set of all locations within the study area boundary line that delimits the spatial sampling frame, from which the sample is actually drawn.

Why is it important to distinguish between target population and sampled population in nonspatial sampling? The CBD revitalization example can illustrate this distinction. Suppose, conceptually, the population being targeted is *all metropolitan area residents,* both in the city and in adjacent suburbs. Operationally, however, this target population is not listed in any single comprehensive source. A sampling frame needs to be constructed that lists the intended target population as completely as possible.

Many possible sources of data could provide an operational definition for this sampling frame. The customer lists from electric utilities would not include all households or all residents at those households listed. People in apartments or other group accommodations would be omitted from the sampling frame. Telephone listings will also not provide a complete enumeration of the target population. Lower-income residents without telephones would be underrepresented in the sample, as would students in dormitories, residents in group quarters, and those with unlisted numbers. The result is that the sampled population often cannot duplicate the target population because the sampling frame used to delineate the sampled population is not complete.

In geographic or spatial sampling problems, it is essential to investigate whether any differences exist between the target area and sampled area. Suppose a geographer wants to study environmental "threats" within the greater Yellowstone ecosystem (figure 6.3). According to officials at Yellowstone National Park, the park has become an ecological island, with boundaries that neither encompass a complete ecological unit nor adequately protect the unique geothermal activity in the area. The target area for the geographic study is the greater Yellowstone ecosystem boundary. The sampled area will almost certainly not correspond exactly with this designated target area. Perhaps entry onto some private ranches and resorts may be prohibited, and access may be denied onto some public lands leased by lumber and energy companies. Will the inability to

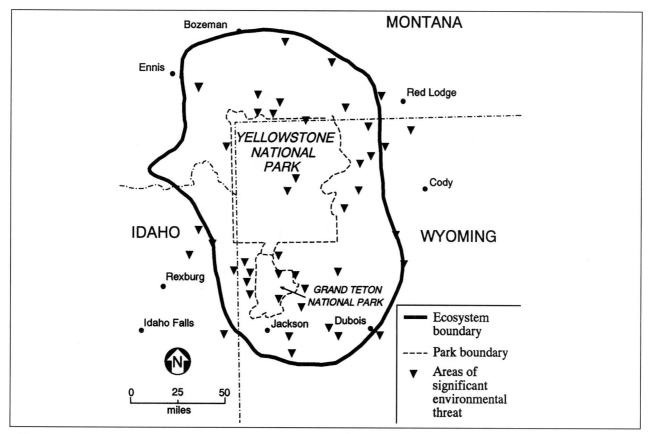

FIGURE 6.3
The Greater Yellowstone Ecosystem and Environmental Threats (*Source:* Redrawn from map in *Washington Post,* June 23, 1985, p.B5, with permission.)

sample from these portions of the target area bias the results and weaken the accuracy of the study? This question must be carefully considered before continuing further.

An important task is to construct effective sampling frames characterized by a target population that matches the sampled population and a target area that matches the sampled area as closely as possible. In the problem concerning CBD revitalization, what is the nature and probable extent of sampling error that will occur if a telephone survey (sampled population) is used to elicit opinions of all metropolitan area residents (target population)? In the Yellowstone example, how much bias will be injected into the sampling procedure if a considerable portion of private land in the region is excluded from analysis? Whenever facing mismatches like these, only qualified or conditional statements are possible, and sources of systematic bias or inaccuracy in the study must be clearly explained.

Step 3: Selecting an appropriate sample design method is crucial to the success of the entire sampling procedure. Sampling design is the way in which individuals or locations are selected from the sampling frame.

A fundamental distinction can be made between **nonprobability** and **probability sampling designs.** Nonprobability sampling is subjective because the selection of sampled individuals or locations is based on personal judgment. Nonprobability samples are, therefore, sometimes called "judgmental" or "purposive" samples. Criteria for sample selection might include

- personal experience and background knowledge;
- convenience or better access to a nonrandom selection of individuals or locations;
- use of only those who volunteer or respond to the survey instrument, such as a questionnaire in a newspaper; and
- selection of a single nonrandom study area or sample region (case study) in which to conduct research.

Excellent descriptive results can sometimes be derived from a nonprobability sampling procedure. For example, the case study approach may provide considerable insight on variable interaction within the study area. However, the underlying problem with all nonprobability approaches to sample selection is that no valid inferences can be made from the sampled elements to any wider group in the population or to any other areas. In other words, nonprobability sampling limits the generalizations that can be drawn from the research findings.

By contrast, the nature of probability sampling is more objective and is closely associated with sci-

entific research. In a probability sample, each individual or location that could be selected from the sampling frame has a known chance (or probability) of being included in the actual sample. The advantage of probability sampling is that the amount of sampling error can be estimated. Because of the importance of probability sampling in geographic research, section 6.2 is devoted entirely to this topic.

Step 4: Many procedural matters have to be resolved when sampling in real-world geographic situations. A key task is the selection and design of an appropriate measurement instrument. Possible formats for data collection include direct observation, field measurement (especially in areas of physical geography), mail questionnaire, personal interview, and telephone interview. To select the most appropriate format, the nature of the research problem must be evaluated carefully. Even after the method of data collection has been selected, problems in instrument design may remain. In survey design, each question must be carefully worded, all possible responses to a question must be considered in advance, and the questions must be sequenced properly. Questionnaire design is complicated; many different considerations are involved. For those interested in further details regarding questionnaire design, see the references at the end of the chapter.

Various other procedural decisions and logistic arrangements may be necessary, depending on the nature of the sampling problem. Preliminary site reconnaissance may be necessary, and interviewers and other personnel may need to be hired and trained. If not anticipated, any number of difficulties can plague the researcher and even create bias in the sample, undermining its inferential power. Discussing the wide variety of circumstances that might arise is not possible, as every geographic problem is likely to have its own peculiarities. However, common problems include the following:

- low response to a survey or questionnaire;
- unexpected response resulting from ambiguity or misunderstanding;
- environmental or political change in the sampled area; and
- unqualified or incompletely trained interviewers.

Step 5: Before extensive time and money are expended on the main data collection effort, a small-scale pretest needs to be conducted to assess both the strengths and deficiencies in the research instrument design and operational plan. In a field study, the pretest should reveal problems with instrument calibration or malfunction, site access problems, and other logistical difficulties. Problems in a survey or questionnaire format may include nonresponse, improperly worded questions, and poorly sequenced questions. A quality pretest can identify and correct difficulties in research design and data collection.

A good pretest or pilot survey can also help determine the size of the sample that needs to be taken. Too small a sample will not yield meaningful results, whereas an unnecessarily large sample will waste time, money, and effort. Pretest information can help determine the sample size needed to achieve a certain level of precision in the main sample. This topic will be discussed in section 7.3.

Step 6: The collection of sample data is a major task, where high levels of quality control must be maintained and well-considered data management procedures carefully followed. Consistency in all aspects of data collection and processing is absolutely essential. In this step, the benefits of the careful design of the research instrument and planning (step 4) will become evident, and the improvements made in pretesting (step 5) will expedite the major data collection effort significantly.

6.2 TYPES OF PROBABILITY SAMPLING

In probability sampling, each individual or item that could be selected from the sampling frame has a known chance (or probability) of being included in the sample. This important advantage occurs because a randomization component is incorporated into the sample design in some known way. If the sample data are to be used in any inferential manner (either for estimation of population parameters or for hypothesis testing), then an element of randomness must be built into the sample design procedure. All types of probability sampling contain this characteristic of randomness and avoid the subjectivity of nonprobability sampling.

Simple Random Sampling

Randomization is fundamental to all probability sampling, and a **simple random sample** is the most basic way to generate an unbiased, representative cross section of the population. In a simple random sample, every individual in the sampling frame has an equal chance of being included in the sample.

To select sample members, a list of random numbers is used. Computer-based random number generators provide sets of random numbers. For demonstration purposes, however, the traditional table of random numbers (appendix, table B) will be used here. This table is a long sequence of integers, and the selection of each integer is independent of all other selections. This usually results in a table with all ten integers (0, 1, 2, . . . , 9) present in roughly equal proportion, with no trend or pattern in their sequencing.

The task of probability sampling is to draw a set of elements (individuals or locations) from the sampling frame. Each element in the sampling frame must be identifiable, so that the randomization device (random number table or computer-based random number generator) can select units to be included in the sample. In a nonspatial sampling frame, individuals are placed in a list, with each member identified by a specific number.

If the sampled population in the sampling frame consists of fewer than 100 members, then a two-digit number could be assigned to each individual (00 could represent the first individual in the list, 01 the second individual, and so on). With a sampled population of 100 members or more, numbers with more digits would be necessary.

Suppose in planning a new student center, university personnel want to determine the detailed opinions of students regarding which activities to provide. Because time is limited, only a small sample of students can be surveyed. From a student body of 8,500, a simple random sample of 25 students is selected. Each student must have an equal chance of being chosen (that probability is 25/8,500 = .0029, or about 3 per 1,000).

Any place in the random numbers table may be used as a starting point, but this position must itself be arbitrarily selected. Suppose the fifth column segment and sixth row in that column segment are selected as the starting position. This number is 10440 (see appendix, table B). Since the target population is 8,500, each member of that population can be assigned a four-digit identification number. Therefore, one digit of each five-digit sequence can be dropped. If the last digit is dropped, this would make the starting sequence 1044. Proceeding down the column segment, the following numbers are found:

1044	3551	6141	
0786	3454	7678	
9351*	5851	1344	
9692*	0000	8483	
9117*	7007	9273*	* (rejected—out of range)
4764	7704	3652	
1952	0079	7964	
7833	2645	5756	
1858	6983	5314	
6988	5201		(desired sample size of 25 obtained)

Note that all numbers greater than 8500 are rejected as being "out of range." The first acceptable four-digit number is 1044, the second is 0786, the third is 4764, and so on. Selection continues until the desired sample size (25) is obtained. If a four-digit sequence repeats, that student has already been selected and should not be chosen again. Another number, the next sampling unit in the table, would be selected instead.

Sometimes the method for operationalizing a simple random sample is awkward or inefficient. For example, the student list might be organized alphabetically or by social security number and not numbered sequentially. Considerable effort might be required to create a sequential list to use with the random numbers table. Depending on circumstances, the geographer might want a less cumbersome and time-consuming sample design.

Systematic Sampling

Systematic sampling is a widely used design that often simplifies the selection process. A **systematic sample** makes use of a regular sampling interval (k) between individuals selected for inclusion in the sample. A "1-in-k" systematic sample is generated by randomly choosing a starting point from among the first k individuals in the sampling frame, then selecting every kth individual from that starting point. For example, if a sample of 25 ($n = 25$) is taken from a population of 500 ($N = 500$), the sampling interval (k) would be N/n ($k = 500/25 = 20$—a "1-in-20" sample). Care must be taken to ensure that no nonrandom pattern or sequencing is present in the list from which every kth individual is selected. Otherwise, bias may be introduced into the sample process.

A systematic method of sampling is generally less cumbersome to operationalize than is simple random sampling. Systematic sampling provides a relatively quick way to derive a large size sample and obtain more information at a reasonable cost. As a result, it is used in many practical contexts. Government agencies (such as the U.S. Census Bureau and Statistics Canada) and political polling firms (such as Gallup or Harris) routinely apply systematic samples when detailed analyses or follow-up studies are needed. To estimate the quality of a product coming off an assembly line, factory management could have a more detailed inspection of every 100th item. Every 20th visitor to a park could be asked to complete a survey regarding park services, or every 50th taxpayer in a city could be asked about alternative funding or planning projects. In market analysis, the editors of a magazine could send a detailed survey or questionnaire concerning existing and proposed features to every 10th subscriber.

Stratified Sampling

In many geographic sampling problems, the target population or target area is separated into different identifiable subgroups or subareas, called **strata.** If sample units in the different strata are expected to provide different results, such "target subdivision" is logical. With **stratified sample** design, the effect of certain possible influences can be controlled. Taking a simple random sample from each class or stratum makes the fullest possible use of available information and increases the precision of sample estimates.

Stratified sample designs may be either **proportional (constant-rate)** or **disproportional (variable-rate).** In a proportional stratified sample, the percentage of the total population in each stratum matches the proportion of individuals actually sampled in that stratum as closely as possible. For example, suppose 20 percent of all residents in a city are apartment dwellers. Proportional representation of the apartment stratum in a stratified sample design would require that 20 percent of the sampled individuals be apartment dwellers. Suppose further that this housing study specifically involves a resident survey on rent control. Because apartment dwellers would be particularly affected by the proposed legislation and their opinions important to decision makers, council members might want their views to be represented more heavily. If such a disproportional (or variable-rate) stratified sampling design is appropriate, the apartment resident stratum would be **oversampled** (with a larger than proportional sample size), and residents not living in apartments would be **undersampled** (with a smaller than proportional sample size). Another practical sampling strategy is to maintain proportional representation of both the apartment and nonapartment strata, but weigh the apartment dweller responses more heavily when calculating sample statistics.

For the opinion survey on the new student center, a simple random sample would probably yield less precise results than would a well-designed stratified sample. A university has a diverse student population, and different student subgroups would likely have varying opinions on activities for a student center. Several stratified sample designs are possible (figure 6.4). For example, those living in campus dormitories may have significantly different priorities than do commuters living off-campus (figure 6.4, case 1). Also, the views of freshmen may differ from those of sophomores, juniors, and seniors (figure 6.4, case 2). Stratifying *both* by place of residence *and* year in school might prove most useful, resulting in a composite sample design structure with eight strata (figure 6.4, case 3).

If the views of each student are considered equally important, a proportional stratification is applied. However, if the views of dormitory residents are considered more important, that group could be oversampled in a disproportional stratified sampling scheme. In both instances, if results are ex-

Case 1:
Stratify by place of residence

Campus dormitory   Off-campus housing

1. ___   1. ___
2. ___   2. ___
3. ___   3. ___
   etc.   etc.

Case 2:
Stratify by year in school

Freshman Sophomore Junior Senior

1. ___   1. ___   1. ___ 1. ___
2. ___   2. ___   2. ___ 2. ___
3. ___   3. ___   3. ___ 3. ___
 etc.   etc.   etc.   etc.

Case 3:
Stratify by place of residence and year in school

	Freshman	Sophomore	Junior	Senior
Campus dormitory	1. ___ 2. ___ 3. ___ etc.	1. ___ 2. ___ 3. ___ etc.	1. ___ 2. ___ 3. ___ etc.	1. ___ 2. ___ 3. ___ etc.
Off-campus housing	1. ___ 2. ___ 3. ___ etc.	1. ___ 2. ___ 3. ___ etc.	1. ___ 2. ___ 3. ___ etc.	1. ___ 2. ___ 3. ___ etc.

Note: Random samples of varying sizes taken from each stratum

FIGURE 6.4
Alternative Stratified Sample Designs: Student Opinion Survey

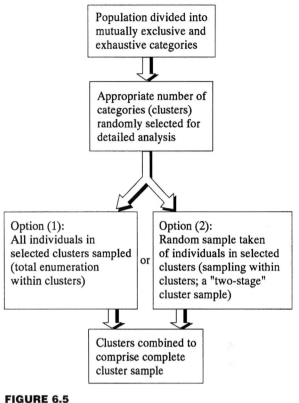

FIGURE 6.5
Steps in Cluster Sampling

pected to vary by stratum, a stratified sample is preferable to a simple random sample of the same size. That is, stratification should be applied if background knowledge or logic suggests that it would be beneficial and practical.

Cluster Sampling

For some geographic problems, **cluster sampling** is most appropriate and may be more efficient or cost-effective than random, systematic, or stratified sampling. A cluster sample is derived by first subdividing the target population or target area into mutually exclusive and exhaustive categories (figure 6.5). An appropriate number of categories (clusters) is selected for detailed analysis through random sampling. Two alternatives exist:

- All individuals within each cluster are included in the sample, making a total enumeration or census within that cluster.

- A random sample of individuals is taken from each cluster.

The latter option is sometimes called a "two-stage" cluster sampling procedure because a random process is used twice—once to select clusters and then again to select sampled individuals within each cluster. The actual approach used depends on the circumstances and practical sampling conditions. The complete cluster sample is the composite of these selected clusters of observations.

Cluster sampling is generally an effective design option when practical or logistic problems make other choices more expensive, difficult, or time-consuming. In some situations, a complete sampling frame is not available, but partial information can be obtained. This might occur in an urban geography study where total enumerations are available for many city blocks, but no complete list of all city residents exists. A cluster could be defined as all (or many) homes on a city block. In other geographic problems, the population may be widely dispersed, resulting in high travel costs and increased logistic problems. In these circumstances, noncluster design options (e.g., simple random sample) would not be practical. The choice of a spatially contiguous or concentrated cluster will keep the costs of obtaining the necessary sample to a minimum and permit an adequate sample size to be generated with reasonable effort.

A total enumeration cluster approach makes it easier to obtain large sample sizes (option 1 in figure 6.5). In the student opinion survey, interviewing all students in a randomly selected number of dormitories or sampling all students in a number of general education classes would result in an appropriately large sample size.

For cluster sampling to be most effective, however, the individuals within each cluster should be as different or heterogeneous as possible. This will make the cluster sample observations representative of the entire sampled population and avoid systematic bias. If the clusters are internally similar or homogeneous in nature, stratified sampling may be a better alternative. Therefore, the appropriateness of clusters in a geographic research problem must be evaluated carefully. In this context, the total enumeration cluster sample of a dormitory or general education class may not be appropriate because these clusters could be too homogeneous.

Combination or Hybrid Sampling Designs

Choosing a sampling design is seldom a simple, straightforward matter. Decision makers should consider cost, time, and convenience, as well as various practical problems unique to the specific situation. Common sense and experience are also important in selecting a sample design.

In many cases, the simplicity of a simple random approach offers an important advantage. However, practical circumstances may make the use of *any* simple type of sampling (random, systematic, stratified, or cluster) difficult or unwieldy. When practical conditions dictate, some combination or **hybrid sampling** design may be most appropriate. The following experience illustrates how practical realities influence the selection of sample design.

A geography department was asked to conduct a survey of air passengers enplaning at the nearby regional public airport (McGrew and Rosing, 1979). The survey had three general objectives: (1) establish the airport's service area; (2) determine selected passenger characteristics; and (3) provide various recommendations on how to improve or expand airport services. Since a complete census of all passengers was impractical, a survey sampling procedure had to be devised. Selection of a proper sample design was critical to ensure that the passengers being surveyed accurately represented the entire population of passengers enplaned at the airport. A number of statistical requirements and practical realities limited the sampling design choice:

- ensuring that each passenger was selected at random with known probability of inclusion
- guaranteeing representative coverage over the different seasons of the year and different days of the week
- working within economic constraints, to keep both the researchers' transportation costs and the number of hours required by airport personnel to administer the survey at a reasonable level
- selecting a convenient sample design that would produce appropriate accuracy, allow airport personnel to closely supervise the procedure, and result in minimal administrative errors

Among the alternative sampling methods available, a combination or hybrid design appeared most able to meet both the statistical requirements and practical constraints. A year-long survey period was chosen, and an initial survey date (cluster) was selected at random. All flights on this calendar day were sampled (total enumeration within clusters) rather than distributing questionnaires to individual passengers on scattered individual flights (sampling within clusters). Succeeding survey dates were systematically spaced every 13th day after the initial date. This sample design produced a total of 28 survey days or clusters, 7 in each season of the year, with each day of the week represented once each quarter. Thus, all seasons were proportionally represented, as were all weekdays and weekend days in each season. By systematically spacing the sample clusters every 13th day, the researchers assumed that no bias was introduced for either the season of the year or day of the week. If a holiday happened (randomly) to be selected as a sample day, it was included in the study. In fact, to exclude a holiday purposely would have introduced an unwanted systematic bias into the analysis.

The overall result of this sample design was a type of *random systematic cluster* sample. Randomness existed because the initial survey date was randomly chosen, surveying passengers every 13th day added a systematic component, and giving all passengers questionnaires on the selected survey dates created a cluster effect.

6.3 SPATIAL SAMPLING

The types of sample design considered to this point have not been explicitly spatial in nature. Under certain circumstances, however, spatial sampling is necessary. Spatial sampling is applied when using a map of a continuously distributed variable (such as vegetative cover, soil type, or pH of surface water) and when a sample of locations is being selected from this map. If a geographer conducting fieldwork must select sample site locations within a defined target area, spatial sampling is also needed.

Spatial sampling from maps or other spatial sampling frames may involve point samples, line samples (traverses), or area samples (quadrats) (figure 6.6). Of these three types of spatial sampling, geographers use point sampling most frequently. For this reason, and because the various concepts involving spatial sampling procedures can be illustrated most easily through point sampling designs, this section focuses on point sampling. Those wanting details on quadrat and traverse sampling are referred to more specialized texts.

For all types of spatial sampling, a spatial sampling frame (such as a map) must be constructed. This frame must include a coordinate system that allows clear identification of locations within the sampled area. For point sampling from a map, designation of the (X, Y) Cartesian coordinate pair will identify a unique location for each point.

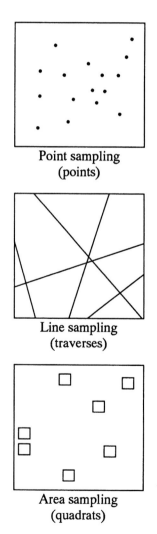

Point sampling
(points)

Line sampling
(traverses)

Area sampling
(quadrats)

FIGURE 6.6
Types of Spatial Sampling

In a **simple random point sample** (figure 6.7, case 1), two numbers (an X and Y coordinate pair) are selected to designate each point. To locate n sample points in a sampled area for a study, 2(n) random numbers must be drawn from a random numbers table or other source. The resulting pattern of points may be uneven, with some portions of the sampled area seemingly underrepresented, and others appearing overrepresented. Locational or spatial unevenness is a natural consequence of the randomization process.

Systematic point sampling is a convenient way to avoid the problems possible with an uneven distribution of points across the study area. Analogous to the regular sampling interval (k) from a list of individuals in nonspatial systematic sampling, a distance interval (figure 6.7, case 2) is used in systematic point sampling. A starting point is randomly selected; then all other points are located using the distance interval to space them evenly across the entire sampled area. Only one point (two numbers from the random numbers table) must be randomly located; the subsequent placement of all other points in the systematic pattern is determined automatically by the distance interval. This approach has the advantage of offering representative, proportional coverage of the entire sampled area. Bias will enter the sample design when a spatial regularity or periodicity exists in the distribution of some phenomenon that happens to match the distance interval. Systematic point sampling is widely used in geographic research, particularly when dealing with environmental and resource problems where data are continuously distributed across an area.

Just as in nonspatial cases, **stratified point sampling** in spatial sample design may be either proportional (constant-rate) or disproportional (variable-rate). The approach selected depends on the circumstances of the problem. Suppose a geographer is studying environmental degradation in a region that includes particularly vulnerable tidal and nontidal wetlands. In figure 6.7 (cases 3 and 4), stratum 1 shows the nonwetland portion of the spatial sampling frame (comprising 60 percent of the total sampled area), while stratum 2 identifies wetlands (the other 40 percent). Suppose proportional representation of both strata is desired. This would dictate that 60 percent of the sample points be placed in nonwetland locations and 40 percent on wetland sites. If a total of 20 sample points is sufficient, 12 points should be in nonwetland locations, and 8 points in wetlands (figure 6.7, case 3). Suppose instead that particular attention needs to be focused on possible environmental problems in the wetlands area. In this situation, the wetlands stratum needs more detailed monitoring or "oversampling," while maintaining

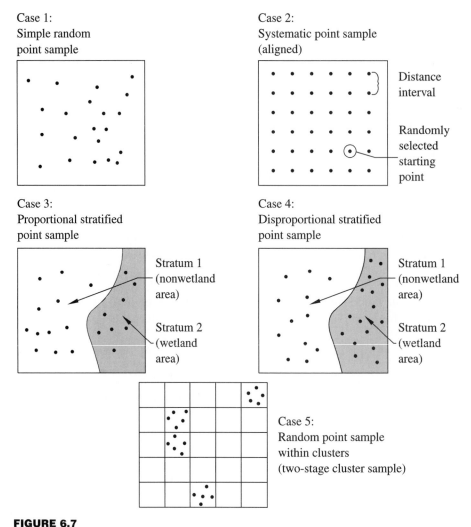

FIGURE 6.7
Types of Spatial Point Sampling

adequate coverage of the nonwetland portion of the sampled area. The disproportional stratified design (figure 6.7, case 4) shows twice the intensity of sample points in the wetland stratum.

In spatial point sampling, a **cluster design** has the important advantage of keeping travel costs and other logistical problems to a minimum because the points of a cluster are located together (figure 6.7, case 5). This feature makes point sampling within clusters particularly attractive in geographic projects having an extensive study area. In many field studies, a cluster approach may be the only practical alternative.

Although two methods of detailed analysis within clusters are available in nonspatial cluster sampling (figure 6.5), only the **two-stage cluster** approach is appropriate in spatial cluster sampling. In the spatial sampling of a variable continuously distributed across an area, an infinite number of points could be selected in each subarea. As a result,

total enumeration within clusters is virtually impossible in a spatial context.

However, a cluster approach to point sampling is not always advisable. If the phenomenon being studied varies from one section of the sampled area to another, it may be necessary to ensure that some point locations are selected from all sections or subareas. Cluster sampling excludes substantial parts of the study area, and such uneven areal representation can be a disadvantage in many geographic problems. As a result, the practical advantages of convenience and reduction in travel cost often have to be balanced against unrepresentative spatial bias.

In addition to simple point sampling, a number of combination or hybrid point sampling designs are also possible (figure 6.8). A composite sample design with its complex procedures is only worthwhile if it is likely to improve the accuracy of the sample in estimating the population.

Case 1:
Stratified systematic
unaligned

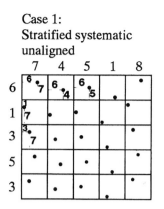

Case 2:
Disproportional stratified
systematic aligned

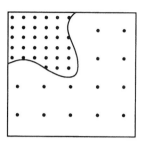

Case 3:
Cluster systematic

Case 4:
Disproportional stratified
cluster

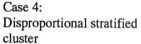

FIGURE 6.8
Selected Types of Composite or Hybrid Point Sampling

Perhaps the most widely used hybrid sampling model is the **stratified systematic unaligned point sample.** First suggested by Berry and Baker (1968), this approach includes the following steps:

1. Place a regular grid system over the entire study area, taking care to ensure that grid size is at an appropriate level of spatial resolution for the problem.
2. Select a random number for each row and column in the gridded area, where the row number specifies the horizontal position of points in that row and the column number specifies the vertical position of points in that column (figure 6.8, case 1).
3. Locate a single point in each grid using the appropriate row and column positions.

This procedure provides proportional representation of all segments of the sampled area, yet avoids possible problems with regularities or periodicities in the spatial pattern that can be encountered with an aligned systematic point sample (figure 6.7, case 2). The needed sample points can also be generated fairly easily. As a result of these advantages, a stratified systematic unaligned point sample is a good choice in many realistic geographic situations.

Other combination or hybrid sampling models can be designed, but should be used only if they are likely to improve the precision of the sampling problem. Geographers try to create models in such areas as climatology, resource management, and migration to replicate or duplicate reality as closely as possible. Similarly, in geographic sampling problems, the sample design must reflect the character of the situation. In this way, precise and accurate estimates can be obtained with reasonable effort. Other hybrid point sample design options include disproportional stratified systematic aligned (figure 6.8, case 2); cluster systematic (figure 6.8, case 3); and disproportional stratified cluster (figure 6.8, case 4).

Are any of these hybrid designs really feasible? Suppose a geographer is concerned with monitoring changes in both the spatial distribution and degree of intensity of nitrogen and phosphorus levels in a bay receiving agricultural runoff. In the narrow estuaries of the bay, runoff problems are likely to be more severe and will need careful monitoring. On the other hand, no portion of the bay should be left totally unmonitored. What type of spatial point sampling procedure should be used when locating monitor buoys in the bay? A disproportional stratified systematic aligned design might be the most practical alternative (figure 6.9). Disproportional stratification places an appropriate density of nutrient monitoring stations in the necessary locations. Systematic

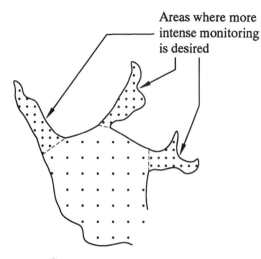

Areas where more
intense monitoring
is desired

· Sample point location

FIGURE 6.9
Location of Nutrient Monitoring Stations on a Bay:
Disproportional Stratified Systematic Aligned Sample

placement allows efficient collection of water samples, since the buoys are aligned regularly for pickup by boat. This sample design is tailor-made to fit the practical circumstances of this geographic problem.

KEY TERMS

basic types of sampling (simple random, systematic, stratified, cluster), 89
cluster sample designs (total enumeration within clusters, two-stage cluster sample), 94
combination or hybrid sampling designs, 92
oversampling and undersampling, 90
population, 82
probability and nonprobability sampling, 88
randomness, 84
sample, 82
sampled population and area, 87
sampling frame, 86
sources of sampling error, 84
strata, 90
stratified sample, 90
stratified sample designs (proportional or disproportional), 90
stratified systematic unaligned point sample, 95
systematic sample, 90
target population and area, 86
types of spatial sampling (simple random point sample, systematic point sample, stratified point sample, cluster point sample), 93

MAJOR GOALS AND OBJECTIVES

If you have mastered the material in this chapter, you should now be able to do the following.

1. Understand the various advantages of sampling in contrast to a complete enumeration or census of an entire population.
2. Identify possible sources of sampling error in a geographic problem or situation.
3. Understand the steps in the general sampling procedure from the conceptual definition of target population and target area to the collection of sample data.
4. Define and distinguish between these sampling terms: target population and target area, sampled population and sampled area, and sampling frame.
5. Explain the characteristics of the important types of probability sampling—including simple random, systematic, stratified, cluster, and combination or hybrid designs.
6. Select an appropriate sampling design (one able to meet both the statistical requirements and practical circumstances) when presented with a geographic situation or practical problem that requires sampling.
7. Explain the characteristics (including both advantages and disadvantages) of the various types of simple and composite point sampling.

REFERENCES AND ADDITIONAL READING

American Statistical Association, Survey Research Methods Section. 1998. *What Is a Survey? How to Plan a Survey, How to Collect Survey Data.* www.amstat.org.

Berry, B. J. L. and A. M. Baker. 1968. "Geographic Sampling," in B. J. L. Berry and D. F. Marble (editors). *Spatial Analysis: A Reader in Statistical Geography.* Englewood Cliffs, NJ: Prentice-Hall, pp. 91–100.

Dixon, C. J. and B. Leach. 1978. *Sampling Methods for Geographical Research.* Norwich, England: Geo Abstracts.

Gregory, S. 1978. *Statistical Methods and the Geographer* (4th edition). London: Longman.

Hanson, M. H. and W. N. Hurwitz (editors). 1993. *Sample Survey Methods and Theory.* New York: John Wiley and Sons.

Hauser, P. 1975. *Social Statistics in Use.* New York: Russell Sage Foundation.

Kish, L. 1995. *Survey Sampling* (Wiley Classics Library). New York: John Wiley and Sons.

McGrew, J. C. and R. A. Rosing. 1979. *Salisbury-Wicomico County Regional Airport Passenger Survey.* Salisbury: Delmarva Advisory Council.

Scheaffer, R. L., W. Mendenhall, and L. Ott. 1996. *Elementary Survey Sampling* (5th edition). Boston: Duxbury.

Thompson, S. K. 1992. *Sampling.* New York: John Wiley and Sons.

Estimation in Sampling

7.1 Basic Concepts in Estimation
7.2 Confidence Intervals and Estimation
7.3 Sample Size Selection

The primary objective of sampling is to make inferences about the population from which a sample is taken. More specifically, sample statistics are used to estimate population parameters such as the mean, total, and proportion. When sample statistics represent a larger population accurately, they are considered unbiased estimators.

This chapter considers both the theory and practice of estimating from samples. The basic terminology and the theoretical concepts that underlie sample estimation are discussed in section 7.1. Distinction is made between point estimation and interval estimation, and the nature of the sampling distribution of a statistic is explained. Also discussed is the importance of the central limit theorem when inferring sample results to a population. Confidence intervals that indicate the level of precision for a population estimate can be determined for any desired sample statistic. Section 7.2 includes a variety of confidence interval equations. Material is organized by type of sample (random, systematic, stratified) and by parameter (mean, total, proportion). A series of examples illustrates the procedure for constructing confidence intervals around each parameter using the different sampling methods.

To save time and effort, geographers often want to know—*before* taking a full sample—how large a sample is needed for a particular research problem. Issues and methods related to determining sample size before taking a full sample are discussed in section 7.3. Examples illustrate how sample size is established for problems using the mean, total, and proportion.

7.1 BASIC CONCEPTS IN ESTIMATION

Point Estimation and Interval Estimation

In statistics, a basic distinction is made between **point estimation** and **interval estimation.** The concept of point estimation is relatively straightforward; a statistic is calculated from a sample and then used to estimate the corresponding population parameter. With probability sampling, the "best" (unbiased) point estimate for a population parameter is the corresponding sample statistic. Therefore, the best point estimate for the population mean (μ) is the sample mean ($\overline{X}$); the best point estimate for the population standard deviation (σ) is the sample standard deviation (s), and so on (table 7.1).

Note that the denominator for calculating sample standard deviation is ($n - 1$) rather than (n). This slight adjustment makes the sample standard deviation a less biased estimator of the population standard deviation, particularly for smaller samples. With larger sample sizes (larger than 30 or so), the difference between dividing by (n) or dividing by ($n - 1$) is insignificant. However, using (n) with a smaller sample would result in an estimate that underrepresents the magnitude of the true standard deviation in the population.

How precise are sample point estimators? How close (or distant) from the true population parameter is the calculated sample statistic? Because probability sampling involves some uncertainty, it is unlikely that a sample statistic will exactly equal the true population parameter. What can be determined, however, is the likelihood that a sample statistic is within

TABLE 7.1

Point Estimators of Population Parameters

Descriptive statistic	Population parameter	Sample statistic*	Calculating formula
Mean	μ	$\bar{X}$	$\dfrac{\sum\limits_{i=1}^{n} X_i}{n}$
Standard deviation	σ	s	$\sqrt{\dfrac{\sum\limits_{i=1}^{n} (X_i - \bar{X})^2}{n-1}}$
Total	τ	T	$N(\bar{X}) = N\dfrac{\sum X_i}{n}$
Proportion**	ρ	p	$\dfrac{x}{n}$

* best point estimate
** x = number of units sampled having a particular characteristic
n = total number of units sampled

a certain range or interval of the population parameter. The determination of this range is the basis for interval estimation. A **confidence interval,** or bound, represents the level of precision associated with the population estimate. Its width is determined by (1) the sample size; (2) the amount of variability in the population; and (3) the probability level or level of confidence selected for the problem. These ideas are now explored in detail, using the mean as the focus of the discussion.

The Sampling Distribution of a Statistic

Suppose a random sample of size n is drawn from a population, and the mean of that sample ($\bar{X}$) is calculated. Now suppose a second sample of size n (totally independent of the first sample) is drawn and its mean calculated. If this process is repeated for many similar-sized independent samples in a population, the frequency distribution of this set of sample means can be graphed (figure 7.1). This curve is referred to as the **sampling distribution of sample means.**

A sampling distribution can be developed for *any* statistic, not just the mean. After many independent samples of size (n) are drawn from a population, the statistic of interest (e.g., mean, total, and proportion) can be calculated for each sample, and the distribution of the sample statistics graphed. The resulting frequency distribution has a shape, a mean, and a certain amount of variability (as reflected by the standard deviation and variance). These general characteristics of sampling distributions are important, but for now the focus is on particular characteristics of sample means.

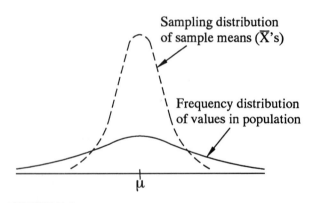

FIGURE 7.1
Sampling Distribution of Sample Means and Frequency Distribution of Population Values

The Central Limit Theorem

When the sample statistic is the mean, its frequency distribution has a particular set of features that are of vital theoretical importance in sampling and general statistical inference. These features are summarized in the **central limit theorem.**

The central limit theorem provides several important insights. Given the effect of randomness in drawing samples, some sample means will be above the population mean, and others will fall below it. If all samples are drawn independently, it seems logical that the mean of a set of sample means is μ. It also follows that a frequency distribution of sample means ($\bar{X}s$) is normally distributed and centered on the population mean (μ). The likelihood that a sample mean differs only slightly from μ is higher than the likelihood that a sample mean differs greatly from μ.

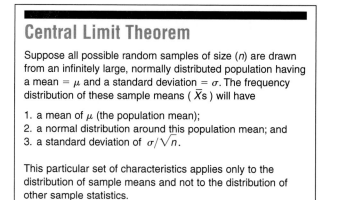

Central Limit Theorem

Suppose all possible random samples of size (*n*) are drawn from an infinitely large, normally distributed population having a mean = μ and a standard deviation = σ. The frequency distribution of these sample means ($\overline{X}$s) will have

1. a mean of μ (the population mean);
2. a normal distribution around this population mean; and
3. a standard deviation of $\sigma/\sqrt{n}$.

This particular set of characteristics applies only to the distribution of sample means and not to the distribution of other sample statistics.

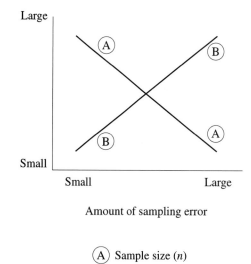

Amount of sampling error

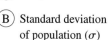

(A) Sample size (*n*)

(B) Standard deviation
of population (σ)

FIGURE 7.2
How the Amount of Sampling Error is Affected by Sample Size and Population Standard Deviation

When a sample is taken, large values in the population are generally counterbalanced by small values. The sample mean is likely to be quite close to the population mean, especially when sample size (*n*) is large. In fact, a large sample is more likely to be closer to the true population mean than a small sample.

The central limit theorem also provides insight into the variability of the sample means. According to this theorem, the standard deviation of the sampling distribution of means is equal to the population standard deviation divided by the square root of the sample size. This measure of standard deviation is called the **standard error of the mean** ($\sigma_{\overline{X}}$):

$$\sigma_{\overline{X}} = \frac{s}{\sqrt{n}} \qquad (7.1)$$

Notice the similar logic between standard deviation and standard error. The standard deviation indicates how much a typical value is likely to differ from the mean of a set of values. In a similar way, the standard error of the mean indicates how much a typical sample mean is likely to differ from the true population mean. Quite simply, standard error is a basic measure of the amount of **sampling error** in a problem.

The sample size and the population standard deviation influence the magnitude of sampling error (figure 7.2). First, the larger the sample size (*n*), the smaller the amount of sampling error ($\sigma_{\overline{X}}$). This has a certain appeal—the mean of a larger sample tends to be closer to the true population mean than the mean of a smaller sample. Second, the larger the standard deviation of the population (σ), the larger the amount of sampling error. This, too, seems logical, because a sample taken from a population containing a large amount of variability should have more error than a similar-sized sample taken from a population containing a small amount of variability.

How can the central limit theorem have such general applicability if the population from which samples are drawn is supposed to be normally distributed and infinitely large? Actually, these appar-

ently restrictive qualifications are not as severe as they may first seem.

In reality, many populations are not normally distributed, and some frequency distributions of interest to geographers are highly skewed and extremely nonnormal. Whatever the shape of the frequency distribution of the underlying population, however, the frequency distribution of sample means is approximately normal if the sample size is large enough. This is a very important result. Generally, a sample size of 30 or more is sufficient to guarantee an approximately normal distribution of sample means, no matter what shape the population distribution takes. Statisticians have established a sample size of 30 as the traditional breakpoint separating large samples from small samples.

Many problems of concern to geographers involve taking a sample from a population of finite size. However, the central limit theorem is completely true only for infinitely large populations. With a finite population, a **finite population correction (fpc)** may be incorporated into the estimation process:

$$\text{fpc} = \sqrt{\frac{N - n}{N - 1}} \qquad (7.2)$$

where fpc = finite population correction
N = population size
n = sample size

This correction factor reduces the amount of sampling error slightly and increases the precision of the sample statistic. When including the finite population correction, the standard error of the mean is calculated as

$$\sigma_{\overline{X}} = \frac{\sigma}{\sqrt{n}} \text{(fpc)} = \frac{\sigma}{\sqrt{n}} \sqrt{\frac{N - n}{N - 1}} \qquad (7.3)$$

The finite population correction is particularly important in the estimation process when the **sampling fraction,** defined as the ratio of sample size to population size (n/N), is large. If a relatively large sample is taken from a relatively small population, the sampling fraction will be large, and the finite population correction should be included in the sampling error formula to provide a more precise estimate. Conversely, if a relatively small sample is taken from a very large population, the sampling fraction will be small, and the fpc will be very close to 1. In the latter situation, the correction factor can be excluded from the standard error formula. If a sample of 25 is taken from a population of 10,000, the finite population correction is calculated as

$$\text{fpc} = \sqrt{\frac{N - n}{N - 1}} = \sqrt{\frac{10,000 - 25}{10,000 - 1}} = .9988$$

Because this fpc is so close to 1.0, multiplying the standard error by an fpc of this magnitude will not change the size of the standard error very much. Consequently, the width of the confidence interval (and the precision of the population estimate) will not change very much either. In general, the fpc value is included in population estimation equations if the sampling fraction is greater than .05. The **fpc rule** states: Include the fpc in the population estimate equations only when the ratio of sample size to population size exceeds 5 percent ($n/N > .05$).

In summary, the central limit theorem provides important information about the frequency distribution of sample means. Such a distribution is considered normal, has a mean of μ (the population mean), and a standard deviation of $\sigma/\sqrt{n}$ (called the standard error of the mean, $\sigma_{\overline{X}}$). However, the practical focus of geographic research is usually on a *single* sample mean ($\overline{X}$) drawn from a population having a mean of μ and standard deviation of σ. Although this single sample mean falls somewhere within the normal frequency distribution of sample means, its actual location is not known.

How precise or exact is this single sample mean? The central limit theorem allows this question to be answered. A confidence interval is placed about the sample mean, and the probability of the true population mean falling within this interval can be calculated. That is, a confidence interval is positioned around the sample mean, with a measurable level of confidence that the true population mean lies within that interval.

7.2 CONFIDENCE INTERVALS AND ESTIMATION

Suppose a geographer wants to place a confidence interval about a sample mean with 90 percent certainty that the interval range contains the actual population mean. The general formula for a confidence interval is

$$\overline{X} \pm Z\sigma_{\overline{X}} \qquad (7.4)$$

where $\overline{X}$ = sample mean
Z = Z-value from the normal table
$\sigma_{\overline{X}}$ = standard error of the mean

Because the central limit theorem establishes the normality of the distribution of sample means, Z-values from the normal table (appendix, table A) provide the proper probabilities for defining the confidence interval. Given a desired certainty of 90 percent, what is the corresponding Z-value or area under the normal curve? That is, what Z-value from the normal table is associated with the confidence interval having a 90 percent likelihood of containing the true population mean? The desired Z-value represents the situation where 90 percent of the total area under the curve is encompassed by the confidence interval and 45 percent of this total area is on either side of the mean (figure 7.3).

Look at the table of normal values in appendix A. According to this table, when the area A equals 4495, the Z-value is 1.64; and when the area A equals 4505, the Z-value is 1.65. Following the convention of rounding up if midway between two numbers, a Z-value of 1.65 is used for this confidence interval estimate. Therefore, one can be 90 percent confident that the true population mean lies within the confidence interval defined by

$$\overline{X} \pm 1.65 \, \sigma_{\overline{X}}$$

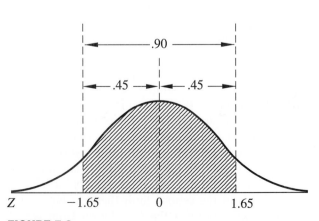

FIGURE 7.3
The Standard Scores (Z-Values) Associated with a 90 Percent Confidence

The confidence interval contains expressions that are added to and subtracted from the mean to define the **upper and lower bounds** of the interval:

$$\text{Upper bound} = \overline{X} + 1.65\,\sigma_{\overline{X}}$$

$$\text{Lower bound} = \overline{X} - 1.65\,\sigma_{\overline{X}}$$

Recall that the entire area under the sampling distribution curve of figure 7.3 represents the set of *all* possible sample means that could be drawn from the original population. The shaded area in the center of the sampling distribution shows the location of 90 percent of all the sample means that could be drawn from the population. Suppose ten actual samples are taken from this population, and the location of each sample mean is plotted (figure 7.4). Of the

ten sample means shown ($\overline{X}_1$ to $\overline{X}_{10}$), nine of them (.90 probability) are within the confidence interval. This result is expected because the interval was constructed with a 90 percent chance of containing μ. Of course, when taking a single sample, the location of μ is not known. However, a .90 probability exists that the interval or bound placed around that single sample mean *does* contain the true population mean. Conversely, a .10 probability exists that the confidence interval placed around the single sample mean *does not* include μ. The probabilities that an unusually large sample mean (well above (μ) or an unusually small sample mean (well below μ could be drawn are represented graphically by the two unshaded tails of the sampling distribution. Note that by chance, the fifth sample mean ($\overline{X}_5$) falls below the lower bound of the confidence interval around μ.

Several terms are used when making interval estimates in sampling. The **confidence level** refers to the probability that the interval surrounding a sample mean *encompasses* the true population mean. This confidence level probability is defined as $1 - \alpha$. The **significance level** refers to the probability that the interval surrounding a sample mean *fails to encompass* the true population mean. The significance level is denoted by α and equals the total sampling error. Because error is equally likely in either direction from μ, the probability of the sample mean falling into either tail of the distribution is $\alpha/2$.

General Procedure for Constructing a Confidence Interval

The best way to learn about the construction of confidence intervals is to present a simple example in some detail. Suppose that a random sample of 50 commuters in a metropolitan area revealed that their average journey-to-work distance was 9.6 miles. Moreover, a recent study has determined that the standard deviation of journey-to-work travel distances for this metropolitan area is approximately 3 miles. What is the confidence interval around this sample mean of 9.6 that guarantees with 90 percent certainty that the true population mean is enclosed within that interval?

The confidence interval for μ is calculated from the following values:

- the sample mean ($\overline{X} = 9.6$)
- the population standard deviation ($\sigma = 3$)
- the sample size ($n = 50$)
- the Z-value associated with a 90 percent confidence level (Z = 1.65)

The single best estimate of the true population mean (μ) is the sample mean ($\overline{X} = 9.6$). This sample mean is the statistic around which the confidence

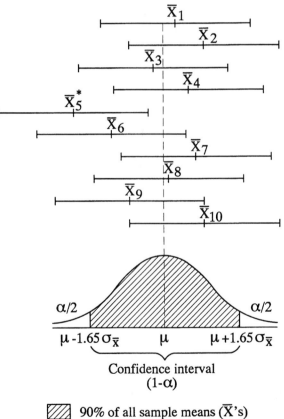

90% of all sample means ($\overline{X}$'s)

Confidence level (1-α) = .90
Significance level (α) = .10
Number of standard errors (Z) = 1.65

*Confidence interval around $\overline{X}_5$ fails to include μ

FIGURE 7.4

Distribution of Sample Means and the Confidence Interval Concept interval is placed. From equation 7.4, the confidence interval is defined as

$$\overline{X} \pm Z\sigma_{\overline{X}}$$

Substituting the value of the standard error, the confidence interval equation becomes

$$\overline{X} \pm Z\frac{\sigma}{\sqrt{n}}$$

Inserting all values for the journey-to-work problem:

$$\overline{X} \pm Z\frac{\sigma}{\sqrt{n}} = 9.6 \pm 1.65\frac{3}{\sqrt{50}} = 9.6 \pm 0.70$$

This interval ranges from a lower bound of 8.90 (9.6 − 0.70) to an upper bound of 10.30 (9.6 + 0.70). It can now be concluded with 90 percent certainty that the mean journey-to-work distance for all commuters is between 8.90 and 10.30 miles.

The finite population correction is not needed in this problem. When a relatively small sample is taken from a relatively large population, the correction is not necessary. Here, a sample of 50 is drawn from a very large metropolitan population, making the sampling fraction very low (definitely less than .05).

When dealing with confidence intervals, various situations arise. Sometimes sample sizes are small and often population standard deviations are unknown. Proper decisions must be made about how to proceed under these circumstances.

What level of confidence should be used? A confidence interval can be created for any desired level of confidence (1 − α). The most commonly accepted and widely used confidence levels are .99, .95, and .90. The corresponding likelihoods of making a sampling error (also called the significance level, α) when using these confidence levels are .01, .05, and .10. These conventional levels of confidence are generally considered rigorous enough to guarantee acceptably low probabilities of sampling error.

The impact of various confidence levels on significance levels, number of standard errors, and con-

fidence interval characteristics are now summarized for the journey-to-work problem (table 7.2). Notice what happens when the confidence level increases from .80 to .99. First, the significance level decreases from .20 to .01, reducing the sampling error for the problem. Also, the number of standard errors (Z) increases from 1.28 to 2.58, thereby widening the confidence interval range and width. Thus, higher levels of confidence result in wider confidence intervals, less precise estimates, but lower sampling error. The investigator has the responsibility for deciding how to trade off confidence level against level of precision.

What if the population standard deviation (σ) and variance (σ²) are unknown? In the previous example, it was assumed that the population standard deviation and variance were known, while the population mean (μ) was unknown. This assumption is unrealistic and impractical; in most situations, if the population mean is unknown, the standard deviation and variance of the population are also unknown. With an unknown population variance, the sample variance (s^2) provides the best estimator and is the statistic inserted into the standard error formula.

Recall the standard error formula, with the finite population correction (fpc) included (equation 7.3):

$$\sigma_{\overline{X}} = \frac{\sigma}{\sqrt{n}}(\text{fpc}) = \frac{\sigma}{\sqrt{n}}\sqrt{\frac{N-n}{N-1}}$$

Squaring both sides to convert standard deviations into variances:

$$\sigma_{\overline{X}}^2 = \frac{\sigma^2}{n}\left(\frac{N-n}{N-1}\right) \tag{7.5}$$

If the population variance (σ²) is unknown, the sample variance (s^2) is substituted:

$$\sigma_{\overline{X}}^2 = \frac{\sigma^2}{n}\left(\frac{N-n}{N-1}\right) = \frac{s^2}{n}\frac{(N-n)}{N} \tag{7.6}$$

To get this expression into an effective form to place confidence intervals (bounds) around point

---

TABLE 7.2

The Relationship between Confidence Level and Confidence Interval for the Journey-to-Work Example

Confidence level (1 − α)	Significance level (α)	Number of standard errors (Z)	Confidence interval ($\overline{X} \pm Z\sigma_{\overline{X}}$)	Confidence interval	
				Range	Width
.80	.20	1.28	9.6 ± .54	9.06 to 10.14	1.08
.90*	.10	1.65	9.6 ± .70	8.90 to 10.30	1.40
.95	.05	1.96**	9.6 ± .83	8.77 to 10.43	1.63
.98	.02	2.33	9.6 ± 1.00	8.60 to 10.60	2.00
.99	.01	2.58	9.6 ± 1.09	8.51 to 10.69	2.18

* Calculations for (1 − α) = .90 are detailed in situation 1.
** A Z-value of 2.00 is often used for 1.96 for convenience.

estimates, one final step is needed—take the square root of both sides.

$$\sigma_{\overline{X}} = \sqrt{\frac{s^2}{n}\left(\frac{N-n}{N}\right)} \qquad (7.7)$$

This standard error formula is the expression used for the many confidence interval equations that follow.

What if the sample size is small? The journey-to-work example incorporated the standard normal deviate (Z) into the confidence interval calculation. However, Z is valid only if the sample size is greater than 30. With smaller samples, the confidence interval equation must be altered, and the standard Z-value must be replaced by the corresponding value from the student's t distribution. This modification is necessary because the sample standard deviation (s) is not always accurate when the sample size is smaller than 30. In some situations, it may be impractical or impossible to have a sample size as large as 30. If the value of s appears to be inaccurate because of the small sample size, logic demands that a wider interval be presented as the confidence interval estimate for μ to avoid introducing bias. This wider interval allows an equivalent amount of confidence in the result that the true population mean lies within the interval.

Whenever the sample size is less than 30, the confidence interval formula is

$$\overline{X} \pm t\sigma_{\overline{X}} \qquad (7.8)$$

Like the normal distribution, the t distribution is symmetric and bell-shaped. The exact shape of the t distribution depends on the sample size; as n approaches 30, the value of t approaches the standard normal (Z) value. Two pieces of information are needed to use the t table (appendix, tables C and D):

- the desired significance level (α)—common levels are $\alpha = .10, .05,$ and .01.
- the number of degrees of freedom (df), where df is defined as one less than the sample size (df = $n - 1$). An explanation of degrees of freedom is included in chapter 8 in the context of inferential statistics.

For smaller sample sizes (30 or less), the t value will always be slightly larger than the Z-value at the corresponding level of significance, resulting in a slightly larger confidence interval, or bound, on the error of estimation. For example, when placing a 95 percent confidence interval ($\alpha = .05$) around a sample statistic when $n = 30$, $Z = 1.96$, while $t = 2.04$. This larger confidence interval for t is to be expected because the smaller the sample, the larger the sampling error and the greater the uncertainty that the sample precisely represents the population from which it is drawn.

All of the components are now in place to calculate various confidence intervals.

However, deciding which one to select depends on two factors:

- Different *sample types* require different confidence interval equations. The following discussion of confidence interval calculation will be limited to three sample types—random, systematic, and stratified
- Different *population parameters* require different confidence interval equations. The following discussion will be restricted to three parameters—mean, total, and proportion

These different sample types and population parameters can be combined to create six basic situations with regard to constructing confidence intervals. Each of these basic situations is illustrated with a geographic example. Other situations focusing on additional sampling designs, such as cluster, composite, or hybrid samples, are discussed in advanced texts that cover sampling more extensively. Interested readers should refer to the readings at the end of the chapter for more information.

Geographic Examples of Confidence Intervals

Nearly ten years have passed since the last census of population was taken in Middletown, and local planners wish to update selected demographic statistics. Tax records show a total of 3,500 households in the community. However, precise estimates of the following variables are desired:

1. The *mean* number of people per household. Planners suspect this figure has declined sharply over the recent intercensal period.
2. The *total* number of people in the community, based on the average number of people per household estimated from item (1).
3. The *proportion* of households with one or more children aged 18 years or younger.

Middletown planners decide to use either a simple random or systematic sample to calculate point estimates and enclose these parameters within confidence intervals. The procedures they followed are illustrated for the sample mean, total, and proportion.

Situation 1
Random or Systematic Sample—Estimate of Population Mean

To estimate and place bounds on the average number of people per household, Middletown planners take a simple random or systematic sample of 25 households. From the sample data, the sample mean and variance are $\overline{X} = 2.73$ and $s^2 = 2.6$. They wish to determine the confidence interval that contains the true average household size of Middletown, with a 90 percent certainty $(1 - \alpha = .90)$. The relevant values for this problem are

$$\overline{X} = 2.73 \quad s^2 = 2.6 \quad N = 3{,}500 \quad n = 25 \quad \alpha = .10$$

Since $n < 30$, t rather than Z should be used in the confidence interval equation (shown in table 7.3). Since $\alpha = .10$ and degrees of freedom $= n - 1 = 24$, the value from the t table is 1.71. Because the sampling fraction (n/N) is less than .05, the finite population correction can be ignored. The confidence interval in this problem is calculated as

$$\overline{X} \pm t\sqrt{\frac{s^2}{n}\left(\frac{N-n}{N}\right)} \tag{7.9}$$

$$2.73 \pm 1.71\sqrt{\frac{2.6}{25}} = 2.73 \pm .552$$

Middletown planners are 90 percent certain that the true mean household size falls within the interval of 2.178 to 3.282.

A confidence interval of this width may not provide enough precision to permit practical policy decisions. One of two strategies is available to narrow the interval—lower the confidence level from .90 or increase the sample size to above 25. The latter strategy requires more effort, but the larger sample size will permit the confidence interval to be narrowed.

Suppose the decision is made to increase the number of households surveyed from 25 to 250. Keeping the level of confidence at .90, a narrower, more precise confidence interval results. Suppose the mean and variance of this larger survey of households are calculated from the sample data as $\overline{X} =$

2.68 and $s^2 = 4.3$. With $n = 250$, two changes occur that help narrow the width of the confidence interval. First, since $n > 30$, Z rather than t is used in the confidence interval formula. With $\alpha = .10$, the value from the normal table becomes $Z = 1.65$. Second, the finite population correction (fpc) is useful here because the sampling fraction n/N is greater than .05. The revised confidence interval is calculated as

$$\overline{X} \pm Z\hat{\sigma}_{\overline{X}} = \overline{X} \pm Z\sqrt{\frac{s^2}{n}\left(\frac{N-n}{N}\right)} \tag{7.10}$$

$$= 2.68 \pm 1.65\sqrt{\frac{4.3}{250}\left(\frac{3{,}500-250}{3{,}500}\right)}$$

$$= 2.68 \pm .209$$

With this larger sample, planners are now 90 percent certain that the true mean household size in Middletown, μ, falls within the narrower interval from 2.471 to 2.889.

Situation 2
Random or Systematic Sample—Estimate of Population Total

Middletown planners can use the estimated average number of people per household to determine the total number of people in their community. The best estimate of the population total $(\hat{\tau})$ is the sample total (T), which is the sample mean $(\overline{X})$ multiplied by the population size (N):

$$\hat{\tau} = T = N\overline{X}$$

For the Middletown problem, the best estimate of the population total is

$$\hat{\tau} = T = 3{,}500\,(2.68) = 9{,}380$$

Using the equations in table 7.4, the confidence interval around the sample total can be determined by

$$T \pm Z\sqrt{N^2\left(\frac{s^2}{n}\right)\left(\frac{N-n}{N}\right)} \tag{7.11}$$

When $\alpha = .10$ and $Z = 1.65$, the confidence interval for the total population in Middletown is calculated as

$$9{,}380 \pm 1.65\sqrt{3{,}500^2\left(\frac{4.3}{250}\right)\left(\frac{3{,}500-250}{3{,}500}\right)}$$

$$= 9{,}380 \pm 729.84$$

Planners are 90 percent certain that the true total number of people in the community, τ, is found within the interval from 8,650.16 to 10,109.84.

TABLE 7.3

Confidence Interval Equations for Random and Systematic Samples: Population Parameter—Mean $(\mu)^*$

Best point estimate:

$$\hat{\mu} = \overline{X} = \frac{\Sigma X_i}{n}$$

Standard error of the point estimate (sampling error):

$$\hat{\sigma}_{\overline{X}} = \sqrt{\frac{s^2}{n}\left(\frac{N-n}{n}\right)} \text{ where}$$

$$s^2 = \frac{\Sigma(X_i - \overline{X})^2}{n-1}$$

Confidence interval (bound) around the point estimate:

$$\overline{X} \pm Z\hat{\sigma}_{\overline{X}} = \overline{X} \pm Z\sqrt{\frac{s^2}{n}\left(\frac{N-n}{N}\right)}$$

---

*Exclude the finite population correction from the confidence interval equations, if $n/N < .05$. Replace Z with the corresponding t if $n < 30$.

TABLE 7.4

Confidence Interval Equations for Random and Systematic Samples: Population Parameter—Total (τ)*

Best point estimate: $\hat{\tau} = T = N\bar{X} = \dfrac{N\Sigma X_i}{n}$

Standard error of
the point estimate $\hat{\sigma}_\tau = \sqrt{N^2 \left(\dfrac{s^2}{n}\right)\left(\dfrac{N-n}{N}\right)}$
(sampling error):

Confidence interval
(bound) around the $T \pm Z\hat{\sigma}_\tau = T \pm Z\sqrt{N^2\left(\dfrac{s^2}{n}\right)\left(\dfrac{N-n}{N}\right)}$
point estimate:

*Exclude the finite population correction from the confidence interval equations if $n/N < .05$. Replace Z with the corresponding t if $n < 30$.

Situation 3
Random or Systematic Sample—Estimate of Population Proportion

The best estimate of the proportion of all households in Middletown with one or more children 18 years of age or younger is the sample proportion. A survey of 250 households in Middletown reveals 105 with children. The best point estimate of the population proportion ($\hat{\rho}$) is the sample proportion (p), which is the number in the sample having the specified characteristic (x) divided by the total sample size (n).

$$\hat{\rho} = p = \frac{x}{n}$$

The best estimate of the proportion of Middletown households having children is

$$\hat{\rho} = p = \frac{x}{n} = \frac{105}{250} = .42$$

Using the equations in table 7.5, the confidence interval around this population estimate of the proportion is calculated as

$$p \pm Z\sqrt{\left(\frac{p(1-p)}{n-1}\right)\left(\frac{N-n}{N}\right)} \qquad (7.12)$$

To enclose the true population proportion with 90 percent confidence ($1 - \alpha = .90$ and $Z = 1.65$), the confidence interval is

$$.42 \pm 1.65\sqrt{\left(\frac{(.42)(.58)}{250-1}\right)\left(\frac{3,500-250}{3,500}\right)}$$

$$= .42 \pm .05$$

Middletown planners can conclude with 90 percent certainty that ρ, the true proportion of households with children, is in the interval from .370 to .470.

Many geographic problems use sampling designs other than simple random or systematic. When geographers work with stratified samples, alternative equations are needed to place confidence intervals around point estimates (tables 7.6, 7.8, and 7.10). Problems follow that show calculation procedures for a sample mean (situation 4), total (situation 5), and proportion (situation 6).

Situation 4
Stratified Sample—Estimate of Population Mean

School officials wish to evaluate the "geographic competencies" of all high school juniors in a large school district. One group of students has completed an introductory geography course taught in several of the district's high schools, whereas another group of students has not. School officials suspect that the geographic competencies of these two groups of juniors are quite different. If a simple random sample is taken of all juniors, either of the groups could be over- or underrepresented in the sample. Therefore, a stratified sample design, with proportional representation of each group of students, is appropriate.

To evaluate students' geographic skills, school officials decide to use the Secondary-Level Geography Test designed by the National Council for Geographic Education (NCGE). This test evaluates a student's basic geographic knowledge in three areas: geographic skills, physical geography, and human

TABLE 7.5

Confidence Interval Equations for Random and Systematic Samples: Population Parameter—Proportion (ρ)*

Best point estimate: $\hat{\rho} = p = \dfrac{x}{n}$

Standard error of the point estimate (sampling error): $\hat{\sigma}_p = \sqrt{\left(\dfrac{p(1-p)}{n-1}\right)\left(\dfrac{N-n}{N}\right)}$

Confidence interval (bound) around the point estimate: $p \pm Z\hat{\sigma}_p = p \pm Z\sqrt{\left(\dfrac{p(1-p)}{n-1}\right)\left(\dfrac{N-n}{N}\right)}$

*Exclude the finite population correction from the confidence interval equations if $n/N < .05$. Replace Z with the corresponding t if $n < 30$.

TABLE 7.6

Confidence Interval Equations for Stratified Samples: Population Parameter—Mean (μ)*

Best point estimate:

$$\hat{\mu} = \bar{X} = \frac{1}{N} \sum_{i=1}^{m} N_i \bar{X}_i = \frac{1}{N} [N_1 X_1 + \ldots + N_m \bar{X}_m]$$

Standard error of the point estimate (sampling error):

$$\hat{\sigma}_{\bar{X}} = \sqrt{\frac{1}{N^2} \sum_{i=1}^{m} N_i^2 \left(\frac{s_i^2}{n_i}\right) \left(\frac{N_i - n_i}{N_i}\right)}$$

$$= \sqrt{\frac{1}{N^2} \left[N_1^2 \left(\frac{s_1^2}{n_1}\right) \left(\frac{N_1 - n_1}{N_1}\right) + \ldots + N_m^2 \left(\frac{s_m^2}{n_m}\right) \left(\frac{N_m - n_m}{N_m}\right) \right]}$$

Confidence interval (bound) around the point estimate:

$$\bar{X} \pm Z\hat{\sigma}_{\bar{X}} = \bar{X} \pm Z\sqrt{\frac{1}{N^2} \sum_{i=1}^{m} N_i^2 \left(\frac{s_i^2}{n_i}\right) \left(\frac{N_i - n_i}{N_i}\right)}$$

*Exclude the finite population correction from the confidence interval equations if $n/N < .05$. Replace Z with the corresponding t if $n < 30$.
m = number of strata; subscript i refers to stratum i, N_i = size of population stratum i, n_i = size of sample from stratum i, $\bar{X}_i$ = sample mean of stratum i, s_i^2 = variance of stratum i, N = total population of all strata ($N = N_1 + N_2 + \ldots + N_m$).

TABLE 7.7

Estimate of Stratified Sample Mean—Situation 4

Task: Estimate the mean test score of all high school juniors on the Secondary-Level Geography Test, based on a stratified sample, and place a 95% ($\alpha = .05$) confidence interval (bounds) around the estimate.

Stratum 1	**Stratum 2**
Students who have taken introductory geography course	Students who have not taken introductory geography course
$N_1 = 480$ $n_1 = 20$	$N_2 = 1680$ $n_2 = 70$
$\bar{X}_1 = 64.36$ $s_1^2 = 65.61$	$\bar{X}_2 = 51.18$ $s_2^2 = 90.25$

The best estimate of the true population mean (μ) is the stratified sample mean ($\bar{X}$):

$$\hat{\mu} = \bar{X} = \frac{1}{N} \sum_{i=1}^{m} N_i \bar{X}_i$$

$$= \frac{1}{2160} [480(64.36) + 1680(51.18)] = 54.11$$

The confidence interval around $\bar{X}$ is

$$\bar{X} \pm Z\hat{\sigma}_{\bar{X}} = \bar{X} \pm Z\sqrt{\frac{1}{N^2} \sum_{i=1}^{m} N_i^2 \left(\frac{s_i^2}{n_i}\right) \left(\frac{N_i - n_i}{N_i}\right)}$$

Exclude the finite population correction $\left(\dfrac{N_i - n_i}{N_i}\right)$ as $(n/N) < .05$ for each stratum:

$$\bar{X} \pm Z\sqrt{\frac{1}{N^2} \sum_{i=1}^{m} N_i^2 \left(\frac{s_i^2}{n_i}\right)}$$

$$= 54.11 \pm 1.96\sqrt{\frac{1}{2160^2}\left[(480)^2 \left(\frac{65.61}{20}\right) + (1680)^2 \left(\frac{90.25}{70}\right) \right]}$$

$$= 54.11 \pm 1.90$$

With 95% certainty, the mean test score of all high school juniors, μ, is within the interval from 52.21 to 56.01.

geography. Of 2,160 total juniors in the school district, 480 have completed the geography course, and 1,680 have not. School officials have only 90 test booklets from NCGE and must restrict their sample to 90 students. The procedure for estimating the mean test score from a stratified sample of students is shown in table 7.7.

Situation 5
Stratified Sample—Estimate of
Population Total

In a county with 430 farms, the agricultural extension agent wants to estimate the total number of uncultivated farm acres in woodlots or fallow. The agent believes that the number of uncultivated acres will vary by size of farm, so a simple random sample could result in certain farm sizes being over- or underrepresented. Taking a stratified sample that represents each of the farm size categories proportionally is a better sample design. In the county, 172 farms are less than 50 acres, 107 farms are 50 to 99 acres, 65 farms are 100 to 299 acres, 42 are 300 to 499 acres, and 44 are more than 500 acres. The procedure for estimating the total number of uncultivated farm acres from a stratified sample of farms is shown in table 7.9. Note that the relatively small sample sizes and the extreme differences in variability within the different strata combine to produce a very wide (imprecise) confidence interval.

Situation 6
Stratified Sample—Estimate of
Population Proportion

An urban park and recreation department wants to investigate the public support for a bond issue to finance various park improvements. They want to estimate the proportion of the population in several neighborhoods adjacent to a city park that favors a proposed tax levy. Three different neighborhoods (with residents likely to have different attitudes about the bond issue) are situated adjacent to the park. Neighborhood 1 contains 155 households, neighborhood 2 has 62 households, and neighborhood 3 has 95. Park and recreation officials decide a stratified sample is appropriate to ensure proportional representation of each neighborhood in the sample. Suppose officials also decide that a total sample size of 80 households is sufficiently large and practical. The procedure for estimating the proportion of residents in favor of the bond issue from a stratified sample of neighborhoods is shown in table 7.11. The very narrow interval around the estimated sample proportion is caused by the relatively large sampling fractions from all strata and the similar estimated proportions from each stratum. A very precise population estimate is highly desirable.

7.3 SAMPLE SIZE SELECTION

In problems using sampling, geographers often want to determine the minimum sample size needed to make sufficiently precise estimates *before* the complete sample is actually taken. Taking a sample larger than necessary wastes both time and effort. Some of the major factors to consider in selecting sample size are

- the type of sample (random, stratified, etc.),
- the population parameter being estimated (mean, total, proportion),
- the degree of precision (width of confidence interval that can be tolerated), and
- the level of confidence to be obtained for the estimate.

As sample size increases, key trade-offs occur. At a particular confidence level (e.g., .95), increasing

---

TABLE 7.8

Confidence Interval Equations for Stratified Samples: Population Parameter—Total (τ)*

Best point estimate:

$$\hat{\tau} = T = \sum_{i=1}^{m} N\bar{X}_i = [N_1\bar{X}_1 + \ldots + N_m\bar{X}_m]$$

Standard error of the point estimate (sampling error):

$$\hat{\sigma}_\tau = \sqrt{\sum_{i=1}^{m} N_i^2 \left(\frac{s_i^2}{n_i}\right)\left(\frac{N_i - n_i}{N_i}\right)}$$

Confidence interval (bound) around the point estimate:

$$T \pm Z\sigma_\tau = T \pm Z\sqrt{\sum_{i=1}^{m} N_i^2\left(\frac{s_i^2}{n_i}\right)\left(\frac{N_i - n_i}{N_i}\right)}$$

---

*Exclude the finite population correction from the confidence interval equations if $n/N < .05$. Replace Z with the corresponding t if $n < 30$.
m = number of strata; subscript i refers to stratum i, N_i = size of population stratum i, n_i = size of sample from stratum i, $\bar{X}_i$ = sample mean of stratum i,
N = total population of all strata ($N = N_1 + N_2 + \ldots + N_m$)

TABLE 7.9

Estimate of Stratified Sample Total—Situation 5

Task: Estimate the total number of uncultivated farm acres in woodlots or fallow in the county, based on a stratified sample, and place a 90% ($\alpha = .10$) confidence interval (bound) around the estimate. The number of uncultivated acres from each sampled farm is individually listed under each stratum.

Stratum 1				Stratum 2			Stratum 3			Stratum 4		Stratum 5	
(< 50 acres)				(50–99 acres)			(100–299 acres)			(300–499 acres)		(≥ 500 acres)	
6	2	17	4	12	27	16	120	40	85	212	170	220	380
10	20	10	26	50	20	40	10	55	32	40	110	60	80
23	16	15	3	4	10								
5	12	8	40										

Stratum 1	Stratum 2	Stratum 3	Stratum 4	Stratum 5
$N_1 = 172$	$N_2 = 107$	$N_3 = 65$	$N_4 = 42$	$N_5 = 44$
$n_1 = 16$	$n_2 = 8$	$n_3 = 6$	$n_4 = 4$	$n_5 = 4$
$\bar{X}_1 = 13.56$	$\bar{X}_2 = 22.37$	$\bar{X}_3 = 57.0$	$\bar{X}_4 = 133.0$	$\bar{X}_5 = 185.0$
$s_1^2 = 102.01$	$s_2^2 = 248.69$	$s_3^2 = 1576.09$	$s_4^2 = 5596.54$	$s_5^2 = 21966.20$

The best estimate of the true population total τ is the stratified sample total (T):

$$\hat{\tau} = T = \sum_{i=1}^{m} N_i \bar{X}_i = 172(13.56) + 107(22.37) + 65(57.0) + 42(133.0) + 44(185.0) = 22156.91$$

The confidence interval about T is

$$T \pm Z\hat{\sigma}_T = T \pm Z\sqrt{\sum_{i=1}^{m} N_i^2 \left(\frac{s_i^2}{n_i}\right)\left(\frac{N_i - n_i}{N_i}\right)}$$

Include the finite population correction $\left(\dfrac{N_i - n_i}{N_i}\right)$ as (n/N) > .05 for each stratum:

$$T \pm Z\sqrt{\sum_{i=1}^{m} N_i^2 \left(\frac{s_i^2}{n_i}\right)\left(\frac{N_i - n_i}{N_i}\right)}$$

$$= 22156.91 \pm 1.65\sqrt{(172)^2 \left(\frac{102.01}{16}\right)\left(\frac{172 - 16}{172}\right) + (107)^2 \left(\frac{248.69}{8}\right)\left(\frac{107 - 8}{107}\right) + \ldots}$$

$$\ldots (65)^2 \left(\frac{1576.09}{6}\right)\left(\frac{65 - 6}{65}\right) + (42)^2 \left(\frac{5596.54}{4}\right)\left(\frac{42 - 4}{42}\right) + (44)^2 \left(\frac{21966.20}{4}\right)\left(\frac{44 - 4}{44}\right)}$$

$$= 22156.91 \pm 6041.33$$

With 90% certainty, the total number of uncultivated farm acres in woodlots or fallow in the county, τ, is within the interval from 16,115.58 to 28,198.24.

TABLE 7.10

Confidence Interval Equations for Stratified Samples: Population Parameter—Proportion (ρ)*

Best point estimate:

$$\hat{\rho} = p = \frac{1}{N}\sum_{i=1}^{m} N_i p_i = \frac{1}{N}\sum_{i=1}^{m} N_i \frac{x_i}{n_i} = \frac{1}{N}\left[N_1\frac{x_1}{n_1} + \ldots + N_m\frac{x_m}{n_m}\right]$$

Standard error of the point estimate (sampling error):

$$\hat{\sigma}_p = \sqrt{\frac{1}{N^2}\sum_{i=1}^{m} N_i^2 \left(\frac{p_i(1 - p_i)}{n_i - 1}\right)\left(\frac{N_i - n_i}{N_i}\right)}$$

Confidence interval (bound) around the point estimate:

$$p \pm Z\hat{\sigma}_p = p \pm Z\sqrt{\frac{1}{N^2}\sum_{i=1}^{m} N_i^2 \left(\frac{p_i(1 - p_i)}{n_i - 1}\right)\left(\frac{N_i - n_i}{N_i}\right)}$$

*Exclude the finite population correction from the confidence interval equations if n/N < .05. Replace Z with the corresponding t if n < 30.
m = number of strata; subscript i refers to stratum i, N_i = size of population stratum i, p_i = sample proportion with characteristic of stratum i; N = total population of all strata ($N = N_1 + N_2 + \ldots + N_m$)

TABLE 7.11

Estimate of Stratified Sample Proportion—Situation 6

Task: Estimate the proportion of all households favoring a bond issue to finance various park improvements, based on a stratified sample, and place a 90% (α = .10) confidence interval (bound) around the estimate.

Stratum 1 (Neighborhood 1)	**Stratum 2** (Neighborhood 2)	**Stratum 3** (Neighborhood 3)
$N_1 = 155$	$N_2 = 62$	$N_3 = 95$
$n_1 = 40$	$n_2 = 16$	$n_3 = 24$
$x_1 = 22$	$x_2 = 12$	$x_3 = 17$
$p_1 = \dfrac{x_1}{n_1} = \dfrac{22}{40} = .55$	$p_2 = \dfrac{x_2}{n_2} = \dfrac{12}{16} = .75$	$p_3 = \dfrac{x_3}{n_3} = \dfrac{17}{24} = .71$

where x_i = number of households from sample stratum i favoring the bond issue

The best estimate of the true population proportion (ρ) is the stratified sample proportion (p):

$$\hat{\rho} = p = \frac{1}{N}\sum_{i=1}^{m} N_i p_i = \frac{1}{312}\left[155(.55) + 62(.75) + 95(.71)\right] = .64$$

The confidence interval about p is

$$p \pm Z\hat{\sigma}_p = p \pm Z\sqrt{\frac{1}{N^2}\sum_{i=1}^{m} N_i^2 \left(\frac{p_i(1-p_i)}{n_i - 1}\right)\left(\frac{N_i - n_i}{N_i}\right)}$$

Include the finite population correction $\left(\dfrac{N_i - n_i}{N_i}\right)$ as $(n/N) > .05$ for each stratum:

$$p \pm Z\sqrt{\frac{1}{N^2}\sum_{i=1}^{m} N_i^2 \left(\frac{p_i(1-p_i)}{n_i - 1}\right)\left(\frac{N_i - n_i}{N_i}\right)} =$$

$$.64 \pm 1.65\sqrt{\frac{1}{(312)^2}\left[(155)^2\left(\frac{(.55)(.45)}{40-1}\right)\left(\frac{155-40}{155}\right) + (62)^2\left(\frac{(.75)(.25)}{16-1}\right)\left(\frac{62-16}{52}\right) + (95)^2\left(\frac{(.71)(.29)}{24-1}\right)\left(\frac{95-24}{95}\right)\right]}$$

$$= .64 \pm 1.65(.0021) = .64 \pm .0035$$

With 90% certainty, the proportion of all households favoring a bond issue to finance park improvements, ρ, is within the interval from .6365 to .6435.

---

the sample size provides greater precision and narrows the confidence interval width around the population estimate. Similarly, at a particular level of precision (e.g., estimating the population proportion within .03 of its true value), increasing the sample size will raise the level of confidence that the estimate is within the selected interval.

Unfortunately, a larger sample generally requires more time and effort. In many practical sampling problems, the investigator has multiple conflicting objectives. He or she must take a sample large enough to achieve the desired precision level and confidence interval width, *but simultaneously* avoid taking too large a sample, which wastes time and effort and provides estimates more precise than necessary. When samples get much larger than needed, the extra effort yields smaller and smaller incremental improvements in precision. In other words, the extra effort is more costly and yields fewer benefits.

This section illustrates how an appropriate sample size is determined from random sampling for these three basic population parameters: mean (μ), total (τ), and proportion (ρ).

Sample Size Selection—Mean

Suppose the task is to determine the minimum sample size needed to place the population *mean* estimate within a desired confidence interval around the true population mean. After choosing a confidence interval having an acceptable range, the issue becomes finding the appropriate sample size needed to place the sample statistic within that range. Simply stated, how large a sample is needed to get within the stated range of the actual population mean?

Recall that the confidence interval for μ is $\overline{X} \pm Z\hat{\sigma}_{\overline{X}}$, where $\hat{\sigma}_{\overline{X}}$ is the standard error of the mean (sampling error). Let E (for **Error**) designate the amount of sampling error one is willing to tolerate. That is, for any particular problem, a maximum acceptable difference separates the sample mean statistic from its population mean:

$$E = Z\hat{\sigma}_{\overline{X}} = Z\sqrt{\frac{\sigma^2}{n}} \qquad (7.13)$$

The magnitude of tolerable sampling error depends on the circumstances of the problem. In some applications, a considerable amount of error can be tolerated, making E a large number. In other instances, only a limited amount of error can be accepted, so E must be a smaller number. For example, a political party taking a survey many months before an election wants only a rough estimate of a candidate's popularity and is willing to accept an E value of large magnitude. Just a few days before the election, however, the party wants a more precise estimate of the expected election outcome; so a small-sized E value is desired.

The tolerable sampling error (E) is equivalent to one-half the width of the confidence interval. After some algebraic manipulation, the sample size (n) needed to estimate μ with a certain level of precision or tolerable error (E), at a chosen level of confidence (Z), can be expressed as

$$E = Z\sqrt{\frac{\sigma^2}{n}}$$

$$E^2 = Z^2\left(\frac{\sigma^2}{n}\right) \qquad (7.14)$$

$$n E^2 = Z^2 \sigma^2$$

$$n = \frac{Z^2 s^2}{E^2} = \left(\frac{Z\sigma}{E}\right)^2$$

If the population standard deviation (σ) is unknown, the sample standard deviation (s) is substituted:

$$n = \left(\frac{Zs}{E}\right)^2 \qquad (7.15)$$

The required minimum sample size is directly related to the desired level of confidence (Z) and variability in the sample (s), but inversely related to the degree of error (E) the investigator is willing to tolerate. In the political example, this relationship suggests that sample size will need to be increased in the second polling taken just a few days before the election to ensure that sampling error (confidence interval range) is reduced to an acceptable level.

In most real-world examples, the population standard deviation is not known. The sample standard deviation value is usually substituted to determine an appropriate sample size. A value for s is best derived from a pretest or preliminary sample. However, s could also be obtained from some previous study, or, if no better alternative exists, s could be an "educated guess" based on past experience.

Since a population parameter is being estimated, at least 30 observations should be included in a pretest or preliminary sample. The geographer may return to the general population to draw additional sample units if needed. These new units can then be combined with the original sample units to create a single larger sample that meets the size requirement. If this "return" procedure is used, it is considered a **two-stage sampling design.**

Sample Size Selection Example—Mean

Middletown planners wish to estimate the *mean* number of people per household and be 90 percent confident that their estimate will be within .3 persons of the true population mean, μ. What is the minimum number of households that must be surveyed to ensure this degree of precision at this selected confidence level?

Suppose a preliminary sample of 30 households is drawn, and s is calculated as 1.25. Since the population standard deviation is unknown, equation 7.15 is used:

$$n = \left(\frac{Zs}{E}\right)^2 = \left(\frac{(1.65)(1.25)}{0.3}\right)^2 = 47.26$$

Therefore, a random sample of at least 48 households should be taken to ensure the desired degree of precision at the 90 percent confidence level. It is always better to include an extra observation or two above the absolute minimum necessary sample size, so rounding up to 48 households is appropriate. Since this result was obtained from a preliminary sample of 30 households in a two-stage sampling design, only 18 additional households need be contacted to complete the study satisfactorily.

Sample Size Selection—Total

The minimum sample size needed to make an interval estimate of a population *total* within a certain tolerable error level (E) can also be determined. The confidence interval for τ is $T \pm Z\hat{\sigma}_\tau$, where $\hat{\sigma}_\tau$ is the sampling error or standard error of the total:

$$E = Z\hat{\sigma}_\tau = Z\sqrt{N^2\left(\frac{\sigma^2}{n}\right)} \qquad (7.16)$$

Algebraic manipulation isolates (n), the minimum sample size needed to estimate τ, with a selected level of tolerable error (E) at a chosen confidence level (Z):

$$n = \left(\frac{NZ\sigma}{E}\right)^2 \qquad (7.17)$$

If the population standard deviation (σ) is unknown, the sample standard deviation (s) is substituted:

$$n = \left(\frac{NZs}{E}\right)^2 \qquad (7.18)$$

Once again, the population standard deviation is seldom known and must be estimated with the sample standard deviation, which generally requires a pretest or preliminary sample.

Sample Size Selection Example—Total

Middletown planners also want to estimate the *total* number of people in the community. Tax records show a total of 3,500 households, but the number of people per household is not known. Suppose the planners wish to estimate total community population within 1,000 people and be 90 percent confident with that level of precision and sampling error (E). What is the minimum number of households that must be surveyed?

From a pretest or preliminary survey of 30 households, the standard deviation of the number of people per household(s) is calculated as 2.05. Because the population standard deviation is not known, the sample size needed to meet the requirements of the problem is calculated from equation 7.18:

$$n = \left(\frac{NZs}{E}\right)^2 = \left(\frac{(3,500)(1.65)(2.05)}{1,000}\right)^2 = 140.16$$

A random sample of at least 141 households should be taken to estimate the total population within 1,000 people and be 90 percent confident in that level of precision. Since 30 households were already surveyed in the preliminary survey, another 111 households should be contacted in the second stage of the survey design.

Sample Size Selection—Proportion

To estimate a population *proportion* within a certain allowable level of error (E), the minimum sample size can also be calculated in advance of full sampling. Again, a pretest or preliminary survey can be used to estimate the population proportion (ρ) from the sample proportion (p). The confidence interval for ρ is $p \pm Z\hat{\sigma}_p$, where $\hat{\sigma}_p$ is the sampling error or standard error of the proportion:

$$E = Z\hat{\sigma}_p = Z\sqrt{\frac{\rho(1-\rho)}{n-1}} \qquad (7.19)$$

The researcher can isolate the minimum sample size (n) algebraically as

$$n = \frac{Z^2\,\rho(1-\rho)}{E^2} \qquad (7.20)$$

If the population proportion (ρ) is unknown, the sample proportion (p) is substituted:

$$n = \frac{Z^2p(1-p)}{E^2} \qquad (7.21)$$

Unlike the mean and total, it is possible to determine the minimum sample size needed to estimate a population proportion *without* taking a pretest or preliminary sample. Note that the numerator of equation 7.21 contains the product $p(1-p)$. Consider the range of values this product can take given different values of p:

p	.1	.2	.3	.4	.5	.6	.7	.8	.9
$p(1-p)$	.09	.16	.21	.24	.25	.24	.21	.16	.09

The maximum product of .25 occurs when $p = .5$ and $(1-p) = .5$. Inserting .25 into equation 7.21, the result is the largest minimum sample size needed under conditions of maximum uncertainty and represents a worst case scenario. This value, .25, will *always* provide a large enough minimum sample size, thereby making this a popular sampling strategy:

$$N = \frac{Z^2p(1-p)}{E^2}$$

$$= \frac{Z^2(.5)(.5)}{E^2}$$

$$= \frac{Z^2(.25)}{E^2} \qquad (7.22)$$

$$= \left(\frac{1}{4}\right)\frac{Z^2}{E^2}$$

Sample Size Selection Example—Proportion

Middletown planners wish to estimate the *proportion* of households having one or more children 18 years of age or younger and want to be 90 percent certain their sample statistic is within .04 (4 percent) of the true population estimate. To determine in advance the minimum sample size needed when estimating a population proportion (ρ), two options are possible:

1. *Preliminary survey taken.* Suppose a preliminary survey reveals that 36 percent of Middletown households contain one or more children. Since the population proportion (ρ) is unknown, $p = .36$ is used with equation 7.21 to determine the necessary sample size:

$$n = \frac{Z^2p(1-p)}{E^2} = \frac{(1.65)^2(.36)(.64)}{(.04)^2} = 392.0$$

Thus, a random sample of 392 households should be taken to estimate the proportion of Middletown households with children within .04 (4 percent) and be 90 percent confident in a result that precise. Since 30 households have already been surveyed, another 362 observations must be taken in the second stage of the survey design.

2. *Preliminary survey not taken.* If no preliminary survey is taken, the maximum product $p(1 - p) = .5(.5) = .25$ may be used in a worst case situation (equation 7.22):

$$n = \frac{Z^2(.5)(.5)}{E^2}$$

$$= \frac{Z^2(.25)}{E^2}$$

$$= \left(\frac{1}{4}\right)\frac{(1.65)^2}{(.04)^2} = 425.4$$

In this case, a random sample of 426 households should be taken to achieve the desired level of precision. A larger required minimum sample size should always be expected with this option.

KEY TERMS

bounds of the interval (upper and lower), 101
central limit theorem, 98
confidence interval, 98
confidence level, 101
E (tolerable sampling error), 109
finite population correction (fpc), 99
fpc rule, 100
interval estimation, 97
point estimation, 97
sampling distribution of sample means, 98
sampling fraction, 100
significance level, 101
standard error of the mean (sampling error), 99
two-stage sampling design, 110

MAJOR GOALS AND OBJECTIVES

If you have mastered the material in this chapter, you should now be able to do the following.

1. Understand the basic concepts in point estimation and interval estimation, including the sampling distribution of a statistic and the central limit theorem.

2. Recognize the conditions when it is appropriate to apply the finite population correction.

3. Explain the terms "confidence interval," "significance level," and "confidence level"—and the relationship between these terms.

4. Understand the procedure for constructing a confidence interval (bounds around the point estimate) when appropriate sample statistics are provided.

5. Understand clearly the factors involved in selecting the exact equation used to place a confidence interval around a population estimate—including type of sample taken and population parameter being estimated by the sample statistic.

6. Know how to apply the appropriate equations to calculate confidence intervals around population estimates for various geographic problems.

7. Identify the factors that need consideration when selecting sample size for a geographic problem and know how to determine the appropriate minimum sample size when presented with a geographic situation.

REFERENCES AND ADDITIONAL READING

American Statistical Association, Survey Research Methods Section. 1998. *What Is a Survey? How to Plan a Survey, How to Collect Survey Data.* www.amstat.org.

Dixon, C. J. and B. Leach. 1978. *Sampling Methods for Geographical Research.* Norwich, England: Geo Abstracts.

Gregory, S. 1978. *Statistical Methods and the Geographer* (4th edition). London: Longman.

Hanson, M. H. and W. N. Hurwitz (editors). 1993. *Sample Survey Methods and Theory.* New York: John Wiley and Sons.

Hauser, P. 1975. *Social Statistics in Use.* New York: Russell Sage Foundation.

Kish, L. 1995. *Survey Sampling* (Wiley Classics Library). New York: John Wiley and Sons.

Scheaffer, R. L., W. Mendenhall, and L. Ott. 1996. *Elementary Survey Sampling* (5th edition). Boston: Duxbury.

Thompson, S. K. 1992. *Sampling.* New York: John Wiley and Sons.

INFERENTIAL PROBLEM SOLVING IN GEOGRAPHY

Elements of Inferential Statistics

The previous chapter introduced various concepts concerning estimation in sampling. Sample estimation involves inference. A primary objective of sampling is to infer some characteristic of the population based on statistics derived from a sample of that population. Sample statistics are used to make point estimates of population parameters, such as the mean, total, and proportion. In addition, to determine the level of precision of these point estimates, a confidence interval, or bound, is placed around the sample statistic, making it possible to state the likelihood that a sample statistic is within a certain range or interval of the population parameter. In this chapter, these ideas are extended to a form of statistical inference known as **hypothesis testing.** Geographers can apply inferential hypothesis testing to help reach conclusions for a wide variety of problems.

The practical application and value of inferential statistics are best understood in the context of the scientific research process. As outlined and discussed in chapter 1 (figure 1.1), the formulation of hypotheses and their testing through inferential statistics play a central role in the development of the science of geography. For example, hypothesis evaluation may lead to the refinement of spatial models and the development of laws and theories. In addition, conclusions from inferential testing often contribute toward the advancement of scientific research.

Methods of hypothesis testing are introduced in this chapter with examples of one-sample differ-

ence tests. A properly created sample is essential in the successful application of inferential statistics. Geographers sometimes need to verify whether a particular sample is truly typical or representative of the population from which it is drawn. Why would a sample need to be analyzed to see whether it is representative of a population? If a sample is used in further analysis or research and *does not* represent the original population accurately, then future results obtained using this sample may be flawed. For example, community leaders sampling the local population want to determine the level of support for constructing a new vocational/technical school. Suppose the views of upper- and lower-income residents differ significantly on this issue. If the sample happens to disproportionately represent upper-income residents, survey results are likely to be flawed because the views of lower-income residents are underrepresented. Therefore, a one-sample difference test needs to be used early in the planning process to verify that the sample opinions are truly representative, thereby avoiding subsequent problems.

In other cases, the geographer might be interested in comparing known statistics from a local sample to population parameters that are available from some external source. The issue here may be to show that the sample taken from the local population differs significantly (or does not differ significantly) from some given external population. For example, a geographer might want to know whether pollution levels from a local sample of sites differ significantly from national or state EPA standards (i.e., the external population) for those pollutants. If the sample of local pollution levels meets the standards, then federal or state penalties may not be imposed on this region.

If the population parameter and the sample statistic are not significantly different, the geographer can be confident that the sample is truly representative of the target population. In such cases, the chosen sample is adequate for further analysis. On the other hand, if the population parameter and sample statistic are significantly different, the sample may be inaccurate or otherwise deficient. Excessive sampling error may produce an unrepresentative sample. Human error may occur in one of the steps taken to produce the sample.

Alternatively, a particular sample may represent the situation described in figure 7.4, where the mean of sample 5 $(\overline{X}_5)$ falls into a tail of the sampling distribution of means. That illustration demonstrates how a sample statistic calculated from a properly drawn random sample has a 10 percent chance of being quite different from the actual population parameter value.

An example helps clarify the need for testing the difference between a sample and population mean. Suppose a study is undertaken to determine the attitudes of community residents toward constructing a neighborhood swimming pool. The advisory committee wants to select a 5 percent sample of families at random for a telephone survey. However, because families with children are more likely to desire this public facility, the committee wants to make sure that their sample accurately represents the family structure in the community.

One way to measure the success of the committee's objective is to determine the number of children in each family selected in the sample. Using these data, the mean family size for the sample could be computed. This sample mean could then be compared to the average family size for the target population found in either recent census data for the community or perhaps in information from a school census. If the mean family size for the sample matches the mean for the target population, then the committee can have confidence that their sample is truly representative of family structure in the area. The critical results of the survey can then be analyzed with the knowledge that the sample is not biased.

In this chapter, two complementary methods of hypothesis testing are presented. The classical/traditional approach (section 8.1) provides a solid, logical foundation for understanding hypothesis testing. The *p*-value, or prob-value, method of hypothesis testing (section 8.2) builds on this solid foundation, yet provides additional useful information concern-

ing the research problem. An example problem from the previous chapter is used to illustrate these methods. A difference of means Z test uses the concepts and procedures developed in chapter 7 on sampling estimation and provides a transition from the placement of confidence intervals around sample estimates to the use of inferential statistics in hypothesis testing.

In section 8.3, some additional topics related to one-sample difference of means tests are presented. Section 8.4 discusses the one-sample difference of proportions test. Finally, in section 8.5, the discussion focuses on the circumstances or geographic problems for which inferential testing is appropriate. In addition, the various issues that influence the selection of the proper statistical test are discussed.

8.1 CLASSICAL/TRADITIONAL HYPOTHESIS TESTING

Classical hypothesis testing involves a formal multi-step procedure that leads from the statement of hypotheses to a conclusive statement (decision) regarding the hypotheses (table 8.1). Hypothesis testing has the general goal of making inferences about the magnitude of one or more population parameters, based on sample statistics estimating those parameters. Hypotheses regarding a population parameter are evaluated using sample information, and a conclusion is reached (at some preselected significance level) about the hypotheses. Because of the nature of sampling, a measurable probability can always be assigned to the conclusions reached through statistical hypothesis testing. This logic should sound very familiar, as it is based directly on the themes on sampling estimation discussed in chapter 7.

Recall in example 7.2, planners in the city of Middletown wanted to update selected demographic statistics. More specifically, community officials estimated the mean number of people per household from two random samples: a smaller sample having only 25 households and a larger sample of 250 households. Both samples were taken from a population of 3,500 Middletown households, and a bound or confidence interval was placed around both sam-

---

TABLE 8.1

Steps in Classical/Traditional Hypothesis Testing

Step 1: State null and alternate hypotheses
Step 2: Select appropriate statistical test
Step 3: Select level of significance
Step 4: Delineate regions of rejection and nonrejection of null hypothesis
Step 5: Calculate test statistic
Step 6: Make decision regarding null and alternate hypotheses

ple estimates. The smaller, 25-household sample used the t table to estimate the width of the confidence interval. This procedure resulted in a 90 percent confidence interval of .552 placed around a mean of 2.73 people per household. In the larger, 250-household sample, the mean was 2.68 people per household, and the associated Z-value for 90 percent confidence produced an interval of .209:

$$\text{smaller sample } (n = 25): \overline{X} \pm t\hat{\sigma}_{\overline{X}} = 2.73 \pm .552$$

$$\text{larger sample } (n = 250): \overline{X} \pm Z\hat{\sigma}_{\overline{X}} = 2.68 \pm .209$$

Suppose that the planners needed additional information. From the 1990 Census of Population, Middletown officials have learned that the average number of people per household in the United States is 2.61 ($\mu = 2.61$), a continuation of the general national trend in declining household size (3.14 in 1970, 2.76 in 1980, and 2.61 in 1990). Middletown planners want to see if household size in their community is typical, or representative, of this national figure. Is mean household size in their community similar to the national mean or significantly different? To answer this question, a hypothesis testing procedure is established to determine how closely the sample of Middletown households compares with the 1990 national average household size of 2.61.

The Middletown mean household size example is used to discuss the steps of the classical hypothesis testing process. Appropriate terminology and concepts regarding statistical testing are introduced as needed.

Step 1: State the Null and Alternate Hypotheses

Two complementary hypotheses of interest are the **null hypothesis** (denoted H_0) and the **alternate or alternative hypothesis** (denoted H_A). Consider the formulation of hypotheses concerning the mean of a population (μ). The typical claim is that μ is equal to some value, μ_H (for **hypothesized mean**). This claim of equality is called the null hypothesis, and takes the general form

$$H_0: \mu = \mu_H$$

The null hypothesis can also be stated

$$H_0: \mu - \mu_H = 0$$

In the latter form, attention is focused on H_0 as a statement of "no difference" between μ and μ_H, which is the case if ($\mu - \mu_H$) equals zero (or null). The null hypothesis statement always includes the equal sign.

The converse of the null hypothesis is the alternate hypothesis, H_A. The alternate hypothesis expresses the conditions under which H_0 is to be re-

jected and can be viewed as a positive statement of difference. The two hypotheses are mutually exclusive, for if H_0 is rejected, H_A is accepted. The alternate hypothesis takes one of two forms, depending on how the research problem is structured. In some problems, the form of H_A is **nondirectional**, while in others it is **directional**, but H_A always consists of an inequality indicating the conditions under which H_0 is rejected. So, for example, each of the following is a valid form of H_A:

$$H_A: \mu \neq \mu_H \text{ (nondirectional)}$$

$$H_A: \mu < \mu_H \text{ (directional)}$$

$$H_A: \mu > \mu_H \text{ (directional)}$$

The selection of a specific form of H_A depends on how the hypothesized difference is stated. The directional and nondirectional formats offer two possibilities. In the Middletown household size example, the alternate hypothesis can be stated in a nondirectional form:

$$H_0: \mu = 2.61$$

$$H_A: \mu \neq 2.61$$

These statements hypothesize that the mean size of Middletown households is no different from the mean national household size of 2.61. If the Middletown mean is close to this population mean, the likely conclusion is that the null hypothesis *should not* be rejected. Conversely, if the Middletown mean is not close to the national population mean, the likely conclusion is that H_0 *should* be rejected. By expressing H_A in this form, the direction of difference between the household size of Middletown and the United States is not important. That is, the Middletown household size could be either greater than *or* less than the national figure. If H_0 is rejected, the *only* conclusion that can be drawn is that the difference in household sizes is significant; no conclusion on the direction of that difference is possible.

For some situations, the alternate hypothesis can provide more specific information about the hypothesized direction of difference:

$$H_0: \mu = 2.61$$

$$H_A: \mu < 2.61$$

$$\text{or}$$

$$H_A: \mu > 2.61$$

In addition to determining whether a significant difference exists, the direction of that difference (expressed in H_A) is also specified. If $H_A: \mu < 2.61$, then rejection of H_0 indicates that the average size of Middletown households is smaller than the national average. This would be a reasonable alternate hypothesis to test if Middletown has many single persons, young couples without children, or elderly residents. If $H_A: \mu > 2.61$, then rejection of H_0 indicates that Middletown household size on average is larger than the national average. This might be a logical alternate hypothesis if many families with children reside in the community.

With classical hypothesis testing, Middletown planners must decide which form of H_A to use. The critical issue is whether a priori knowledge exists about any direction of difference between Middletown and household sizes nationwide. If Middletown planners have no preconceived rationale for believing their household size is larger or smaller than the national average, the nondirectional format would be appropriate.

In classical hypothesis testing, the conclusion is to either reject or not reject the null hypothesis (i.e., choose between the null and alternate hypotheses). *Because this decision is based on a single sample,* there is a measurable chance or probability of making an incorrect decision or reaching a wrong conclusion. Error comes from two possible sources (table 8.2):

Type I Error   A decision could be made to reject the null hypothesis as false when, in fact, it is true. For the Middletown example, it could be concluded that the difference between the national average household size and the sample of Middletown households is significant, when actually no difference exists. The likelihood of this sort of error, known as a **Type I error,** is equivalent to the significance level (α), discussed in chapter 7 (see figure 7.4).

Type II Error   Conversely, a decision could be made not to reject the null hypothesis when it is actually false. For the Middletown example, it could be concluded that their household size does not differ from the national average, when a significant difference really exists. The likelihood of this error, known as a **Type II error,** is beta (β).

Hypothesis testing operates on much the same principle as judicial decision making in a court of law (table 8.3). In court, a defendant is presumed innocent until proven guilty. Similarly, in hypothesis testing, the null hypothesis is presumed correct until rejected or proven otherwise. In the judicial system, convicting a person who is truly not guilty is considered a more serious error than freeing a guilty person. Similarly, in hypothesis testing, rejecting H_0 as false when it is actually true (a Type I error) is considered more serious than not rejecting H_0 when it is actually false (a Type II error). In a court of law, when reasonable doubt exists about the guilt of a defendant, the jury reaches a verdict of not guilty. The defendant is *not* declared "innocent," but rather "not guilty." Similarly, if reasonable doubt exists about rejecting the null hypothesis, the researcher should not reject it. For this reason, the significance level (α) of most problems is kept relatively low (at a level such as .05 or .01), to minimize the chances of a serious Type I error.

Step 2: Select the Appropriate Statistical Test

The second step in classical hypothesis testing is selecting the most appropriate statistical test to examine the research problem. As will be explained toward the end of this chapter, a logical and convenient way to categorize the many statistical tests available is by the type of question asked and the assumptions that are met. At this point, however, it is sufficient to state that the appropriate statistical test for the Middletown household size problem is a **one-sample difference of means test**:

$$Z \text{ or } t = \frac{\overline{X} - \mu}{\sigma_{\overline{X}}} = \frac{\overline{X} - \mu}{\sigma/\sqrt{n}} \qquad (8.1)$$

where Z or t = test statistic
$\overline{X}$ = sample mean
μ = population mean
$\sigma_{\overline{X}}$ = standard error of the mean
σ = population standard deviation
n = sample size
(select Z if $n \geq 30$; select t if $n < 30$)

Because a single sample mean is being compared to a population mean, this test is sometimes referred to

---

TABLE 8.2

Possible Decisions in Classical/Traditional Hypothesis Testing

Decision from hypothesis testing	Null hypothesis in reality	
	True	**False**
Reject H_0 as false	Type I error (prob. = α)	Correct decision (prob. = $1-\beta$)
Do not reject H_0	Correct decision (prob. = $1-\alpha$)	Type II error (prob. = β)

TABLE 8.3

Possible Decisions in a Court of Law

Decision from jury	True situation (unknown)	
	Not guilty	**Guilty**
Guilty	Incorrect decision	Correct decision
Not guilty	Correct decision	Incorrect decision

as a *one-sample* difference of means test to differentiate it from tests that compare two or more samples for differences.

In chapter 5 on probability, the concept of a Z-score or standard score was introduced. The one-sample difference of means test is structurally similar to the Z-score equation (table 8.4). The standardized Z-score of a value in a set of data may be interpreted as the number of standard deviations a value lies above or below the mean. Using similar logic, the Z-value or t-value difference of means test statistic measures the number of standard errors a sample mean lies above or below the hypothesized population mean.

For the Middletown household size example, the population standard deviation (σ) is not known. Recall from chapter 7 (table 7.1) that the sample standard deviation, s, is a proper estimator of σ, but slight adjustment may be needed if the sample size is less than 30. Thus, substituting s for σ in equation 8.1 gives

$$Z = \frac{\overline{X} - \mu}{\sigma_{\overline{X}}} = \frac{\overline{X} - \mu}{s/\sqrt{n}} \text{ (if } n \geq 30) \qquad (8.2)$$

or

$$t = \frac{\overline{X} - \mu}{\sigma_{\overline{X}}} = \frac{\overline{X} - \mu}{s/\sqrt{n - 1}} \text{ (if } n < 30) \qquad (8.3)$$

Which of these equations should be chosen? For illustrative purposes, suppose Middletown officials decide to use the larger, 250-household sample ($\overline{X} = 2.68$ persons per household) to make the comparison, since it is more precise than the smaller sample of 25 households, and the data have presumably already been collected. Therefore, the Z test statistic will be used here, and the one-sample t test example will be discussed later in the chapter. Notice in the boxed inset labeled "one-sample difference of means tests" that the same set of assumptions and requirements must be met for both Z and t. A random sample must be taken from a normally distributed population and the data measured at the interval/ratio scale.

TABLE 8.4

Structure of Z-Score of a Data Value and Difference of Means (Z and t) Tests

Z-score of a value (i) in a set of data	Z-value or t-value of a sample mean in a frequency distribution of sample means
$Z_i = \dfrac{X_i - \overline{X}}{s}$	$Z \text{ or } t = \dfrac{\overline{X} - \mu}{\sigma_{\overline{X}}}$ where $\sigma_{\overline{X}} = \dfrac{s}{\sqrt{n}}$

> ## One-Sample Difference of Means Z or t Test
>
> **Primary Objective:** Compare a random sample mean to a population mean for difference
>
> **Requirements and Assumptions:**
>
> 1. Random sample
> 2. Population from which sample is drawn is normally distributed
> 3. Variable is measured at interval or ratio scale
>
> **Hypotheses:**
>
> $H_0: \mu = \mu_H$ (where μ_H is the hypothesized mean)
> $H_A: \mu \neq \mu_H$ (two-tailed)
> $H_A: \mu > \mu_H$ (one-tailed) or
> $H_A: \mu < \mu_H$ (one-tailed)
>
> **Test Statistic:**
>
> $$Z \text{ or } t = \frac{\overline{X} - \mu}{\sigma_{\overline{X}}}$$
>
> If sample size ≥ 30, use Z; if sample < 30, use t

Step 3: Select the Level of Significance

The next task in classical hypothesis testing is to place a probability on the likelihood of a sampling error. As mentioned earlier, committing a Type I error and rejecting a null hypothesis as false when it is actually true is generally considered serious. Therefore, the usual procedure is to select a fairly low significance level (α) such as .05 or .01. Then, the conclusion is specified in terms of the level of significance of the result. In classical hypothesis testing, a null hypothesis may be rejected at the .05 level, which is the same as saying the statistical test is significant at the .05 level. This would mean that there is a 5 percent chance that a Type I error has occurred, and it is only 5 percent likely (1 chance in 20) that the null hypothesis has been improperly rejected because of random sampling error.

For many geographic research problems, a highly demanding or extremely stringent significance level may not be necessary. In general, the choice of significance level depends on the nature of the problem and the effects of the decision. With some geographic problems, the consequences of sampling error may be severe, and in these instances, a very low significance level is required. With the Middletown household size example, using a commonly accepted significance level such as $\alpha = .05$ provides sufficient precision.

Step 4: Delineate Regions of Rejection and Nonrejection of Null Hypothesis

Once a significance level has been selected, this value is used to create the regions of **rejection** and

nonrejection of the null hypothesis (figure 8.1). The total area in which H_0 is rejected, as represented by the significance level, encompasses 5 percent (α = .05) of the area under the curve. This area of rejection can be distributed in one of two ways. In case 1, the alternate hypothesis is nondirectional, so the shaded rejection area of H_0 is distributed equally between the two tails of the curve. With this **two-tailed** format and α = .05, 2.5 percent ($\alpha/2$ = .025) of the total area is in each of the rejection regions or tails of the distribution. In case 2, the alternate hypothesis is directional, so the shaded rejection region of H_0 is placed entirely on one tail of the distribution. In a **one-tailed** format shown in case 2, the rejection region happens to be on the right tail, but the placement of the rejection region depends on the form of H_A. In both the two-tailed and one-tailed cases, the unshaded area under the curve delineates test statistic values where H_0 is not rejected. This area of nonrejection is 95 percent ($1 - \alpha$ = .95) of the total area under the curve.

The next task is to determine the critical Z-values that delimit the boundaries separating the rejection and nonrejection regions. When α = .05 and H_A is nondirectional, each of the two tails contains

Case 1:
Two-tailed (nondirectional) format

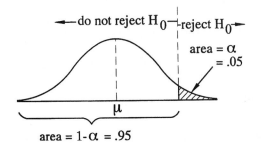

Case 2:
One-tailed (directional) format

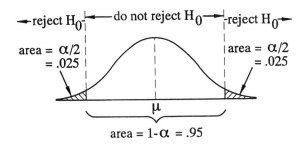

FIGURE 8.1
General Regions of Rejection and Nonrejection of Null Hypothesis: Significance Level (α) = .05

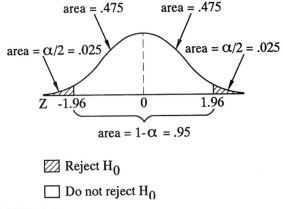

☒ Reject H_0

☐ Do not reject H_0

FIGURE 8.2
Normal Table Probability Values Associated with a Significance Level (α) = .05: Two-Tailed Case

$\alpha/2$, or .025, of the area under the curve (figure 8.2). This leaves .475 of the area on *each* side of the distribution ($1 - \alpha$ = .95 in total) in the nonrejection region. In the normal curve table (appendix, table B), the Z-value that corresponds with the .475 probability for the upper side of the distribution is 1.96. Therefore, in this problem, a Z-value of 1.96 defines the boundary between the rejection and nonrejection regions, and a calculated Z-value (test statistic) must be less than or equal to 1.96 to keep the null hypothesis from being rejected. Given the symmetrical nature of a normal curve, a similar-magnitude Z-score of -1.96 is the critical value delineating the boundary separating H_0 and H_A at the lower tail.

The following decision rule identifies and summarizes the regions of rejection and nonrejection of the null hypothesis for α = .05 and H_A nondirectional:

Decision rule:

If $Z < -1.96$ or if $Z > 1.96$, reject H_0.
Conversely, if $-1.96 \leq Z \leq 1.96$, do not reject H_0.

Step 5: Calculate the Test Statistic

At this step of the hypothesis testing procedure, sample data are evaluated using the test statistic. In the Middletown sample, mean household size was 2.68 persons, with a sample standard deviation of 2.07 persons. Substituting these sample statistics into the difference of means Z test gives the following:

$$Z = \frac{\overline{X} - \mu}{s/\sqrt{n}} = \frac{2.68 - 2.61}{2.07/\sqrt{250}} = 0.53$$

The Middletown sample mean is .53 standard errors *above* the U.S. average. If household size in Middletown had been less than the national norm of 2.61, the calculated test statistic would have been negative. This calculated value is now compared with the critical Z-values determined in step 4 to reach a final decision.

Step 6: Make Decision Regarding Null and Alternate Hypotheses

All the information needed to make a decision regarding rejection or nonrejection of the null hypothesis is now available. The calculated test statistic is $Z = .53$. The critical Z-values are -1.96 and 1.96. Since the test statistic lies between the critical values, $(-1.96 \leq Z \leq 1.96)$, the null hypothesis should *not* be rejected. The conclusion is that mean household size in Middletown does not differ from the mean household size nationally. The steps in the classical hypothesis testing procedure for the Middletown example are summarized in table 8.5.

8.2 *P*-VALUE, OR PROB-VALUE, HYPOTHESIS TESTING

The formal multistep procedure of classical hypothesis testing provides a logical basis and excellent theoretical underpinning for inferential decision making. However, the usefulness of the results from classical analysis is limited in some important ways. First, a specific significance level must be selected to delineate the regions of rejection and nonrejection of the null hypothesis. This a priori selection of α is often arbitrary and may lack a clear theoretical basis. A significance level of .05 or .01 is often chosen because they are the conventional probabilities commonly provided in statistical tables. Second, the final decision regarding the null and alternate hypotheses is binary in nature: either H_0 is rejected or not rejected at that arbitrary significance level. This type of conclusion provides only limited information about the calculated test statistic. Geographers rarely use classical hypothesis testing today for statistical problem solving; instead, the *p*-value, or prob-value, approach is commonly employed in geographic research.

The more flexible *p*-value method of hypothesis testing provides additional valuable information. With this approach, the *exact* significance level associated with the calculated test statistic value is determined. That is, the ***p*-value** is the exact probability of getting a test statistic value of a given magnitude, *if* the null hypothesis of no difference is true. This can generally be interpreted as the probability of making a Type I error. Preselection of a "standard" significance level (α) is avoided, and decisions to reject or not reject H_0 at that arbitrary level need not be made.

In the Middletown example, the decision rule was to reject the null hypothesis if $Z < -1.96$ or $Z > 1.96$ and to not reject H_0 if $-1.96 \leq Z \leq 1.96$. The calculated test statistic value of $Z = .53$ led to the conclusion that the mean size of Middletown households was not different from the mean national household size, and the null hypothesis was *not* rejected.

This conclusion is of limited use. All that has been decided is that the Middletown household size is not different from the national average household size. However, knowing the exact significance level associated with the specific sample mean or calculated test statistic value of 2.68 would be more informative. If the null hypothesis had been rejected on the basis of the 2.68 sample mean, what would be the exact significance level and likelihood that a Type I error had been made? The probability of making a Type I error cannot be determined with the classical approach.

The Middletown sample can be viewed in a more informative way using the *p*-value approach (figure 8.3). The critical values that separate the regions of rejection and nonrejection of the null hypothesis are now based on the location of the particular sample mean $\overline{X} = 2.68$ relative to the population mean ($\mu = 2.61$). The unshaded region of nonrejection

---

TABLE 8.5

Summary of Classical Hypothesis Testing: Middletown Household Size Example

Step 1: H_0: $\mu = 2.61$ and H_A: $\mu \neq 2.61$
Step 2: One-sample difference of means Z test selected as test statistic
Step 3: $\alpha = .05$
Step 4: If $Z < -1.96$ or if $Z > 1.96$, reject H_0
If $-1.96 \leq Z \leq 1.96$, do not reject H_0
Step 5: Calculate Z (from random sample) = .53
Step 6: Since $-1.96 \leq Z \leq 1.96$, do not reject H_0

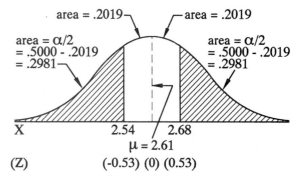

p-value (α) = 2(.2981) = .5962

FIGURE 8.3
P-Value in Two-Tailed Hypothesis Test: Middletown Mean Household Size $(\overline{X})$

is centered on $\mu = 2.61$, and the difference between $\overline{X}$ and μ ($\overline{X} - \mu = .07$) is used to establish the width of the nonrejection interval on either side of μ. Thus the nonrejection region has an upper bound of 2.68 ($\mu + .07$) and a lower bound of 2.54 ($\mu - .07$). The rejection regions occupy the extremes of the distribution, outside the upper and lower bound (shaded areas).

In the p-value approach to hypothesis testing, the area within the rejection region(s) represents the p-value. This area, or p-value, is calculated in four steps:

1. The test statistic for Z is calculated, as in the classical approach.
2. The probability or relative area under the normal curve is determined for that Z-value.
3. The shaded rejection area is determined by subtracting the probability (of step 2) from .5000.
4. This area is doubled if a nondirectional (two-tailed) alternate hypothesis is used.

The resulting probability is the p-value.

This procedure is illustrated with the Middletown example of mean household size. The difference of means Z test statistic is calculated as

$$Z = \frac{\overline{X} - \mu}{s/\sqrt{n}} = \frac{2.68 - 2.61}{2.07/\sqrt{250}} = .53$$

The resulting Z-value is then used to determine a probability or area under the normal curve. When $Z = .53$, the table of normal values lists a probability of .2019. This probability represents the nonrejection or unshaded area between $Z = 0$ and $Z = +.53$ (figure 8.3). The area in the shaded *upper* tail of the normal distribution is found by subtraction: $.5000 - .2019 = .2981$. Since this problem is nondirectional (two-tailed), a second rejection region of area $= .2981$ is located at the *lower* tail below $\overline{X} = 2.54$. Thus, the total area in the rejection region is twice .2981, or .5962. This result defines the p-value associated with the null hypothesis ($p = .5962$).

How can this p-value be interpreted? The significance level is the total area in the rejection region and indicates the likelihood of making a Type I error. If the decision is made to reject H_0, the significance level equals the p-value, or .5962, representing a 59.62 percent chance of a Type I error. In this situation, the likelihood of making an error is very high, so the clear and logical decision is not to reject the null hypothesis.

The p-value provides a measure of the belief or conviction that the decision not to reject the null hypothesis is correct. For example, a p-value relatively close to 1 indicates a high degree of trust in the validity of the null hypothesis, whereas a p-value rela-

tively close to 0 suggests that little faith should be placed in the null hypothesis. In the Middletown example, the p-value of .5962 indicates that the null hypothesis should not be rejected, and the Middletown mean household size is not different from the nationwide average household size.

Suppose the null and alternate hypotheses for the Middletown household size example are

$$H_0: \mu = 2.61$$

$$H_A: \mu > 2.61$$

This is a one-tailed or directional test that asks whether the mean household size of Middletown is greater than the national average. This alternate hypothesis would be appropriate if planners thought that Middletown has an atypically large number of households with children. The p-value associated with $\overline{X} = 2.68$ is .2981 ($p = .2981$) (figure 8.4). In the earlier two-tailed example, this value was doubled because H_0 was rejected if $\overline{X}$ was different from μ, regardless of the direction of that difference. In this example, no such doubling is necessary. The critical region of rejection of H_0 is located entirely in one tail of the normal distribution (in this case, the upper or right tail).

The p-value from a one-tailed test is interpreted in much the same way as a two-tailed p-value. That is, if the null hypothesis is rejected in this problem, the chance of a Type I error is 29.81 percent. Since this p-value is not at all close to any conventional significance level (such as .05), the decision should be against rejecting the null hypothesis. The conclusion in this one-tailed test is that mean household size in Middletown is not larger than the nationwide average household size.

Note how the one-tailed p-value reduced the likelihood of making a Type I error by one-half. Therefore, whenever a rationale exists for doing so, a one-

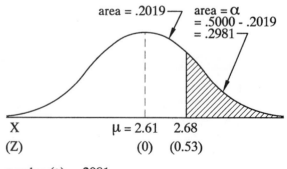

area = .2019

area = α
= .5000 - .2019
= .2981

| X | $\mu = 2.61$ | 2.68 |
| (Z) | (0) | (0.53) |

p-value (α) = .2981

FIGURE 8.4
P-Value in One-Tailed Hypothesis Test: Middletown Mean Household Size ($\overline{X}$)

tailed (as opposed to two-tailed) hypothesis testing format should be applied.

The p-value method provides more information than the classical method. The exact significance level and probability of making a Type I error is calculated for any test statistic value. The geographer can then evaluate the particular situation revealed by the sample data and report the p-value results without having to reach a decision to reject or not to reject a null hypothesis based on an arbitrary significance level. Since virtually all statistical analysis in geography is now done with the aid of statistical computer packages, p-values are calculated automatically. Furthermore, the p-value method contains all of the information derived from classical testing. The decision to reject the null hypothesis when $p < .05$ is equivalent to a classical test using a significance level of .05.

The ease of deriving multiple p-values also permits geographic studies to emphasize comparative, investigative analysis. Through multiple applications of a statistical test, for example, trends in spatial data over time can be examined or different regions can be compared with one another at a particular time. In other instances, the application of different test statistics to the same set of spatial data may provide new geographic insights.

Use of the p-value approach, however, does not provide an excuse to interpret results subjectively or avoid making decisions. Although the computer can produce multiple p-values quickly, the researcher must still be responsible for interpreting the statistical and geographic meaning of the p-values in a consistent and rational way. P-values report what the sample data reveal about the credibility of a null hypothesis, but still demand the same stringent rules of inference as required in classical hypothesis testing.

Since the p-value method offers many advantages over classical hypothesis testing, *all statistical tests presented in the remainder of this text will use the p-value method*. In the statistical analysis of the geographic problems that follow, all associated p-values will be reported without showing their manual calculation. This approach provides all necessary information to reach conclusions in a succinct and consistent manner. Subsequent chapters emphasize investigative analysis and the comparative use of multiple p-values.

8.3 ONE-SAMPLE DIFFERENCE OF MEANS TEST: SMALL SAMPLE

In the example from the previous section, the sample size was fairly large ($n = 250$), and the Z-score one-sample difference test was appropriate. In some situations, however, a large sample is neither practical nor feasible. For a sample size less than 30, the stu-

dent's t distribution is used instead of the normal distribution (Z).

The t distribution was used in chapter 7 to estimate confidence intervals or bounds around sample estimates for small samples. Compared with the standard normal distribution, applying the t distribution to confidence interval estimation slightly widened the bound on the error of estimation. The smaller sample resulted in greater uncertainty about the precision of the estimate. Similar logic is applied to an inferential difference test. If the sample size is small ($n < 30$), the difference of means test is adjusted. Either of the following options is procedurally correct:

$$t = \frac{\overline{X} - \mu}{s/\sqrt{n-1}} \text{ if } s = \sqrt{\frac{\Sigma(X - \overline{X})^2}{n}} \qquad (8.4)$$

or

$$t = \frac{\overline{X} - \mu}{s/\sqrt{n}} \text{ if } s = \sqrt{\frac{\Sigma(X - \overline{X})^2}{n-1}} \qquad (8.5)$$

With either of these small-sample options, the results of the difference of means t test are comparable to Z.

The one-sample t test is illustrated with another example from Middletown. Suppose Middletown contains a new planned unit development (PUD) that is predominantly residential. The development includes condominiums, townhouses, and apartments, but also contains some detached single-family homes. Planners in Middletown believe that average household size in this PUD is not representative of national household size ($\mu = 2.61$), but is instead significantly smaller. This hypothesis is reasonable if planners believe this development has been particularly attractive to older singles and couples (including retirees) or to young professionals without children. In this case, a directional or one-tailed alternate hypothesis is logical:

$$H_0: \mu = 2.61$$

$$H_A: \mu < 2.61$$

Rejection of H_0 will indicate that PUD household size is significantly smaller than the national average.

Because time and personnel are limited, only 25 households in the PUD will be sampled. Since $n < 30$, the appropriate difference test is the one-sample t test, not the Z test. Suppose the sample of 25 households in this development had a mean household size of 2.03 persons and a standard deviation of 1.50 persons. The resulting test statistic (t) is

$$t = \frac{\overline{X} - \mu}{s/\sqrt{n-1}} = \frac{2.03 - 2.61}{1.50/\sqrt{25-1}} = -1.90$$

The degrees of freedom in the problem equal $n - 1$, or 24. The probability associated with $t = -1.90$ at 24 degrees of freedom is .4652 (appendix, table C), and the corresponding p-value is .0348 (.5000 − .4652 = .0348), as shown in figure 8.5.

In this example, if the decision is made to reject the null hypothesis and conclude that mean household size in the Middletown PUD is less than mean household size nationally, the probability that a Type I error has been made is only .0348 (or 3.48 percent). With a p-value this close to 0, the decision to reject H_0 is probably correct, since the chance of making an error when rejecting H_0 is so small. Middletown planners should feel confident when concluding that PUD household size is less than the national average.

8.4 ONE-SAMPLE DIFFERENCE OF PROPORTIONS TEST

Hypotheses can also be formed to study the difference between a sample proportion and a population proportion. The test statistic for a difference of proportions test is sometimes called the Z test for proportions. Like the difference of means Z test, the normal distribution is used.

Middletown planners estimated the proportion of all households in the community with one or more children less than 18 years of age (situation 3, section 7.2). Of the 250 families sampled, 105, or 42 percent, had children, for a sample proportion (p) and 90 percent confidence interval ($Z\hat{\sigma}_p$) of

$$p \pm Z\hat{\sigma}_p = p \pm Z\sqrt{\frac{p(1 - p)}{n - 1}\left(\frac{N - n}{N}\right)}$$

$$= .42 \pm .05$$

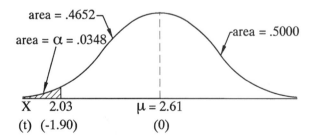

At t = -1.90 and df = 24, table value for t = .4652
p-value (α) = .5000 - .4652 = .0348

FIGURE 8.5
P-Value in Small Sample, One-Tailed Hypothesis Test: Mean Household Size ($\overline{X}$) in Middletown PUD

It was 90 percent likely that ρ, the true population proportion of all Middletown families with children, is in the interval $.37 < \rho < .47$.

This example is now developed further to create a hypothesis test for differences. According to the 1990 census, 48.6 percent of all families in the United States have one or more children (that is, $\rho_H = 0.486$). Is Middletown typical or representative of the national norm, or is Middletown's proportion different? That is:

$$H_0: \rho = \rho_H$$

$$H_A: \rho \neq \rho_H \text{ (nondirectional)}$$

and for this example:

$$H_0: \rho = 0.486$$

$$H_A: \rho \neq 0.486$$

With a nondirectional two-tailed test, H_0 could be rejected if the Middletown proportion is either much greater than or much less than the national proportion.

The appropriate test statistic is the one-sample difference of proportions test:

$$Z = \frac{p - \rho}{\sigma_p} \tag{8.6}$$

where σ_p = standard error of the proportion.

The standard error of the proportion (σ_p) is the standard deviation of the sampling distribution of proportions:

$$\sigma_p = \sqrt{\frac{p(1 - p)}{n}} \tag{8.7}$$

In the Middletown example, the one-sample difference of proportions test statistic is calculated as

$$\sigma_p = \sqrt{\frac{p(1 - p)}{n}} = \sqrt{\frac{(.42)(.58)}{250}} = .0312$$

$$Z = \frac{p - \rho}{\sigma_p} = \frac{.42 - .486}{.0312} = \frac{-.066}{.0312} = -2.12$$

If the null hypothesis is rejected when $Z = -2.12$, the exact associated significance level and p-value is .034 (figure 8.6). This indicates a 3.4 percent chance that a Type I error has been made if H_0 is rejected. This fairly low p-value suggests that the null hypothesis should probably be rejected, and the conclusion made that the proportion of Middletown families with children is indeed different from the national proportion.

The various one-sample difference tests are related, which can be seen in the similar structure of the one-sample test statistics (table 8.6). A general

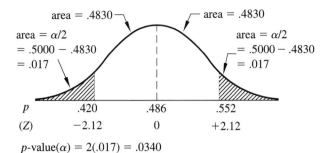

$p\text{-value}(\alpha) = 2(.017) = .0340$

FIGURE 8.6
P-Values in One-Sample Difference of Proportions Test: Proportion (*p*) of Middletown Families with Children

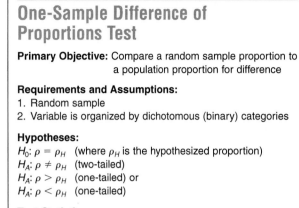

One-Sample Difference of Proportions Test

Primary Objective: Compare a random sample proportion to a population proportion for difference

Requirements and Assumptions:
1. Random sample
2. Variable is organized by dichotomous (binary) categories

Hypotheses:
H_0: $\rho = \rho_H$ (where ρ_H is the hypothesized proportion)
H_A: $\rho \neq \rho_H$ (two-tailed)
H_A: $\rho > \rho_H$ (one-tailed) or
H_A: $\rho < \rho_H$ (one-tailed)

Test Statistic:

$$Z = \frac{p - \rho}{\sigma_p}$$

TABLE 8.6

The Similar Structure of One-Sample Difference Tests

Difference test	Test statistic
Difference of means* large sample ($n > 30$)	$Z = \dfrac{\overline{X} - \mu}{\sigma_{\overline{X}}} = \dfrac{\overline{X} - \mu}{s/\sqrt{n}}$
Difference of means* small sample ($n \leq 30$)	$t = \dfrac{\overline{X} - \mu}{\sigma_{\overline{X}}} = \dfrac{\overline{X} - \mu}{s/\sqrt{n-1}}$
Difference of proportions	$Z = \dfrac{p - \rho}{\sigma_p} = \dfrac{p - \rho}{\sqrt{p(1-p)/n}}$

---

*Population standard deviation (σ) is unknown.

Best estimate of $\sigma = s = \sqrt{\dfrac{\Sigma(X_i - \overline{X})^2}{n-1}}$

General Format of One-Sample Difference Test

Z or $t = \dfrac{\text{sample statistic} - \text{hypothesized population parameter value}}{\text{standard deviation of sample statistic (standard error)}}$

---

format is common to each of these difference tests. Note that the numerator of each test statistic is the difference between the sample statistic and the hypothesized population parameter value. The denominator is the standard deviation of the sampling distribution, which is also referred to as the standard error of the sample statistic.

8.5 ISSUES IN INFERENTIAL TESTING AND TEST SELECTION

So far in this chapter, attention has been focused on a related set of one-sample difference tests. However, there are numerous other types of inferential tests

that can be applied to help solve a wide variety of geographic problems. Before continuing, however, three basic questions need to be addressed: First, in nonmathematical terms, what is meant by degrees of freedom? Second, under what general circumstances are inferential statistics appropriate? Third, how does one determine which of the many available inferential techniques is appropriate and should be selected for a particular geographic problem?

Degrees of Freedom

In some inferential procedures, a test statistic is calculated with the sample size (n) as one of the critical components. If a particular problem has a sample size of n, the problem can be thought of as starting with n **degrees of freedom.** Whenever a parameter must be estimated to calculate a test statistic, one degree of freedom is lost. For example, in the one-sample difference of means t test, one degree of freedom is lost because only one population parameter, μ, is being estimated. This explains why the denominator in equation 8.4 contains $n - 1$ rather than n. Looking ahead one chapter, a two-sample test is used to estimate and compare two population means (μ_1 and μ_2). In this example, two degrees of freedom are lost (one for each parameter); so the degrees of freedom equal $n - 2$. Degrees of freedom will appear in other contexts when other inferential procedures (ANOVA, chi-square, correlation, and regression) are discussed.

Sampling Issues and Inferential Testing

All inferential techniques have certain general characteristics and assumptions in common. As stated earlier, the general goal of hypothesis testing is to make inferences about the magnitude of one or more

TABLE 8.7

Organizational Structure for Selection of Appropriate Inferential Test

Level of measurement	Type of question Questions about differences			
	One sample	Dependent sample (matched-pairs)	Two samples	Three or more (k) samples
Nominal scale (categorical variables)				
Ordinal scale				
Interval/ratio scale	One sample difference of means (Z or t)			
	One sample difference of proportions test (Z)			

population parameters, based on sample estimates of those parameters. When applying any inferential test statistic, it is assumed that a *random sample* has been drawn from a population. Often, however, other unbiased types of sampling (such as systematic sampling) are also valid for inferential hypothesis testing. Recall from chapters 6 and 7 that an element of randomness must be included in the sample procedure, whatever specific type of sample is selected. If multiple samples are required for a particular problem, it is always assumed that *each sample is drawn separately and independently.* Suppose a geographer wishes to compare the average size of pebbles on two beaches. This would be done by taking a random sample of pebbles from one beach and a separate and independent random sample of pebbles from the other beach. (An exception to the assumption of independence occurs if a matched-pairs, or dependent-sample, difference test is used, examples of which will be seen in chapter 9).

Beyond these general characteristics, geographers have actively debated the circumstances for appropriate use of inferential testing in problem solving. The discussion centers on the difference between artificial and natural sampling. In an **artificial sample,** the investigator draws an unbiased, representative sample from a statistical population and then is able to infer certain characteristics about the population based on the sample data. No one questions the appropriateness of inferential techniques in geographic problems or situations in which a proper artificial sample has been taken.

Opinion differs on whether inferential statistics should be applied on geographic data sets not obtained from artificial sampling. Geographers often wish to analyze spatial patterns or data sets that comprise complete enumerations or total populations. In these situations, are inferential procedures ever appropriate? Some geographers suggest that inferential statistics may be permitted when using a data set considered a "natural sample."

In **natural sampling,** the "natural" or real-world processes that produce the spatial pattern under analysis contain random components. Suppose a geographer wishes to analyze a data set showing the pattern of all hurricane landfalls along a coastline during the last century. The landfall site of a hurricane is partly a function of prevailing global water currents and wind patterns (nonrandom or systematic influences), but is also affected by a complex variety of meteorological processes associated with that particular hurricane (random influences). It is possible to argue, then, that the observed pattern of hurricane landfall sites is actually a natural sample from the population of all possible landfall locations that could have occurred.

The pattern of state-level 1980–90 population change (shown in figure 1.2) is another example of a natural sample. This choropleth map pattern is partly the result of such nonrandom or systematic factors as climate and job opportunities related to the location of natural resources and partly the result of numerous individual and family decisions to migrate from one state to another (some of which could be random).

TABLE 8.7 *(CONTINUED)*

Level of measurement	Type of question	
	Questions about strength of relationship between two variables	Questions about form or nature of relationship between variables
Nominal scale (categorical variables)		
Ordinal scale		
Interval/ratio scale		

If inferential procedures are applied to natural samples such as these, *the results must be interpreted with extreme caution.* Descriptive summaries of natural samples are appropriate, but the researcher must take care to avoid making improper inferential statements. Issues regarding the application of inferential statistics to natural samples are quite complex and continue to generate considerable discussion and controversy in applied statistics. Those interested in pursuing these arguments further (particularly from the geographer's perspective) are directed to the references at the end of this chapter.

Inferential Test Selection

A geographer needs to know more than the general characteristics of a particular problem to select the most appropriate inferential test. Because of the idiosyncratic nature of geographic problems, selecting the single best statistical procedure is often difficult. To complicate matters, the same geographic problem can be structured or organized in different ways, or data can be collected in various ways. Very often, determining the structure of a problem and organizing data in that problem direct the geographer to a particular technique or set of appropriate techniques. To help make the proper selection, a logical organizational framework is needed in which to place each of the many inferential techniques.

The fundamental dimensions of such an organizational framework are shown in table 8.7. The only specific inferential tests covered so far in the text are the one-sample difference tests presented earlier in this chapter, and they are placed in the appropriate cells of the table. Many other inferential tests are presented in the following chapters, and this table is reproduced again in the final (epilogue) chapter, with all specific inferential tests covered in the text placed in their proper location.

The columns of the table represent the *type of question* being asked or being investigated. At the most basic level, statistical inference is concerned with differences between a sample (or set of samples) and a hypothesized value of a population parameter (or set of parameters). However, problems are organized or structured differently according to the type of difference being examined.

The first four columns of table 8.7 refer to various types of questions about differences. For example, the problems considered earlier in this chapter were concerned with the magnitude of difference between a single sample and a hypothesized population parameter. These problems are collectively referred to as one-sample difference tests (first column of table 8.7). In certain special cases, the data set under investigation consists of one-sample of observations collected for two or more different variables or at two or more different time periods. In these circumstances, a matched-pairs (dependent-sample) difference test is appropriate (second column of table 8.7). Other inferential tests consider the magnitude of difference between two or more independently drawn random samples, and hence these tests are known as two-sample, three-sample, or *k*-sample difference tests (columns 3 and 4 of the table). Difference tests can be structured in various ways. Some geographic problems ask a question about differences, but the data are in the form of frequency distributions or frequency counts by category, and special purpose categorical tests are used. In other circumstances, data may be explicitly spatial—in the form of a point or area pattern. Specially designed one-sample difference tests are available to analyze the point pattern or area pattern to determine if it is random or significantly different from random (more clustered or more dispersed than random). Clearly, there are various types of difference tests. Nearly all of the tests discussed in chapters 8–12 are examples of difference tests.

The fifth and sixth columns of table 8.7 refer to questions about the strength of relationship between two variables and questions about the form of relationship between two variables, respectively. Notice this is a fundamentally different approach. Rather than asking if a sample is significantly different from a population, or if two or more samples are significantly different from one another, the focus is on how strong or weak the relationship is between two variables, and on a description of the form or nature of that relationship. If the strength of relationship between two variables is a concern (column 5), then either some form of contingency analysis (section 11.2) or some test of correlation (chapter 13) is appropriate. Alternatively, if the primary concern is with the form or nature of relationship between two variables (column 6), then regression analysis is appropriate (chapter 14).

The rows of table 8.7 represent different *levels of measurement*. Depending on the nature of the geographic problem under investigation, data may be scaled at the nominal (categorical), ordinal, or interval/ratio level of measurement (rows 1, 2, and 3, respectively), and selection of an appropriate inferential test depends in part on this level of measurement.

Statisticians traditionally divide inferential techniques into two categories: parametric and nonparametric. Inferential tests that require knowledge about population parameters and make certain assumptions about the underlying population distribution are termed **parametric tests.** For example, in the one-sample difference of means tests (Z or t) and the one-sample difference of proportions test (Z), population parameters such as μ, σ, and ρ were included in the formulas for the test statistic. A commonly applied assumption is that the population be normally distributed with mean μ and standard deviation σ. Other assumptions that apply for some multisample parametric techniques will be discussed as needed.

Another group of statistical tests requires no such knowledge about population parameter values and has fewer restrictive assumptions concerning the nature of the underlying population distribution. This group of tests is termed **nonparametric** or **distribution-free**.

Parametric tests *require* sample data measured on an interval/ratio scale, because their test statistics use parameters such as the mean or standard deviation—descriptive statistics that can be calculated only from interval/ratio data. A mean or standard deviation cannot be calculated from nominal or ordinal data. Nonparametric tests, on the other hand, do not require such interval/ratio statistics. Some nonparametric tests are specifically designed to be used with ordinal data, whereas others are designed to be applied effectively to nominal or categorical data.

For data measured at a nominal or ordinal scale, *only* a nonparametric test can be applied. Different strategies may be possible, however, with data at the interval/ratio scale:

1. *Run only a parametric test.* This approach is appropriate if there is virtually no doubt that the requirements and assumptions needed to use the test have all been met.
2. *Run only a nonparametric test.* If there is reason to believe that one or more of the parametric test assumptions is moderately or severely violated, then the results from running the parametric test are likely to be invalid or inaccurate. Therefore, the set of interval/ratio data may be converted to an ordinal or nominal scale, and a corresponding appropriate nonparametric test can be run. This process of converting an interval/ratio variable to nominal or ordinal is termed "downgrading."
3. *Run both a parametric and nonparametric test.* If there is uncertainty about the degree to which requirements and assumptions are violated or an assumption is only slightly violated (e.g., a sample is drawn from an underlying population distribution that is only slightly nonnormal), then it may be appropriate to run both a parametric and a comparable nonparametric test. Additional insights may be gained by running both tests on the same data and comparing the resulting p-values. This "pairing" strategy may be particularly useful if conducting a comparative or investigative geographic analysis. This idea is demonstrated several times in later chapters.

The number of geographic problems that can be examined using inferential statistics is limitless. In the remainder of the text, inferential techniques are applied to a number of real-world geographic situations. Each inferential technique or set of techniques is presented using a common format:

1. Presentation of the rationale, purposes, and objectives of the technique. A set of appropriate geographic problems that could be solved using the technique are listed, and an overview of the key assumptions and required conditions is provided.
2. Presentation of basic formulas and computations associated with the technique.
3. Discussion of the geographic problem under examination and the application of the inferential technique.
4. Evaluation of the use of the technique for a particular geographic problem and discussion of both inferential issues and geographic factors that might have affected the test results.

KEY TERMS

MAJOR GOALS AND OBJECTIVES

If you have mastered the material in this chapter, you should now be able to do the following.

1. Explain null and alternate hypotheses, directional and nondirectional hypothesis testing formats, Type I and Type II errors, rejection and nonrejection regions, significance levels, and test statistics.
2. Recognize when it is appropriate to use a directional or nondirectional alternate hypothesis.
3. Understand and apply the steps necessary to test hypotheses using the classical/traditional approach.
4. Understand and apply the p-value approach to hypothesis testing and evaluate its advantages over the classical approach.
5. Know when to apply the difference of means Z or t test or the difference of proportions Z test in comparing a sample statistic to a population parameter.
6. Recognize the difference between artificial and natural sampling approaches.
7. Understand the issues involved in selecting the appropriate inferential statistical test. Issues include the type of question investigated, the level of measurement used, and whether a parametric and/or nonparametric test is appropriate.

REFERENCES AND ADDITIONAL READING

Burt, J. E. and G. M. Barber. 1996. *Elementary Statistics for Geographers*. (2nd edition). New York: Guilford Press.

Cliff, A. D. 1973. "A Note on Statistical Hypothesis Testing." *Area* 5, No. 3:240.

Coakley, C. W. 1996. "Suggestions for Your Nonparametric Statistics Course." *Journal of Statistics Education* 4, No. 2. www.stat.ncsu.edu/info/jse/v4n2/coakley.html.

Court, A. 1972. "All Statistical Populations Are Estimated from Samples." *Professional Geographer* 24:160–61.

Gould, P. R. 1970. "Is Statistix Inferens the Geographical Name for a Wild Goose?" *Economic Geography* (Supplement) 46:439–50.

Hays, A. 1985. "Statistical Tests in the Absence of Samples: A Comment." *Professional Geographer* 37:334–38.

Marzillier, L. F. 1990. *Elementary Statistics*. Dubuque, IA: Wm. C. Brown.

Maxwell, N. P. 1994. "A Coin-Flipping Exercise to Introduce the P-Value." *Journal of Statistics Education* 2, No. 1. www.stat.ncsu.edu/info/jse/v2n1/maxwell.html.

Meyer, D. R. 1972a. "Geographical Population Data: Statistical Description Not Statistical Inference." *Professional Geographer* 24:26–28.

———. 1972b. "Samples and Populations: Rejoinder to 'All Statistical Populations Are Estimated from Samples.'" *Professional Geographer* 24:161–62.

Ott, R., C. Rexroat, R. Larson, and W. Mendenhall. 1992. *Statistics: A Tool for the Social Sciences* (5th edition). Boston: PWS-Kent Publishing Co.

Siegel, S. and N. J. Castellan. 1988. *Nonparametric Statistics for the Behavioral Sciences* (2nd edition). New York: McGraw-Hill.

Silk, J. 1986. *Statistical Concepts in Geography*. London: Allen and Unwin.

Summerfield, M. 1983. "Population, Samples, and Statistical Inference in Geography." *Professional Geographer* 35:143–49.

CHAPTER 9

Two-Sample and Matched-Pairs (Dependent-Sample) Difference Tests

Chapter 8 introduced the process of hypothesis testing using both the classical and *p*-value approaches. In that chapter, the focus was on the application of *one-sample* difference tests, where a sample statistic was compared to a population parameter using both means and proportions. Geographers encounter many other situations where the objective is to determine whether significant differences exist between *two samples*. If the sample differences are significant, statistical inference permits the researcher to conclude that the samples were drawn from truly different populations. However, if sample differences are not significant, it may be concluded that the samples have been drawn from a single population. The hypothesis testing procedure discussed in the preceding chapter is now extended to applications of **two-sample difference tests**.

This chapter is organized according to the type of statistic being compared (mean or proportion) as well as by the nature of relationship between the samples (independent or dependent). Sections 9.1 and 9.2 expand on the methods discussed in chapter 8 to test for differences between two *independent* samples. Independent samples occur when the items collected in the first sample do not relate to (are independent of) the items collected in the second sample. In section 9.1, differences between two sample means are tested for statistical significance using both parametric and nonparametric procedures. Section 9.2 considers the two-sample difference of proportions test, which is designed to compare two sample proportions for significant difference.

In other problems, the samples may not be independent. For example, a geographer might measure each subject or location twice: *before* and *after* some type of treatment or event. The before and after measurements constitute a *matched pair of observations* and are considered *dependent samples*. In other geographic situations, two variables or indicators for the same sample of observations or locations are compared to see if the samples of matched pairs of observations are significantly different. Parametric and nonparametric methods for comparing differences between dependent samples are examined in section 9.3.

9.1 TWO-SAMPLE DIFFERENCE OF MEANS TESTS

Geographers use two primary methods to compare and test means from two independent samples for significant difference. The basic distinction separating the two methods is the requirements for parametric and nonparametric tests. If the sample data and population distributions meet the requirements of parametric testing, the appropriate independent-sample difference of means procedure would be either the Z or t test. For these methods, the sample data must be measured on an interval/ratio scale, and the samples must be drawn from normally distributed populations. Conversely, if the sample data are measured at the ordinal scale or if the samples are drawn from populations that are clearly not normal, a nonparametric procedure is required. For two-sample difference of means problems that do not meet parametric requirements, the Wilcoxon rank sum W test or Mann-Whitney U test is used.

Two-Sample Difference of Means *Z* or *t* Test

Suppose a geographer wants to compare surface water runoff levels in two different environmental settings: (1) a natural landscape with little or no modification by human activity, and (2) an exurban setting that has been partially modified with the development of several residential subdivisions and their access roads. The geographer might hypothesize that the presence of artificial surfaces like concrete and asphalt affects the volume of surface water runoff, making the amount different in the exurban setting versus the natural setting. To test for significant differences in runoff, spatial random samples could be taken from both settings and surface runoff levels measured. The question being asked is whether the mean volume of runoff from the natural locations differs significantly from the mean volume of runoff at the exurban locations. If the two sample means are significantly different, it can be logically inferred that the samples have been taken from two distinct populations.

In another possible application, suppose a study is conducted to determine if current home values differ between two adjacent residential subdivisions built at the same time and with similar construction costs, but with very different design principles. One of the subdivisions used cluster zoning of homes on smaller lots, including such amenities as a tennis court, baseball diamond, playing field/village common, and some open space. The other subdivision used a conventional "cookie-cutter" approach to design with larger private lots and no community recreational amenities or open space. Twenty years later, if a random sample of homes is taken from each subdivision, a research study could determine if the properties in the cluster-zoned development appreciated in value at a faster rate than those in the conventional subdivision.

As another example, suppose a geographer wants to compare fertility patterns in two different Chinese provinces. One is mostly an urban coastal province that has been heavily exposed to the one-child policies of the government over recent years, whereas the second is a predominantly rural interior province with a large ethnic population that has experienced less pressure from central government officials regarding the one-child policy. Suppose independent random samples of women aged 30 to 34 are chosen from each of the two provinces, and the number of births for each woman is recorded. The question can then be asked: Is there a significant dif-

ference between these two provinces in the mean number of children per woman?

The null hypothesis for problems testing two independent sample means for significant differences has the form $H_0: \mu_1 = \mu_2$, which is equivalent to $H_0: \mu_1 - \mu_2 = 0$. The alternate hypothesis can be stated in two ways: $H_A: \mu_1 \neq \mu_2$, when the direction of difference is not hypothesized (two-tailed format), or as $H_A: \mu_1 > \mu_2$ (or $H_A: \mu_1 < \mu_2$), when the direction of difference is hypothesized (one-tailed format).

Which of the two forms should be used in the preceding examples? In the water runoff problem, a one-tailed (directional) format seems appropriate because natural surfaces in an undeveloped landscape logically absorb more water, allowing more moisture to soak directly into the ground, resulting in a *lower* volume of water runoff. Conversely, precipitation falling on the artificial surfaces in the exurban setting might not be absorbed as readily, resulting in *higher* volumes of runoff into gutters and drains along residential streets. In the subdivision design problem, a two-tailed (nondirectional) format is suggested if no logical, preestablished basis exists for hypothesizing that either design type would produce a faster appreciation rate. In the problem examining patterns of Chinese fertility, a one-tailed format seems appropriate because a particular direction of difference makes sense: it would be logical to hypothesize that women in the coastal province have *fewer* children on average than women in the interior province.

Similar to the one-sample difference of means procedure, the parametric two-sample difference of means test can take different forms depending on the size of the samples and nature of the population variances. If the populations are normally distributed with known variances and the sample sizes are large (n_1 and n_2 greater than 30), the sampling distribution for the difference of means follows the normal (Z) distribution, and the test statistic is

$$Z = \frac{\overline{X}_1 - \overline{X}_2}{\sigma_{\overline{X}_1 - \overline{X}_2}} \qquad (9.1)$$

where $\overline{X}_1$ = mean of sample 1

$\overline{X}_2$ = mean of sample 2

$\sigma_{\overline{X}_1 - \overline{X}_2}$ = standard error of the difference of means

and

$$\sigma_{\overline{X}_1 - \overline{X}_2} = \sqrt{\frac{\sigma_1^2}{n_1} + \frac{\sigma_2^2}{n_2}} \qquad (9.2)$$

Following the general format for difference tests, the numerator of the two-sample test statistic in equation 9.1 shows the actual or observed difference between the two sample means. Recall from chapter 7 that $\overline{X}$, the sample mean, is the best estimate of μ, the population mean. Similarly, in this situation involving two samples, $(\overline{X}_1 - \overline{X}_2)$ is the best estimate of the *difference* of population means $(\mu_1 - \mu_2)$.

The denominator of equation 9.1 represents the standard error of the difference of means. This standard error expression is an estimate of how much difference between $\overline{X}_1$ and $\overline{X}_2$ is expected to occur by sampling and can be considered a measure of the expected sampling error. Think of it this way: if two independent random samples were taken from the very same population, the two sample means would likely be somewhat different, just because of the random effects of sampling. Thus, the standard error of the difference of means estimates the expected difference that should have occurred simply by chance.

What factors are influencing the size of the test statistic in equation 9.1? The magnitude of the *actual* or *observed* difference in sample means (the numerator) is divided by the magnitude of the *expected* difference in sample means (the denominator). If the actual difference in sample means is considerably greater than the expected difference in sample means, then the resultant test statistic Z will be large (either positive or negative) and not very close to zero. This leads to the inferential conclusion that the two sample means are more different than could have happened by chance if both were taken from the same population. This also leads to the complementary conclusion that the two sample means must have been taken from two distinctive (different) populations, supporting the likelihood of the alternate hypothesis.

Conversely, if the actual difference in sample means (numerator of equation 9.1) is not much greater than the expected difference in sample means (denominator), then the resultant test statistic Z will be relatively small. This leads to the conclusion that the two sample means must have been taken from the same population and supports the validity of the null hypothesis. It is also possible for the actual difference in sample means to be smaller than the expected difference in sample means, also resulting in the conclusion that the two sample means have been taken from the same population.

In most geographic applications of a two-sample difference of means test, the variances of the two populations are not known, and therefore the test statistic for Z in equation 9.1 cannot be used. In these situations, the standard error must be estimated from the *sample* variances (s^2), the difference of means test will follow the *t* distribution, and the test statistic is

$$t = \frac{\overline{X}_1 - \overline{X}_2}{\sigma_{\overline{X}_1 - \overline{X}_2}} \qquad (9.3)$$

Although equation 9.3 has the same general appearance as equation 9.1, the equations differ in the way the standard error (denominator) is estimated. In equation 9.1, the standard error is derived directly from the known variances of the two populations using equation 9.2. In equation 9.3, the standard error can be estimated only from the variances calculated for the samples taken from the two populations. In addition to its usefulness for problems with unknown population variances, the *t* distribution also provides more accurate results than Z when one or both sample sizes are small (*n* less than 30).

Two methods exist for using sample data to estimate the standard error in the denominator of equation 9.3. Selection of the appropriate method depends on the assumed relationship between the two population variances. In the first instance, when the population variances are assumed to be *equal*, $\sigma_1^2 = \sigma_2^2$, a **pooled variance estimate (PVE)** is calculated as the weighted average of the two sample variances:

$$\text{PVE} = \sqrt{\frac{s_1^2 (n_1 - 1) + s_2^2 (n_2 - 1)}{n_1 + n_2 - 2}} \qquad (9.4)$$

The denominator of equation 9.4 represents the degrees of freedom in the problem. When the two population variances are unknown but assumed equal, the standard error estimate for equation 9.3 can be written as

$$\sigma_{\overline{X}_1 - \overline{X}_2} = \text{PVE}\sqrt{\frac{1}{n_1} + \frac{1}{n_2}} \qquad (9.5)$$

In problems where the population variances are assumed to be *unequal* (i.e., if $\sigma_1 \neq \sigma_2$), the pooled estimate is not appropriate. In such cases, the sample variances are substituted directly into the standard error portion of equation 9.2 as best estimates of the respective population variances. A **separate variance estimate (SVE)** is then calculated:

$$\text{SVE} = \sigma_{\overline{X}_1 - \overline{X}_2} = \sqrt{\frac{s_1^2}{n_1} + \frac{s_2^2}{n_2}} \qquad (9.6)$$

Since geographers rarely know population parameters (like μ and σ), they must rely heavily on sample data to estimate population characteristics. This task supports one of the central goals of inferential statistics discussed earlier—the use of samples to produce unbiased estimates of population parameters. Thus, the method chosen to estimate variance—pooled versus separate—depends on whether the population variances are *assumed* to be equal. In practice, researchers usually decide equality or inequality of population variances by testing the corresponding sample data. Sample variances are considered the best estimators of population variances.

To determine whether the variances of the two samples are significantly different, a test for equality of variances is needed. A widely used approach is the F statistic for the Levene test for equality of variances, derived by computing a one-way analysis of variance (ANOVA) on the absolute deviations of each observation from its sample mean. Details on the application of analysis of variance and the associated F test are discussed in the next chapter. If the two sample variances are found to be different, the population variances are assumed to be unequal, and the separate variance method is used to test for the difference in the means. However, if the sample variances are not significantly different, the pooled variance estimate is applied. Most computer programs will provide the result of an equality of variance test, as well as the t test statistic and p-values for both situations (variances assumed to be equal and not equal).

Wilcoxon Rank Sum *W* (Mann-Whitney *U*) Tests

For some geographic problems testing the difference between two independent samples, the parametric two-sample difference of means tests (Z or t) may not be appropriate. Sometimes data are available only in ordinal or ranked form, making it impossible to calculate sample means and sample variances. In other situations, a geographer may be working with data measured at the interval/ratio level, but has reason to believe that the samples are from populations that are not normally distributed. When populations exhibit moderate-to-severe deviation from normality, the use of a difference of means test raises serious questions about the validity of such a parametric procedure. In these cases, a nonparametric test for two independent samples is a better alternative.

The most widely used nonparametric alternatives for the two-sample difference of means test are the Wilcoxon rank sum W test and the directly related Mann-Whitney U statistic. Like the t test, the Wilcoxon and Mann-Whitney tests examine two independent samples for differences. Rather than using parameters like mean and variance, however, these techniques use the ranks of sample observations to measure the magnitude of the differences in the ranked positions or locations between the two sets of sample data. Although no form of distribution is assumed for the two populations, the procedure requires that the two distributions be similar in shape. This characteristic makes the Wilcoxon and Mann-Whitney tests especially useful for problems with small samples drawn from populations that are not necessarily normal.

To keep the explanation of the nonparametric two-sample difference tests simple, the discussion here will focus on the Wilcoxon test procedure, which uses a difference test format very similar to the two-sample difference of means Z and t tests. The discussion includes a brief explanation of the direct relationships of the Mann-Whitney and Wilcoxon test statistics.

In the Wilcoxon rank sum W test, the data from the two samples are combined and placed in a single ranked set. When two or more values are tied for a particular rank, the average rank value is assigned to each position. The samples are then considered separately and the sum of ranks (W) calculated for each sample. Suppose the sum of ranks for sample 1 is called W_1, and the sum of ranks for the second sample W_2. If the two samples are drawn from the same population, the ranks should be randomly mixed between the two samples, and the sum of ranks for each sample should be roughly equal when the sample sizes are equal. That is, if $n_1 = n_2$, then W_1 should be roughly equal to W_2. If the sample sizes are not equal, but drawn from the same population, then their respective sum of ranks should be proportionate to their respective sample sizes. These findings would all confirm the null hypothesis of no significant difference between the two samples. However, if the sum of ranks for the first sample (W_1) is quite different from the sum of ranks for a second sample of similar size (W_2), it is more likely that the two samples have been drawn from different populations, making confirmation of the null hypothesis less likely.

Two-Sample Difference of Means *Z* or *t* Test

Primary Objective: Compare two independent random sample means for difference

Requirements and Assumptions:

1. Two independent random samples
2. Each population is normally distributed
3. Variable is measured at interval or ratio scale

Hypotheses:

$H_0: \mu_1 = \mu_2$
$H_A: \mu_1 \neq \mu_2$ (two-tailed)
$H_A: \mu_1 > \mu_2$ (one-tailed) or
$H_A: \mu_1 < \mu_2$ (one-tailed)

Test Statistic:

$$Z \text{ or } t = \frac{\overline{X}_1 - \overline{X}_2}{\sigma_{\overline{X}_1 - \overline{X}_2}}$$

If sample sizes ≥ 30 and population variances are known, use Z
If sample sizes < 30 and population variances are unknown, use t

The Wilcoxon test uses a variation of the Z test to see if the sum of sample ranks is significantly different from what it should be if the two samples are actually drawn from the same population. The test statistic (Z_W) for the two-sample Wilcoxon procedure is

$$Z_W = \frac{W_i - \overline{W}_i}{s_W} \qquad (9.7)$$

where W_i = sum of ranks for sample i

$$\overline{W}_i = \text{mean rank of } W_i = n_i\left(\frac{n_1 + n_2 + 1}{2}\right) \quad (9.8)$$

s_W = standard deviation of W

$$= \sqrt{n_1 n_2\left(\frac{n_1 + n_2 + 1}{12}\right)} \qquad (9.9)$$

$\overline{W}_i$ and s_W represent the theoretical mean and standard deviation of W_i, respectively, and are determined totally by the sample sizes. As shown in equation 9.9, only one standard deviation exists for W. However, because rank sums (W_1 and W_2) can be determined for each sample, two means can also be calculated using equation 9.8—one for sample 1 ($\overline{W}_1$) and another for sample 2 ($\overline{W}_2$).

In the Wilcoxon rank sum procedure, two complementary test statistics (Z_{W_1} and Z_{W_2}) can be calculated. Z_{W_1} uses the rank sum and the mean from the first sample, while Z_{W_2} uses the corresponding statistics for the second sample. The two test statistics differ only in sign: one will be negative and the other positive.

The Wilcoxon statistic, W, is simply the value or magnitude of the sum of ranks of the group with the *smaller* sample size. The Mann-Whitney statistic, U, which complements the Wilcoxon statistic, is determined by the number of times an observation from the group with the smaller sample size ranks lower than an observation from the group with the larger sample size. Because the Mann-Whitney calculation can be more cumbersome than the Wilcoxon calculation, only the Wilcoxon worktable is included. Statistical software packages usually calculate both Wilcoxon and Mann-Whitney test statistics. The two test statistic values are equivalent, in that they always provide the same significance level (p-value) when applied to the same set of data.

Application to Geographic Problem

Many countries have been experiencing profound changes in development as the result of an increasingly integrated global economy. A variety of economic, social, political, and demographic changes are taking place as less-developed countries attempt to advance through economic development and modernization. This is a continuing and complicated process, involving literally hundreds of different development indicators, as people across the globe seek to improve their standard of living and quality of life.

This multitude of changes offers geographers many opportunities for research. Geographic questions can be proposed for any development indicator. Suppose the percentage of labor force that is female is selected as a target indicator for investigation. In the rapidly changing global economy, is the involvement of women in the labor force changing in any way? Does the participation level of women in the labor force vary by the economic level of the country in which they live? Do females constitute a higher percentage of the labor force in low-income economies or in high-income economies? What are the recent trends?

According to the World Bank, as found in *World Development Indicators 1997*, the percentage of the world's total labor force that is female has been slowly increasing, from 38 percent in 1980 to 40 percent in 1995. The World Bank data on development indicators list 133 countries with a population of 1 million or more. Of these 133 countries, 90 are classified as "low-income" and "low-middle-income" economies (with GNP per capita of $3,035 or less), whereas 43 are classified as "upper-middle-income" and "high-income" economies (with GNP per capita of $3,036 or more).

A random sample of 30 countries is selected from the two "lower" categories of income, and a second independent random sample of 14 countries is taken from the "higher"-income categories. Using these independent samples, a two-sample difference of means test is applied to the data on percentage of females in the labor force for 1980 as well as for 1995. Table 9.1 shows the 44 nations selected in the two independent samples, the economic classification to which each nation belongs, and the data on females in the labor force from 1980 and 1995.

First, the two-sample difference of means t test is selected for the analysis. This test statistic seems appropriate since the data are interval/ratio, the population variances are unknown, and both sample sizes are 30 or less.

The calculations in table 9.2 summarize the testing of lower-income countries against higher-income countries according to percentage of labor force that is female, using the two-sample difference of means t test. Note that for the 1980 data, the countries with lower income have a greater mean female labor force participation rate than the countries with higher income ($\overline{X}_1 = 38.33$ percent, as compared to $\overline{X}_2 = 30.93$ percent). The central question is whether these two sample means differ statistically. The calculated

TABLE 9.1

Percentage of Labor Force Female, 1980 and 1995, for Samples of Lower-Income Countries and Higher-Income Countries

Country	Economic classification*	Percentage of females in labor force	
		1980	1995
Tanzania	Lower	50	49
Chad	Lower	43	44
Nepal	Lower	39	40
Madagascar	Lower	45	45
Vietnam	Lower	48	49
Mali	Lower	47	46
Cambodia	Lower	56	53
Togo	Lower	39	40
India	Lower	34	32
Nicaragua	Lower	28	36
Angola	Lower	47	46
Mauritania	Lower	45	44
Guinea	Lower	47	47
China	Lower	43	45
Albania	Lower	39	41
Sri Lanka	Lower	27	35
Egypt	Lower	26	29
Moldova	Lower	50	49
Philippines	Lower	35	37
Papua New Guinea	Lower	42	42
Guatemala	Lower	22	26
Romania	Lower	46	44
Algeria	Lower	21	24
Paraguay	Lower	27	29
Colombia	Lower	26	37
Russia	Lower	49	49
Costa Rica	Lower	21	30
Panama	Lower	30	34
Estonia	Lower	51	49
Venezuela	Lower	27	33
Mexico	Higher	27	31
Brazil	Higher	28	35
Malaysia	Higher	34	37
Oman	Higher	7	14
Argentina	Higher	28	31
Korea, Rep.	Higher	39	40
New Zealand	Higher	28	36
Kuwait	Higher	13	27
Australia	Higher	39	43
Finland	Higher	46	48
Netherlands	Higher	31	40
Singapore	Higher	35	38
Germany	Higher	40	42
Japan	Higher	38	41

*In the World Bank publication *World Development Indicators 1997*, the main criterion used to classify economies and broadly distinguish stages of economic development is GNP per capita. The GNP per capita cutoff levels used here are combinations of those used by the World Bank to classify economies. In this table, "lower" is defined as GNP per capita of $3,035 or less, and "higher" is defined as GNP per capita of $3,036 or more.

test statistic ($t = 2.187$) has an associated two-tailed p-value or significance level of .034. From this calculation, it may be concluded that the mean percentage of labor force population that is female in lower-income countries differs significantly from that of higher-income countries. With a p-value of .034, researchers can be 96.6 percent confident ($1 - .034 = .996$) that the correct conclusion has been reached using only the sample data.

Using a similar difference of means procedure for the 1995 data, $t = 1.618$ and $p = .113$. It appears that women make up an increasing percentage of the total labor force in many countries, but the sharpest increases (at least from 1980 to 1995) occur in the countries with higher income ($\overline{X}_2 = 30.93$ for 1980, whereas $\overline{X}_2 = 35.93$ for 1995). With this second t-test result, it is probably better to conclude that the mean percentage of labor force that is female is not significantly different between lower- and higher-income countries. The null hypothesis should not be rejected because the likelihood of committing a Type I error is 11.3 percent, a rather high and usually unacceptable level of error.

Following these calculations, the Wilcoxon rank sum W and Mann-Whitney U tests are applied. "Pairing" these nonparametric tests with the parametric t test can be justified. It is not known whether the two populations from which samples have been taken (lower- and higher-income economies) are normally distributed. This is an important consideration, since normality of the two populations being tested is one of the assumptions associated with the parametric two-sample difference tests (Z and t). See the boxed inset for these tests.

The worktable for the Wilcoxon rank sum W test (table 9.3) summarizes the application of these tests on the worldwide female labor force data, with calculations shown for the 1980 information. In 1980,

Wilcoxon Rank Sum Test

Primary Objective: Compare two independent random sample rank sums for difference

Requirements and Assumptions:

1. Two independent random samples
2. Both population distributions have the same shape
3. Variable is measured at ordinal scale or downgraded from interval/ratio scale to ordinal

Hypotheses:

H_0: The distribution of measurements for the first population is equal to that of the second population

H_A: The distribution of measurements for the first population is not equal to that of the second population (two-tailed)

H_A: The distribution of measurements for the first population is larger (or smaller) than that for the second population (one-tailed)

Test Statistic:

$$Z_W = \frac{W_i - \overline{W}_i}{s_W}$$

TABLE 9.2

Worktable for Difference of Means t Test: Percentage of Labor Force Female, 1980 and 1995

	Sample statistics: 1980					
	n	$\bar{X}$	s	$\sigma_{\bar{x}}$	F^*	p-value of F
Lower-income countries (sample 1)	30	38.33	10.42	1.90		
					.506	.481
Higher-income countries (sample 2)	14	30.93	10.56	2.82		

*Levene test for equality of variances. A significance level this far from zero ($p = .481$) allows equality of population variances to be assumed (that is, $\sigma_1^2 = \sigma_2^2$), thereby making the pooled variance estimate (*PVE*) appropriate:

from equation 9.4:

$$PVE = \sqrt{\frac{s_1^2(n_1 - 1) + s_2^2(n_2 - 1)}{n_1 + n_2 - 2}} = \sqrt{\frac{(10.42)^2(30 - 1) + (10.56)^2(14 - 1)}{30 + 14 - 2}} = 10.46$$

from equation 9.5:

$$\sigma_{\bar{x}_1 - \bar{x}_2} = PVE\sqrt{\frac{1}{n_1} + \frac{1}{n_2}} = 10.46\sqrt{\frac{1}{30} + \frac{1}{14}} = 3.39$$

therefore, from equation 9.3:

$$t = \frac{\bar{X}_1 - \bar{X}_2}{\sigma_{\bar{x}_1 - \bar{x}_2}} = \frac{38.33 - 30.93}{3.39} = 2.187 \text{ with two-tailed } p = .034$$

	Sample statistics: 1995					
	n	$\bar{X}$	s	$\sigma_{\bar{x}}$	F^*	p-value of F
Lower-income countries (sample 1)	30	40.13	7.88	1.44		
					.229	.635
Higher-income countries (sample 2)	14	35.93	8.35	2.33		

After similar calculations as above for 1995: $t = 1.618$ with two-tailed $p = .113$

the sample statistics indicate the mean rank for the lower-income economy sample is 24.88, as compared with a mean rank of 17.39 for the higher-income country sample. Results from the Wilcoxon test suggest with moderate certainty that the mean rank for the lower-income sample is too different from the mean rank for the higher-income sample to conclude that the two samples have been drawn from the same population. The associated two-tailed p-value of .071 suggests that the difference in female labor force participation between the two groups of countries is statistically significant. Viewed another way, the confidence level associated with concluding that the two samples have been drawn from two different populations is $(1 - .071 = .929)$, or 92.9 percent.

The conclusion that can be drawn from the 1995 data is not as strong statistically. The mean ranks are not as different as they are in 1980 (24.53 for the lower-income economies versus 18.14 for the higher-income economies), and the Wilcoxon test statistic value has an associated p-value of .124. If it is con-

cluded that female labor force participation rates are different for the lower- and higher-income country populations (i.e., rejecting the null hypothesis), the chance of making a Type I error is 12.4 percent. In this situation, the null hypothesis should not be rejected. For the 1995 data, female labor force participation rates are not different enough to be statistically significant.

Summary Evaluation

The statistical results from the t test and from the Wilcoxon (Mann-Whitney) test are very similar for both the 1980 and 1995 data. This numerical reinforcement from different statistical tests can only increase the confidence that the results are correct. Actually, the p-values associated with the t tests are just slightly lower (stronger statistically) than the comparable Wilcoxon (Mann-Whitney) p-values. This finding is expected, given the slightly stronger statistical power of the parametric t test over its nonparametric equivalents.

TABLE 9.3

Worktable for Wilcoxon Rank Sum W Test: Percentage of Labor Force Female, 1980 and 1995

	Sample statistics: 1980		
	n	Mean rank	Sum of ranks
Lower-income countries (sample 1)	30	24.88	746.5
Higher-income countries (sample 2)	14	17.39	243.5

Wilcoxon W is the sum of ranks of the group with the *smaller* sample size ($W_2 = 243.5$)

Mean rank of sample 2: $\quad \overline{W}_2 = n_2\left(\dfrac{n_1 + n_2 + 1}{2}\right) = 14\left(\dfrac{30 + 14 + 1}{2}\right) = 315$

Standard deviation of W: $\quad s_w = \sqrt{n_1 n_2\left(\dfrac{n_1 + n_2 + 1}{12}\right)} = \sqrt{(30)(14)\left(\dfrac{30 + 14 + 1}{12}\right)} = 39.69$

Test statistic for sample 2: $\quad Z_{W_2} = \dfrac{W_2 - \overline{W}_2}{s_W} = \dfrac{243.5 - 315}{39.69} = -1.805 \quad p = .071$ (two-tailed)

Corresponding Mann-Whitney U (1980) = 138.50, $p = .071$ (two-tailed)

	Sample statistics: 1995		
	n	Mean rank	Sum of ranks
Lower-income countries (sample 1)	30	24.53	736.0
Higher-income countries (sample 2)	14	18.14	254.0

After similar calculations for 1995: $Z_W = -1.540$, $p = .124$ (two-tailed)

These results invite considerable geographic speculation. Why are more women seemingly in the labor force in countries with lower incomes? What is the cause for the changes in higher-income countries? Are more women becoming primary wage earners, or is this increase the result of more two-income households? The sharpest increases in female labor force participation have occurred in such lower-income countries as Nicaragua, Sri Lanka, Colombia, and Costa Rica and in such higher-income countries as New Zealand, Kuwait, and the Netherlands (table 9.1). What geographic factors can explain this rather complex pattern?

The operational definitions of "low-income," "low-middle income," "upper-middle income," and "high-income" used by the World Bank may have influenced the results of this analysis. Certainly the countries of the world have very different incomes positioned over a continuum. As a result, the classification of each country as either "lower" or "higher" income is potentially misleading and may be concealing a more complex reality. It is certainly valid to ask if a two-sample difference test is the best statistical method to use in this problem. Perhaps a procedure that allows differences in female labor force participation to be examined using more than two samples would be more appropriate.

9.2 TWO-SAMPLE DIFFERENCE OF PROPORTIONS TEST

Geographers often work with proportions as well as means. Just as there are two-sample difference of means tests, there is also a two-sample difference of proportions test. Inferences concerning the difference of two population proportions can be made by comparing the difference of two sample proportions, using a similar logic (and similar test structure) to the difference of means tests. For example, a political geographer might want to determine if differences exist in the president's approval rating among the voting population in two neighboring states. Random samples of voters in each state could be polled and the proportion of potential voters giving the president a favorable rating tabulated. A two-sample difference of proportions test would then assess the likelihood that a statistically significant difference in approval ratings exists. A recreation planner could ask visitors at a regional park whether they approve of a change in operating hours. If the park visitors were divided into two samples (weekday visitors versus weekend visitors), it would then be possible to apply a two-sample difference of proportions test to determine if the proportion of those who approve of the new hours differs between weekday and weekend visitors.

In these situations, where a dichotomous (binary) variable has only two possible outcomes or responses, the two-sample difference of proportions test is appropriate. For many problems, the variable being analyzed fits this requirement of exactly two categories or classes, such as male–female, approve–disapprove, or yes–no. In situations with more than two possible outcomes or responses, geographers can sometimes reduce three or more categories to two, so that the difference of proportions test is appropriate across the entire data set. For example, a public survey measuring attitudes toward capital punishment might include the following possible responses: strongly favor, mildly favor, mildly oppose, strongly oppose. If these four possible responses were collapsed to two—favor and oppose, for example—then the data would be structured in a way that would allow a single two-sample difference of proportions test to be applied across the entire set of data. For example, one could then test whether the proportion of elderly (age 65 and above) favoring capital punishment differs from the proportion of young adults (age 20–34) sharing that opinion.

In problems testing for significant differences with a dichotomous (binary) variable, one of the two variable categories is selected as the focus for the analysis. The proportion of the sample in the focus category is termed p, while the other proportion is termed q, where $q = 1 - p$. The objective of the difference test is to determine whether the proportion of *population 1* (ρ_1) having the focus attribute differs significantly from the corresponding proportion of *population 2* (ρ_2).

The null hypothesis for the problems is $H_0: \rho_1 = \rho_2$. As in other difference tests, the alternate hypothesis (H_A) can take two forms. If no direction is hypothesized for difference in proportion between the two samples, a two-tailed procedure is applied where $H_A: \rho_1 \neq \rho_2$. This would be the case if one wished to test if elderly and young adult views differ with regard to capital punishment. On the other hand, if one of the sample proportions is expected to exceed the other, a one-tailed test is appropriate. This would

be the situation for a hypothesis that a greater proportion of elderly favored capital punishment over young adults. In these instances, the alternate hypothesis is $H_A: \rho_1 > \rho_2$ or $H_A: \rho_1 < \rho_2$.

The test statistic for the difference of proportion procedure (Z_p) has a form similar to other two-sample difference tests:

$$Z_p = \frac{p_1 - p_2}{\sigma_{p_1 - p_2}} \qquad (9.10)$$

where p_1 = proportion of sample 1 in the category of focus

p_2 = proportion of sample 2 in the category of focus

$\sigma_{p_1 - p_2}$ = standard error of the difference of proportions

Just as the sample proportion, p, is the best estimate of the population proportion, (ρ), the sample difference of proportions, ($p_1 - p_2$), is the best estimate of the population difference of proportions, ($\rho_1 - \rho_2$). The denominator of equation 9.10 represents the standard error of the difference in proportions and can be estimated as follows:

$$\sigma_{p_1 - p_2} = \sqrt{\hat{p}(1 - \hat{p})\left(\frac{n_1 + n_2}{n_1 n_2}\right)} \qquad (9.11)$$

where $\hat{p}$ = pooled estimate of the focus category for the population

The pooled estimate, $\hat{p}$, is the proportion in the focus category if the two samples were combined into one sample. Operationally, the pooled estimate is the weighted proportion from the two samples:

$$\hat{p} = \frac{n_1 p_1 + n_2 p_2}{n_1 + n_2} \qquad (9.12)$$

Two-Sample Difference of Proportions Test

Primary Objective: Compare two independent random sample proportions for difference

Requirements and Assumptions:

1. Two independent random samples
2. Variable is organized by dichotomous (binary) categories

Hypotheses:

$H_0: \rho_1 = \rho_2$
$H_A: \rho_1 \neq \rho_2$ (two-tailed)
$H_A: \rho_1 > \rho_2$ (one-tailed) or
$H_A: \rho_1 < \rho_2$ (one-tailed)

Test Statistic:

$$Z_p = \frac{p_1 - p_2}{\sigma_{p_1 - p_2}}$$

Application to Geographic Problem

The two-sample difference of proportions test can be illustrated using information collected from the 1995 General Social Survey (GSS). Conducted by the National Opinion Research Center (NORC) at the University of Chicago, the GSS is administered annually in the United States to a nationwide sample of adults. The GSS sample selection method is complex and sequential, involving first a random sample of cities and counties, then random samples of neighborhoods, households, and finally an adult from within a particular household. Although complicated, NORC technical notes indicate that this procedure may be considered statistically equivalent to a random sample of the adult population of the United States.

A series of questions from the GSS investigates attitudes toward a variety of controversial issues, including whether those sampled are in favor of or oppose gun permits. The question can be asked: Does the proportion of those in favor of gun permits differ from one type of place to another? For example, do attitudes differ between residents of large cities, suburbs of large cities, unincorporated areas/smaller cities, and small towns/rural areas? To answer these questions and have sufficient sample sizes to conduct two-sample difference of proportions tests, the original "Expanded NORC Size Code" classification (with ten different categories) is collapsed to four combined categories. Table 9.4 lists the actual GSS data from each of the ten original "Expanded NORC Size Code" categories and combines these data into four "Summary Categories."

A series of two-sample difference of proportions tests is run, a separate test for each possible pair of the four combined categories. In this situation, a total of six difference of proportions tests are run for the six possible pairs. Table 9.5 illustrates the calculation, contrasting the proportion of "large-city" respondents in favor of gun control with the proportion in favor of gun control from "suburbs." The test statistic and p-values for the other five difference of proportions tests are calculated in a similar way and summarized in table 9.6.

The largest difference of proportions in table 9.6 occurs when attitudes of large-city residents are contrasted with attitudes of people living in rural areas (for that pair, $Z_p = 3.38$ and $p = .0008$). Not surprisingly, perhaps, the sample of large-city residents is most supportive of gun control (with a proportion of .890 in favor of requiring gun permits), whereas the sample of those living in small towns and rural areas is least in favor (with a proportion of only .720). The associated p-value of .0008 very strongly suggests that these sample proportions are indicative of true differences in attitude between these two populations within the United States. Perhaps residents of larger cities see more violence and crime, associate this crime with guns, and believe that stricter gun control measures, such as requiring gun permits, will somehow reduce crime and violence. Alternatively, perhaps those living in small places and rural areas don't see as much gun-related violence and crime, might be more likely to associate guns with noncriminal activities such as hunting, and therefore do not advocate stringent gun controls.

TABLE 9.4

Attitude toward Requiring Gun Permits, by National Opinion Research Center (NORC) Size Code Category

Expanded NORC size code category	Number of residents sampled	Number (proportion) in favor of gun permits	Number (proportion) opposed to gun permits
City > 250,000	82	75	7
City 50,000–250,000	45	38	7
Summary category: "Large city"	127	113 (.890)	14 (.110)
Suburb, large city	106	93	13
Suburb, medium city	46	34	12
Summary category: "Suburb"	152	127 (.836)	25 (.164)
Unincorporated, large city	32	24	8
Unincorporated, medium city	52	40	12
City, 10,000–49,999	31	26	5
Summary category: "Small city"	115	90 (.783)	25 (.217)
Town > 2,500	51	40	11
Smaller areas	42	27	15
Open country	25	18	7
Summary category: "Rural"	118	85 (.720)	33 (.280)

TABLE 9.5

Worktable for Difference of Proportions Test Contrasting Two NORC Samples on Attitude toward Requiring Gun Permits

$H_0: \rho_1 = \rho_2$
$H_A: \rho_1 \neq \rho_2$

where ρ_1 = Proportion in favor of requiring gun permits, large/medium city residents
ρ_2 = Proportion in favor of requiring gun permits, suburban residents of large/medium city

Summary NORC category	Sample statistics: (from NORC)		
	Total number of residents sampled (n)	Number (proportion) in favor of gun permits (p)	Number (proportion) opposed to gun permits (q)
"Large city"	127	113 (.890)	14 (.110)
"Suburbs"	152	127 (.836)	25 (.164)

from equation 9.12: $\hat{p} = \dfrac{n_1 p_1 + n_2 p_2}{n_1 + n_2} = \dfrac{127(.890) + 152(.836)}{127 + 152} = .861$

from equation 9.11: $\sigma_{p_1 - p_2} = \sqrt{\hat{p}(1 - \hat{p})\left(\dfrac{n_1 + n_2}{n_1 n_2}\right)} = \sqrt{.861(.139)\left(\dfrac{127 + 152}{127(152)}\right)} = .04158$

therefore, from equation 9.10: $Z_p = \dfrac{p_1 - p_2}{\sigma_{p_1 - p_2}} = \dfrac{.890 - .836}{.04158} = 1.30 \qquad p = .1936 \text{ (two-tailed)}$

Summary Evaluation

This problem provides an opportunity to discuss two related validity issues of general concern. First, as mentioned in chapter 1, if the operational definitions are not appropriate, subsequent statistical analysis may be invalid. In this example, the four combined NORC categories of size and type of place may not be particularly valid for distinguishing attitudes regarding gun control issues, at least in the context of a national sample barely exceeding 500 people. Further studies could examine differences in attitude by region of the country or by the respondents' income and education level.

A second validity issue concerns the level of spatial aggregation. Does a national survey provide the most valid insights regarding differences in attitude? Perhaps, for example, if differences in attitude were explored in a single metropolitan area and its surrounding suburbs and rural hinterland, other locational differences of opinion might be uncovered. However, this alternative strategy is not possible using the GSS data.

9.3 MATCHED-PAIRS (DEPENDENT-SAMPLE) DIFFERENCE TESTS

Geographers often want to determine if two sets of values defined for one group of individuals or spatial locations differ. For example, suppose one wishes to compare the number of migrants moving into a set of counties with the number of migrants leaving the same set of counties. The number of in-migrants and the number of out-migrants could be determined for a sample of geographic areas and the differences tested for statistical significance. In another example, residents of an area could be sampled to determine the miles traveled per week for purposes other than commuting. Following a major rise in the price of gasoline, which can occur during times of increased political tension in the Persian Gulf, the same individuals could be surveyed a second time to determine changes in their discretionary travel behavior. Has the change in gasoline prices led to de-

TABLE 9.6

Difference of Proportions Test Results: Attitude toward Requiring Gun Permits

	"Suburbs"	"Small city"	"Rural"
"Large city"	$Z_p = 1.30$ $p = .1936$	$Z_p = 2.26$ $p = .0238$	$Z_p = 3.38$ $p = .0008$
"Suburbs"		$Z_p = 1.10$ $p = .2714$	$Z_p = 2.30$ $p = .0214$
"Small city"			$Z_p = 1.11$ $p = .2670$

creased automobile use for drivers within a particular region? In another research problem, the same sample of coastal zone residents could be surveyed both before and after a severe winter storm (nor'easter) to determine if attitudes toward coastal zone management have changed significantly. As a final example, suppose that average male life expectancy *and* average female life expectancy data are collected from a single random sample of villages in Mexico. Are male and female life expectancies in these Mexican villages different?

In each of these examples, the data consist of one set of observations (locations or individuals) collected for two different variables or at two different time periods. At first, the difference of means *t* or Wilcoxon rank sum *W* may seem most appropriate to test differences for significance. However, as discussed in section 9.1, these two difference tests require two **independent samples.** In the examples from the previous paragraph, only *one* sample is drawn. In two of those examples, data are collected for two variables at one time period (in- and out-migrants; male and female life expectancy). In the other two examples, data are collected for one variable at two time periods (miles traveled before and after a price increase for gasoline; attitudes toward managing the coast before and after a winter storm). Thus, neither the two-sample *t* test nor the Wilcoxon *W* procedure is appropriate for these problems.

When two sets of data are collected for one group of observations, the samples are termed **dependent,** and a **matched-pairs difference test** is the proper inferential procedure for examining differences between such samples. As the name implies, each observation or sample member has two values, known as a **matched pair.** The differences in the set of matched pairs are tested for statistical significance and results inferred to the population from which the dependent samples are drawn. In the example of life expectancy in the Mexican villages, two equal-sized samples are drawn—one male and the other female. However, the samples are not independent. In fact, average male and female life expectancies from a particular village are directly related to one another, since men and women at the same location are likely affected by many of the same economic, social, and environmental factors. The male and female life expectancies from each village are a matched pair, and it is appropriate to compare a sample of such matched pairs of observations to see if they differ significantly.

Two common inferential procedures for testing differences in dependent samples are the matched-pairs *t* test and the Wilcoxon matched-pairs signed-ranks test. The matched-pairs *t* test is a parametric test requiring interval/ratio-level data and a normally distributed population. Like the Wilcoxon *W* two-sample difference test, the related matched-pairs signed-ranks test is a distribution-free, nonparametric method that uses either ranked data directly or interval/ratio data downgraded to its ordinal equivalent.

Matched-Pairs *t* Test

The matched-pairs *t* test considers the difference between the values for each matched pair. The greater this difference (*d*), the more dissimilar the results of the two values within the matched pair. The mean of the difference values ($\bar{d}$) is determined for the set of all matched pairs in the sample. If the differences within the matched-pairs values are small, this average difference value will be close to zero. However, in a problem where the matched-pairs differences are large, $\bar{d}$ will also be large, suggesting significant differences between the two sets of data being studied.

The null hypothesis in the matched-pairs problem states that the mean difference for all matched pairs in the population (δ) equals zero. The best estimate of the population matched-pairs mean difference (δ) is the sample matched-pairs mean difference ($\bar{d}$). The alternative hypothesis can be either nondirectional or directional, depending on a priori information about the expected matched-pairs differences.

The test statistic for the matched-pairs *t* (t_{mp}) is defined as follows:

$$t_{mp} = \frac{\bar{d}}{\sigma_d} \qquad (9.13)$$

where $\bar{d}$ = mean of matched-pairs differences (*d*)
  σ_d = standard error of the mean difference

The numerator of equation 9.13 is the mean of matched-pairs differences (*d*):

$$\bar{d} = \frac{\Sigma d_i}{n} \qquad (9.14)$$

where d_i = difference for matched-pair *i*
  n = number of matched pairs

The difference (d_i) is found by subtracting the corresponding paired values of the second variable (*Y*) from those of the first variable (*X*):

$$d_i = X_i - Y_i \qquad (9.15)$$

The mean of the difference values ($\bar{d}$) can also be calculated from the means of the two variables:

$$\bar{d} = \overline{X} - \overline{Y} \qquad (9.16)$$

As in other difference tests, the denominator of equation 9.13 contains a standard error measure. In this case, the standard error refers to the mean difference in matched pairs:

$$\sigma_d = \frac{s_d}{\sqrt{n}} \qquad (9.17)$$

where

$$s_d = \sqrt{\frac{\Sigma\,(d_i - \bar{d})^2}{n - 1}} \qquad (9.18)$$

Standard deviation of the matched-pairs differences can also be derived from the computational formula (table 3.3):

$$s_d = \sqrt{\frac{\Sigma d_i^2 - \dfrac{(\Sigma d_i)^2}{n}}{n - 1}} \qquad (9.19)$$

Wilcoxon Matched-Pairs Signed-Ranks Test

In some geographic problems, a matched-pairs (dependent-sample) test may be appropriate, but the sample data for analysis are measured at the ordinal level. In other situations, the sample data may not be drawn from a normally distributed population. In the first instance, the parametric matched-pairs t test cannot be applied because the measurement scale is not appropriate. In the second case, the parametric test may produce biased results. The Wilcoxon signed-ranks test is the nonparametric equivalent for matched-pairs or dependent sample problems and is the appropriate procedure in these situations.

The Wilcoxon signed-ranks test uses matched-pair differences ranked from lowest (rank 1) to highest. The matched-pairs data can come either from direct ordinal measurement or from interval/ratio

differences downgraded to ranks. The *absolute* difference between the two variables (rather than the positive or negative difference) is used to determine the rank for each matched pair. When the difference for any matched pair is zero, the data are ignored and the sample size reduced accordingly. When differences for matched pairs are tied for a particular rank position, the average rank is assigned to each such pair. The null hypothesis for the signed-ranks test states that the matched-pairs differences (in ranks) for the population from which the sample is drawn equals zero.

Two sums can be calculated from the set of ranked matched pairs: T_p, the sum of ranks for positive differences (variable one greater than variable two), and T_n, the sum of ranks for negative differences (variable two greater than variable one). If the two variables measured for the single sample show very little difference, T_p should be approximately equal to T_n. However, for a problem in which the differences between the two variables are large, the disparity between T_p and T_n will also be large. In these situations, one of the rank sums (either the positive or negative differences) will be large and the other small.

The Wilcoxon test for dependent samples uses only one of the two possible rank sum values. The decision of which rank sum to test depends on whether the alternate hypothesis (H_A) is directional (one-tailed) or nondirectional (two-tailed). If no direction of difference between the two variables is hypothesized, a two-tailed test is applied, and the *smaller* of T_p and T_n is chosen. In this instance, the hypothesis concerns only a difference between the two variables under study and not which variable is the largest.

The second possibility involves a directional hypothesis and a one-tailed procedure. In this case, the hypothesis states that either the positive or negative differences for the matched pairs are expected to dominate. The value of T corresponding to the *smaller number* of hypothesized differences (either positive or negative) is selected for testing. Thus, if more differences are expected to be positive, T_n, the sum of the negative differences is used.

When the number of matched pairs exceeds ten, the rank sum (T) can be converted to a Z statistic (Z_W) and tested using the distribution of normal values:

$$Z_W = \frac{T - \dfrac{n(n + 1)}{4}}{\sqrt{\dfrac{n(n + 1)(2n + 1)}{24}}} \qquad (9.20)$$

where n = number of matched pairs ($n > 10$)
T = rank sum

Matched-Pairs t Test

Primary Objective: Compare matched pairs from a random sample for difference

Requirements and Assumptions:

1. Random sample
2. Data are collected for two different samples or at two different time periods
3. Population is normally distributed
4. Variable(s) is (are) measured at interval or ratio scale

Hypotheses:

H_0: $\delta = 0$
H_A: $\delta \neq 0$ (two-tailed)
H_A: $\delta > 0$ (one-tailed) or
H_A: $\delta < 0$ (one-tailed)

Test Statistic:

$$t_{mp} = \frac{\bar{d}}{\sigma_d}$$

Wilcoxon Matched-Pairs Signed-Ranks Test

Primary Objective: Compare matched pairs from a random sample for difference

Requirements and Assumptions:

1. Random sample
2. Data are collected for two different variables or at two different time periods
3. Variable(s) is (are) measured at ordinal scale or downgraded from interval/ratio scale to ordinal

Hypotheses:

H_0: The ranked matched-pair differences of the population are equal
H_A: The ranked matched-pair differences of the populations are not equal (two-tailed)
H_A: The ranked matched-pair differences of the populations are positive or negative (one-tailed)

Test Statistic:

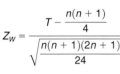

$$Z_W = \frac{T - \dfrac{n(n+1)}{4}}{\sqrt{\dfrac{n(n+1)(2n+1)}{24}}}$$

when $n > 10$

Application to Geographic Problem

One of the most critical geographic issues recently has been climatic change and the corresponding theory of global warming. Scientists have theorized that temperatures in the earth's atmosphere should be rising because carbon dioxide and particulates have increased following the burning of fossil fuels used as energy sources. These by-products from the combustion of natural gas, liquid gasoline, fuel oil, and coal are nonnatural ingredients, which intensify the so-called greenhouse effect. The process is common in all locations that are experiencing increased energy demand resulting from rising levels of development. The result is an artificial "blanket" within the lower levels of the earth's atmosphere that blocks escaping long-wave solar energy and contributes to rising temperatures near the surface.

A way to test for global warming would be to compare recorded temperatures at a single set of locations over a particular time period. Because this problem uses only one variable collected at two different time periods, a matched-pairs or dependent-sample difference test is most appropriate.

To illustrate the matched-pairs t and Wilcoxon signed-ranks procedures, a sample of 28 weather stations in a key agricultural region of the United States was selected at random for analysis. For each of these stations located generally within the American

Corn Belt, the average annual temperature is calculated for a two-year period in the mid-1970s and for a second two-year period in the mid-1990s. Table 9.7 presents the average annual temperatures (in degrees Celsius) for the two periods, as well as the difference in temperature (d) and the rank value of this difference for each location. Since one sample of weather stations was selected to record average temperatures for two different time periods, a matched-pairs procedure is applied.

Of the 28 stations being analyzed, 18 showed warmer average temperatures ($d > 0$) over the 20-year period, 7 showed colder average temperatures ($d < 0$), and 3 stations showed no change in temperature. For the sample of weather stations, the average annual temperature for the 1970 period was 10.046°C, whereas in the 1990 period, the temperatures averaged about 0.2 of a degree higher at 10.243°C. From basic descriptive statistical analysis of the sample

TABLE 9.7

Average Annual Temperature Data for Weather Stations in the American Corn Belt: Matched-Pairs Tests

| Weather station | Average annual temperature (°C) | | d (X−Y) | Rank* |
	1994–95 (X)	1974–75 (Y)		
Cairo	14.5	15.2	−0.7	21.5
Peoria	10.8	10.4	0.4	13.0
Moline	10.1	10.1	0.0	nd
Rockford	8.9	8.9	0.0	nd
Springfield	11.8	11.7	0.1	1.5
Evansville	14.2	13.8	0.4	13.0
Fort Wayne	10.4	10.0	0.4	13.0
Indianapolis	11.7	11.5	0.2	4.0
South Bend	9.8	10.5	−0.7	21.5
Burlington	10.9	10.4	0.5	19.0
Des Moines	9.9	10.3	−0.4	13.0
Dubuque	8.3	7.9	0.4	13.0
Sioux City	9.0	9.4	−0.4	13.0
Waterloo	8.5	8.1	0.4	13.0
Topeka	12.5	12.4	0.1	1.5
Kansas City	12.4	11.9	0.5	19.0
St. Louis	14.1	12.8	1.3	24.0
Lincoln	10.6	10.4	0.2	4.0
Omaha	10.5	10.7	−0.2	4.0
Norfolk	9.7	9.4	0.3	7.0
Sioux Falls	8.0	7.7	0.3	7.0
Huron	7.1	7.5	−0.4	13.0
Minneapolis	7.5	7.0	0.5	19.0
Rochester	6.6	6.6	0.0	nd
Milwaukee	9.5	8.1	1.4	25.0
Madison	8.3	7.9	0.4	13.0
Toledo	10.1	9.3	0.8	23.0
Dayton	11.1	11.4	−0.3	7.0

* Rank of the absolute differences (d).
nd = no data; rank values where difference (d) = 0 have been eliminated.
Source: National Climatic Data Center, U.S. Dept. of Commerce.

data, it would appear that the study region had experienced rising temperatures from the mid-1970s to the mid-1990s. However, was this difference in temperature statistically significant?

The null hypothesis for this problem states that the temperatures recorded in the mid-1970s were not significantly different from the temperatures recorded 20 years later in the mid-1990s. Table 9.8 shows the calculation of the matched-pairs t procedure to test the null hypothesis of no significant temperature change. Since a priori investigation of this issue would logically suggest an increase in temperature over time—the result of global warming—a directional (one-tailed) format seems most logical here. As shown in table 9.8, the calculated matched-pairs t value is 2.07, which leads to a corresponding one-tailed p-value of .024. Given a p-value of this magnitude and a likelihood of a Type I error at 2.4 percent, most researchers would reject the null hypothesis and conclude that the temperature change is significant. For the temperature patterns shown in this problem, we can be more than 97 percent confident that the differences in temperatures are the result of legitimate occurrences as opposed to being produced by sampling fluctuation.

An alternative procedure to analyze the paired temperatures for significant difference is the Wilcoxon signed-ranks test. The absolute values of all nonzero differences are ranked (see table 9.7), and then the ranks are summed, treating the positive and negative differences separately. Note that all observations where the difference in temperature is zero are eliminated first from the analysis. The 18 positive differences produce a rank sum (T_p) of 232, whereas the 7 negative differences produce a rank sum (T_n) of 93 (table 9.9). Since this problem contains more than 10 nonzero differences, the rank sum can be converted to Z_w using equation 9.20.

TABLE 9.8

Worktable for Matched-Pairs (Dependent-Sample) t: Corn Belt Temperatures*

H_0: $\delta = 0$

H_A: $\delta > 0$

Calculate the average difference ($\bar{d}$):

$\Sigma d = 5.50$ $\Sigma d^2 = 7.87$ $n = 28$ $\bar{X} = 10.243$ $\bar{Y} = 10.046$

where d = difference in average annual temperature [(1994–95) − (1974–75)]*

$\quad\quad X$ = temperature (1994–95)

$\quad\quad Y$ = temperature (1974–75)

$$\bar{d} = \frac{\Sigma d_i}{n} = \frac{5.50}{28} = .196$$

$$\bar{d} = \bar{X} - \bar{Y} = 10.243 - 10.046$$

Calculate the standard error of d:

$$s_d = \sqrt{\frac{\Sigma d_i^2 - \frac{(\Sigma d_i)^2}{n}}{n-1}} = \sqrt{\frac{7.87 - \frac{5.50^2}{28}}{27}} = \sqrt{\frac{7.87 - 1.08}{27}} = \sqrt{\frac{6.79}{27}} = .501$$

Calculate the test statistic (t_{mp}) and p-value:

$$t_{mp} = \frac{\bar{d}}{s_d/\sqrt{n}} = \frac{.196}{.501/\sqrt{28}} = \frac{.196}{.095} = 2.07 \quad\quad p\text{-value} = .024 \text{ (one-tailed)}$$

---

*Temperature data are in degrees Celsius; see table 9.7 for temperature data.

TABLE 9.9

Worktable for Dependent-Sample Wilcoxon Signed-Ranks Test: Corn Belt Temperatures

$$Z_W = \frac{T - \dfrac{n(n + 1)}{4}}{\sqrt{\dfrac{n(n + 1)(2n + 1)}{24}}}$$

where T_n = sum of negative ranks = 93
T_p = sum of positive ranks = 232
$n = 25$

$$Z_W = \frac{93 - \dfrac{25(25 + 1)}{4}}{\sqrt{\dfrac{25(25 + 1)(2(25) + 1)}{24}}} = \frac{93 - \dfrac{650}{4}}{\sqrt{\dfrac{33150}{24}}} = \frac{93 - 162.5}{\sqrt{1381.25}} = -1.870$$

p-value = .031 (one-tailed)

Note: Data values where difference (d) = 0 have been eliminated. This has reduced n from 28 to 25. See table 9.7 for matched-pairs ranks.

Because a directional alternate hypothesis is used and the number of negative differences in temperature is less than the number of positive differences, the negative rank sum (T_n) is selected for the computation. As shown in table 9.9, the Z_W value is calculated as -1.870, and the resulting one-tailed p-value is .031. These results correspond closely with those generated by the previous inferential test for the matched-pairs t. Both matched-pairs procedures suggest that the differences in temperature for the set of weather stations over the 20-year period are statistically significant. One can be approximately 97 percent confident that rejection of the null hypothesis is appropriate.

Summary Evaluation

Although the results from both matched-pairs tests support the conclusion that there are significant temperature changes, numerous other factors could influence the results. Perhaps the particular years selected to investigate temperature change in the agricultural heartland of the United States biased the results. Different conclusions may result by changing the years for which data are collected or lengthening the period over which the average temperatures are calculated.

The Corn Belt is typical of major geographic regions; it has a boundary that is difficult to define. Different geographers have drawn different boundary lines for this region. Moreover, environmental, economic, and other factors work to shift the location of agriculture over time. This suggests that the Corn Belt boundary may not be stable over the 20

years studied. As discussed in section 3.5, changes in the boundary of the study region can influence the results produced in any statistical analysis. Further investigation of temperature patterns over additional time periods and varying geographic regions is necessary before valid conclusions can be drawn about changing climates in the American Midwest.

KEY TERMS

MAJOR GOALS AND OBJECTIVES

If you have mastered the material in this chapter, you should now be able to do the following.

1. Recognize those geographic problems or situations for which application of a two-sample difference test is appropriate.
2. Explain the basic difference between independent and dependent samples.
3. Understand the procedure for selecting the proper two-sample difference test based on level of data measurement (interval/ratio, ordinal, nominal). Choices include two-sample difference of means test, Wilcoxon rank sum W test, and two-sample difference of proportions test.
4. Explain the types of geographic problems where a matched-pairs (dependent-sample) difference test is appropriate.

Three-or-More-Sample Difference Tests: Analysis of Variance Methods

10.1 Analysis of Variance (ANOVA)
10.2 Kruskal-Wallis Test

Many geographic problems involve comparison of three or more independent samples for significant differences. The logic applied to these problems is an extension of the reasoning used with two-sample difference tests. If multiple (three or more) samples are taken from the *same* population, their sample means are expected to vary somewhat from one another, just because of sampling variation. However, if multiple sample means vary considerably more than what is expected from sampling variability, then it can reasonably be concluded that the samples were drawn from *at least two different* populations.

Consider again the geographic problem used in section 9.1 to test for differences between two-samples. In that application, a random sample of "lower"-income countries was compared to a second independent sample of "higher"-income countries, with regard to the percentage of labor force that is female. Suppose that the research design of this two-sample problem is expanded. In *World Development Indicators, 1997*, the World Bank divided the 133 countries with a population of 1 million or more into four categories: (1) "high-income" economies (GNP per capita of $9,386 or more); (2) "upper-middle-income" economies (GNP per capita of $3,036 to $9,385); (3) "lower-middle-income" economies (GNP per capita of $766 to $3,035); and (4) "lower-income" economies (GNP per capita of $765 or less). If this four-group classification scheme is used to test for difference of means, two-sample difference of means tests, such as the *t* test or Wilcoxon rank sum *W* test, are no longer appropriate.

Early in the previous chapter, another possible application of two-sample difference of means tests

was suggested. In that example, a geographer was interested in comparing surface water runoff levels in two different environments: a natural setting with little or no human landscape modification and an exurban setting that had been partially modified by development, such as residential subdivisions. If this problem is expanded to three distinct landscapes whose mean surface water runoff levels are compared (e.g., totally natural setting, slightly modified landscape, and urban area substantially modified by human activity), then, once again, *t* and Wilcoxon *W* are not applicable.

The statistical procedures used to test three or more (multiple) samples for differences are **analysis of variance (ANOVA)** and the **Kruskal-Wallis test.** ANOVA is a parametric test that requires interval/ratio data drawn from normally distributed populations. In addition, to use analysis of variance, the variances in all groups in the population are assumed equal. The equivalent nonparametric procedure is the Kruskal-Wallis one-way analysis of variance by ranks test. The Kruskal-Wallis test uses ordinal data directly or interval/ratio data downgraded to ordinal if either the assumption of normality or equal variance is badly violated.

10.1 ANALYSIS OF VARIANCE (ANOVA)

Recall that variance is a descriptive statistic that measures the total amount of variability about the mean. Analysis of variance (ANOVA) involves the separation or partitioning of the total variance found in three or more groups or samples into two distinct components: (1) variability *between* the group or category means themselves; and (2) variability of the observations *within* each group around its group mean. Even though the structure of ANOVA uses variation as the key descriptive statistic, the means of the samples are compared for significant differences, making ANOVA a three-or-more-sample difference of means test.

ANOVA can be explained in this way. If the variability *between* the group means is relatively large as contrasted with a relatively small amount of variability *within* each group around its group mean, then the statistical conclusion will likely be that the different groups have been drawn from different populations. As a result, the null hypothesis would be rejected. The null hypothesis for ANOVA takes the form $H_0: \mu_1 = \mu_2 = \ldots = \mu_k$ (where k is the number of independent random samples), so the null hypothesis asserts that all samples are drawn from the same population. Conversely, the alternate hypothesis takes the form $H_A: \mu_1 \neq \mu_2 \neq \ldots \neq \mu_k$, asserting that *at least one sample* is drawn from a different population than the other samples.

The interpretation of ANOVA becomes more apparent when viewed graphically (figure 10.1). The null hypothesis states that independent samples have been drawn from the *same population,* whereas the alternate hypothesis asserts that the different samples are from *at least two separate and distinct populations.* In case 1 of figure 10.1, only a small difference separates the three sample means $(\bar{X}_1, \bar{X}_2, \bar{X}_3)$, and when ANOVA is applied, the likely decision is to *not reject* the null hypothesis. That is, the three sample means are inferred to be no more different than would be expected with three independent samples drawn from the same population. Conversely, in case 2, the seemingly large difference between the three sample means leads to the inferential conclusion that the different samples have been taken from more than one separate and distinct population. The proper decision in this case is to *reject* the null hypothesis. Note that these decisions are all nondirectional or two-tailed, since the concern is to find a difference between three or more samples. The presence of more than two

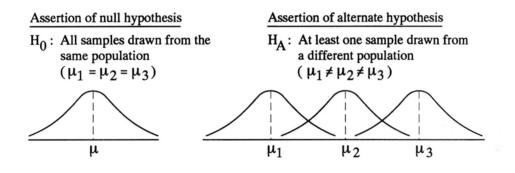

Assertion of null hypothesis

H_0 : All samples drawn from the same population
$(\mu_1 = \mu_2 = \mu_3)$

μ

Assertion of alternate hypothesis

H_A : At least one sample drawn from a different population
$(\mu_1 \neq \mu_2 \neq \mu_3)$

$\mu_1 \quad \mu_2 \quad \mu_3$

Case 1:
Small apparent difference between sample means
Likely decision: do not reject H_0

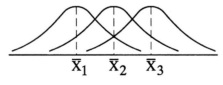

$\bar{X}_1 \quad \bar{X}_2 \quad \bar{X}_3$

Case 2:
Large apparent difference between sample means
Likely decision: reject H_0

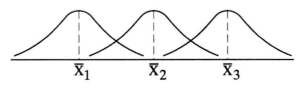

$\bar{X}_1 \quad \bar{X}_2 \quad \bar{X}_3$

FIGURE 10.1
Null and Alternate Hypotheses in Analysis of Variance (ANOVA)

samples makes it impossible to conduct a directional (one-tailed) test of difference.

Between-group variability focuses on how the sample mean of each group differs from the total or overall mean when all categories are grouped together. Quite simply, the total or overall mean is the weighted average of the individual group or sample means. Therefore, if the individual group or sample means differ significantly from the total or overall mean, the between-group variability is large. Sometimes the between-group variation is called the "explained variation" because variability attributed to differences between group or sample means is a measure of how much variation is explained by (or statistically dependent on) the group structure itself.

Within-group variability measures the variation of observations in each group or sample about the mean of that group or sample. That is, each sample has a mean and variance, and when these estimates of variability are totaled for all groups, the result is the total within-group variability. Internal variation within each of the groups is not "explained" by the grouping or categorization scheme and thus is considered "unexplained" or "residual" variation.

Many different testing structures are available for ANOVA, but only a single-variable model called "one-way analysis of variance" is examined here. This testing model is called "one-way" because observations fall into different groups or samples based on their values for *one* variable. Other, more complicated ANOVA techniques incorporate two or more variables simultaneously in the statistical model. In the problem regarding the percentage of labor force that is female, for example, it might be useful to see if another variable such as national literacy rate also varies with female labor force participation. Perhaps literacy should be included with the income variable in the analysis.

More complex ANOVA models with multiple variables are often used in agricultural, medical, and psychological research, but less often in geography. In part, this is because geographers infrequently work with controlled experimental research designs containing equal sample sizes and other restrictive assumptions required by the more complex models. ANOVA testing structures having multiple variables are generally discussed only in nonintroductory statistics texts that examine multivariate statistical techniques.

The basic goal of the ANOVA test is to determine which is more dominant or pronounced: between-group variability or within-group variability. The test statistic for ANOVA (*F*) incorporates these two sources of variation:

$$F = \frac{MS_B}{MS_W} \qquad (10.1)$$

where MS_B = between-group mean squares
MS_W = within-group mean squares

Calculating Between-Group Variability

To calculate between-group mean squares (the numerator in equation 10.1), a three-step procedure is followed:

Step 1: Calculate the total or overall mean ($\overline{X}_T$), the weighted average of the individual group or sample means ($\overline{X}_1, \overline{X}_2, \ldots, \overline{X}_k$):

$$X_T = \frac{\sum_{i=1}^{k} n_i \overline{X}_i}{N} \qquad (10.2)$$

where $\overline{X}_i$ = mean of sample i
n_i = number of observations in sample i
k = number of groups or samples
N = total number of observations in all samples

Step 2: Calculate the between-group sum of squares (SS_B):

$$SS_B = \sum_{i=1}^{k} n_i (\overline{X}_i - \overline{X}_T)^2 \qquad (10.3)$$

$$= \left(\sum_{i=1}^{k} n_i (\overline{X}_i^2) \right) - N(\overline{X}_T^2) \qquad (10.4)$$

Equation 10.3 is the definitional formula of the between-group sum of squares, and equation 10.4 is the computational formula. Notice what is happening in equation 10.3. Within the parentheses, the total or overall mean is subtracted from each group mean. If any of these differences between total mean and individual group mean is large, it will make the entire expression (the between-group sum of squares) a larger value. This makes sense, for if an individual group mean is very different from the total or overall mean, it suggests a large amount of variability *between* the group means. In equation 10.3, each of these differences is squared, then multiplied by the number of observations in that group (to give that group its proper weight). Finally, these weighted differences are summed across all k groups.

Step 3: Calculate the between-group mean squares (MS_B):

$$MS_B = \frac{SS_B}{df_B} = \frac{SS_B}{k-1} \qquad (10.5)$$

where df_B = between-group degrees of freedom

Note that the between-group mean squares is simply the between-group sum of squares divided by the between-group degrees of freedom (df_B), where df_B depends on the number of groups or samples in the problem.

Calculating Within-Group Variability

The calculation of within-group mean squares (the denominator in equation 10.1) also involves multiple steps:

Step 1: Calculate the within-group sum of squares (SS_W):

$$SS_W = \sum_{i=1}^{k}(n_i - 1)s_i^2 \qquad (10.6)$$

Interpretation of equation 10.6 is straightforward. The variance in each group or sample (s_i^2) is multiplied or "weighted" by its sample size ($n_i - 1$). Then all of the weighted variances are summed to obtain the total within-group sum of squares for the groups.

Step 2: Calculate the within-group mean squares (MS_W):

$$MS_W = \frac{SS_W}{df_W} = \frac{SS_W}{N - k} \qquad (10.7)$$

The within-group mean squares is simply the within-group sum of squares divided by the within-group degrees of freedom (df_W), where df_W equals the total number of observations in all samples (N) minus the number of groups or samples (k).

When equation 10.1 is examined, the ANOVA test statistic, F, defines the ratio of the between-group mean squares (equation 10.5) to the within-group mean squares (equation 10.7). How can the magnitude of this ratio of mean squares be interpreted?

Even if a set of independent samples is drawn from the same population, the set of sample means will fluctuate somewhat. This is to be expected because of the nature of sampling and is measured by the standard error of the estimate. Now consider what should occur if the null hypothesis is correct: the group or sample means should vary no more than expected if they are a set of independent random samples drawn from a single population. In this situation, the between-group variance would approximately equal the within-group variance. As a result, the F statistic will have a magnitude of about one.

However, if H_0 is incorrect and the population means are significantly different, then the sample means estimating these population means will vary more than expected from simple random fluctuations of multiple samples from the same population. This will cause the between-group variance to be significantly larger than the within-group variance, and the F statistic will be greater than one.

10.2 KRUSKAL-WALLIS TEST

The Kruskal-Wallis test is the nonparametric equivalent of ANOVA. Kruskal-Wallis may be the most appropriate technique in cases where assumptions re-

---

Analysis of Variance (ANOVA)

Primary Objective: Compare three or more (k) independent random sample means for differences

Requirements and Assumptions:

1. Three or more (k) independent random samples
2. Each population is normally distributed
3. Each population has equal variance
4. Variable is measured at interval or ratio scale

Hypotheses:

$H_0: \mu_1 = \mu_2 = \ldots = \mu_k$
$H_A: \mu_1 \neq \mu_2 \neq \ldots \neq \mu_k$ (not all μ_k equal)

Test Statistic:

$$F = \frac{MS_B}{MS_W}$$

---

quired for the use of the parametric ANOVA test (such as normality and equal population variances) are not fully met. Kruskal-Wallis may be considered the nonparametric extension of the Wilcoxon rank sum W test to problems with three or more samples.

In this test, values from all samples are combined into a single overall ranking (as in the Wilcoxon test). The rankings from each sample are summed, and the mean ranks of each sample are then calculated. For example, in sample one, the sum of ranks is R_1 and the mean rank is R_1/n_1. In sample two, the sum of ranks is R_2 and the mean rank is R_2/n_2, and so on. Because of the random nature of sampling, the sample mean ranks should differ somewhat, even if the samples are drawn from the same population. The Kruskal-Wallis test examines whether the mean rank values are significantly different.

If the multiple (k) samples are from the same population, as asserted by the null hypothesis, their mean ranks should be approximately equal. The best estimate of the mean rank of population i is the sample mean rank (R_i/n_i). Thus, if the null hypothesis is correct:

$$\frac{R_1}{n_1} = \frac{R_2}{n_2} = \ldots = \frac{R_k}{n_k}$$

On the other hand, if the mean ranks differ more than is likely with chance fluctuations, it may be concluded that at least one of the samples comes from a different population, and the null hypothesis is rejected.

The Kruskal-Wallis H test statistic is

$$H = \frac{12}{N(N+1)}\sum_{i=1}^{k}\frac{R_i^2}{n_i} - 3(N+1) \qquad (10.8)$$

where N = total number of observations or values in all samples

$\quad\quad = n_1 + n_2 + \ldots + n_k$

$\quad n_i$ = number of observations or values in sample i

$\quad R_i$ = sum of ranks in sample i

In some geographic data sets, the ranks for a number of observations across the k samples may be tied. If this affects more than 25 percent of the values, a correction factor should be included in the Kruskal-Wallis test statistic. The correction factor for ties is

$$1 - \frac{\Sigma T}{N^3 - N} \quad\quad (10.9)$$

where $T = t^3 - t$

$\quad t$ = number of tied observations in the data

$\quad N$ = total number of observations or values in all k samples

With the correction factor for ties, the H statistic is

$$H = \frac{\frac{12}{N(N + 1)} \sum_{i=1}^{k} \frac{R_i^2}{n_i} - 3(N + 1)}{1 - \frac{\Sigma T}{N^3 - N}} \quad\quad (10.10)$$

This modification for ties increases the value of H, reduces the p-value, and improves the likelihood of discovering significant differences between samples.

The Kruskal-Wallis H distribution approximates the commonly used chi-square (χ^2) distribution, when $k > 3$ and/or at least one sample has a size (n_i) > 5. Thus, to determine a p-value for a calculated Kruskal-Wallis statistic, the researcher would first convert H into chi-square and generate the p-value from this distribution. The sampling distribution of H is chi-square, where degrees of freedom = $k - 1$. The chi-square distribution will be discussed in section 11.1.

Application to Geographic Problem

Suppose an urban geographer is interested in determining whether home purchase prices differ between four adjacent neighborhoods in a particular section of a city. If a random sample of home purchase prices from each neighborhood is collected, a multisample difference test, such as ANOVA or Kruskal-Wallis, could be applied to answer this question. Table 10.1 shows the home purchase prices of the 22 homes randomly selected from the four neighborhoods.

First, an analysis of variance (ANOVA) test is applied. ANOVA seems appropriate since this is a "difference test" type of problem with four samples, and the data are measured on a ratio scale (with the

---

Kruskal-Wallis Test

Primary Objective: Compare three or more (k) independent random sample mean ranks for difference

Requirements and Assumptions:

1. Three or more (k) independent random samples
2. Each population has an underlying continuous distribution
3. Variable is measured at ordinal scale or downgraded from interval/ratio scale to ordinal

Hypotheses:

H_0: The populations from which the three or more (k) samples have been drawn are all identical

H_A: The populations from which the three or more (k) samples have been drawn are not all identical

Test Statistic:

$$H = \frac{\left[\frac{12}{N(N + 1)} \sum_{i=1}^{k} \frac{R_i^2}{n_i}\right] - 3(N + 1)}{1 - \frac{\Sigma T}{N^3 - N}}$$

---

purchase price of each home expressed in thousands of dollars). Before applying ANOVA, it must be determined if the samples are drawn from normally distributed populations and if each population has an equal variance. Preliminary testing of the sample data confirmed that the samples are normally distributed and have equal variances.

Table 10.2 summarizes the procedures for the ANOVA calculation. Note that the average home purchase price varies from a low of $144,600 in neighborhood 3 to a high of $189,250 in neighborhood 4, with a total or overall mean purchase price of $168,680. The statistical question concerns the likelihood or probability that such observed differences in purchase price could occur if indeed no *real* difference exists in home purchase prices. If these observed differences in mean purchase price are significantly more than expected

---

TABLE 10.1

Random Sample of Home Purchase Prices by Neighborhood (in Thousands of Dollars)

Neighborhood			
1	2	3	4
175	151	127	174
147	183	142	182
138	174	124	210
156	181	150	191
184	193	180	
148	205		
	196		

TABLE 10.2

Worktable for Analysis of Variance (ANOVA): Home Purchase Prices by Neighborhood

$H_0: \mu_1 = \mu_2 = \mu_3 = \mu_4$

$H_A: \mu_1 \neq \mu_2 \neq \mu_3 \neq \mu_4$

	Sample statistics		
	n	**$\overline{X}$**	**s**
Neighborhood 1	6	158.00	17.83
Neighborhood 2	7	183.29	17.61
Neighborhood 3	5	144.60	22.49
Neighborhood 4	4	189.25	15.48
Total	22	168.68	24.85

Calculate between-group variability:

Step 1: Total or overall mean ($\overline{X}_T$):

$$\overline{X}_T = \frac{\sum_{i=1}^{4} n_i \overline{X}_i}{N} = \frac{6(158.00) + 7(183.29) + 5(144.60) + 4(189.25)}{22} = 168.68$$

Step 2: Between-group sum of squares (SS_B):

$$SS_B = \left(\sum_{i=1}^{4} n_i(\overline{X}_i^2) \right) - N(\overline{X}_T^2)$$

$$= (6(158.00)^2 + 7(183.29)^2 + 5(144.60)^2 + 4(189.25)^2) - 22(168.68)^2$$

$$= 6769.394$$

Step 3: Between-group mean squares (MS_B):

$$MS_B = \frac{SS_B}{df_B} = \frac{SS_B}{k-1} = \frac{6769.394}{3} = 2256.465$$

Calculate within-group variability:

Step 1: Within-group sum of squares (SS_W):

$$SS_W = \sum_{i=1}^{4} (n_i - 1)s_i^2 = 5(17.83)^2 + 6(17.61)^2 + 4(22.49)^2 + 3(15.48)^2 = 6193.379$$

Step 2: Within-group mean squares (MS_W):

$$MS_W = \frac{SS_W}{df_W} = \frac{SS_W}{N-k} = \frac{6193.379}{22-4} = 344.077$$

Calculate test statistic (F):

$$F = \frac{MS_B}{MS_W} = \frac{2256.465}{344.077} = 6.558 \ (p = .003)$$

ANOVA summary table					
Source of variation	**Sum of squares**	**df**	**Mean squares**	**F ratio**	**p-value**
Between group	6769.394	3	2256.465	6.558	.003
Within group	6193.379	18	344.077		
Total	12962.8	21			

from random sampling, then the result will be a large F statistic and a p-value close to zero.

The calculated F statistic of 6.558 has an associated p-value or significance level of .003. The logical conclusion is that at least one of the four neighborhoods has home purchase prices that differ from the other neighborhoods. We can be 99.7 percent confident $(1 - .003 = .997)$ that rejecting the null hypothesis and concluding that home purchase prices differ by neighborhood is indeed the correct conclusion.

The next logical question of concern is which sample or samples differ from one another. The ANOVA significance level of .003 strongly indicates that significant differences exist *somewhere* in the grouping structure of four neighborhoods, but does not indicate where. To determine the location of these differences, a series of two-sample difference of means t tests could be applied. The descriptive statistics from ANOVA show neighborhood 3 with the lowest mean home purchase price $(\bar{X}_3 = \$144,600)$ and neighborhood 4 with the highest mean purchase price $(\bar{X}_4 = \$189,250)$. Therefore, the most significant t test result (largest difference) should occur when neighborhood 3 is paired with neighborhood 4.

The results (table 10.3) confirm that a significant t value exists when neighborhoods 3 and 4 are contrasted $(t_{34} = -3.363, p = .012)$. However, the *most significant* housing price difference occurs when neighborhoods 2 and 3 are compared $(t_{23} = 3.352, p = .007)$. Of all neighborhood pairs, we are most confident that neighborhoods 2 and 3 have the greatest statistical difference in home purchase prices. Why isn't the comparison of neighborhoods 3 and 4 most significant statistically? The answer to this apparent contradiction can be found by examining neighborhood sample sizes. Neighborhoods 3 and 4 have mean home purchase prices that differ the most, but their sample sizes are the smallest (table 10.2). By comparison, mean home purchase prices of neighborhoods 2 and 3 don't differ as much, but their sample sizes are larger. Quite simply, the smaller sample sizes in neighborhoods 3 and 4 reduce the *level of statistical confidence* that their home purchase prices are truly different.

The ANOVA results validate the existence of significant differences in home purchase prices by neighborhood. Furthermore, the normality and equal variance requirements regarding the application of ANOVA have been met. However, in any problem with very small sample sizes, application of powerful parametric tests like ANOVA must be evaluated cautiously. An alternative strategy often available is to run the equivalent nonparametric test, since this class of tests is designed to work more effectively with small samples. In this case, the equivalent nonparametric test is Kruskal-Wallis.

The Kruskal-Wallis procedure is now applied to the same home purchase price data (table 10.4). With Kruskal-Wallis, the values from all samples are combined into a single overall ranking, the rankings from each sample are summed, and the mean rank of each sample is calculated. The mean ranks of the four samples are quite different (8.67, 15.50, 5.40, and 16.38); so it is not surprising that the Kruskal-Wallis test statistic indicates significant differences in home purchase prices by neighborhood $(H = 10.47$ and $p = .015)$. These results clearly confirm the ANOVA findings that somewhere in the four samples significant differences exist.

Analogous to the ANOVA strategy to determine the locations of the most significant home price differences between samples, Wilcoxon and Mann-Whitney tests are run on each pair of neighborhoods as a follow-up to the Kruskal-Wallis testing. The Wilcoxon/Mann-Whitney results identify several significant differences in home purchase prices, findings that closely match the results obtained earlier from the corresponding parametric t tests (table 10.5). The most significant difference statistically occurs when home purchase prices in neighborhoods 2 and 3 are compared $(W_{23} = -2.517$ and $p = .012)$, and the next most significant difference is between neighborhoods 3 and 4 $(W_{34} = -2.205$ and $p = .027)$.

Summary Evaluation

Statistically, the ANOVA and Kruskal-Wallis test results are remarkably consistent. Both tests confirm that highly significant differences in home purchase prices exist among the four neighborhoods. The somewhat lower p-value associated with ANOVA $(p = .003)$ is expected given the greater statistical strength of this parametric technique, as contrasted with the nonparametric Kruskal-Wallis test $(p = .015)$.

Geographically, why do home purchase prices vary significantly among the four neighborhoods? Why do neighborhood pairs 2–3 and 3–4 differ the most? Since the results from ANOVA and Kruskal-Wallis analyses generate questions such as these, they should be viewed as intermediate steps in the overall research process. For example, a geographer

---

TABLE 10.3

Two-Sample Difference of Means t Test Results: Home Purchase Prices by Neighborhood

t-Test number	Neighborhood pair	t	df	p-value
1	1 and 2	−2.566	11	.026
2	1 and 3	.104	9	.298
3	1 and 4	−2.850	8	.021
4	2 and 3	3.352	10	.007
5	2 and 4	−.562	9	.588
6	3 and 4	−3.363	7	.012

TABLE 10.4

Worktable for Kruskal-Wallis: Home Purchase Prices by Neighborhood

Neighborhood 1		Neighborhood 2		Neighborhood 3		Neighborhood 4	
Price	Rank	Price	Rank	Price	Rank	Price	Rank
				124	1		
				127	2		
138	3						
				142	4		
147	5						
148	6						
		151	8	150	7		
156	9						
		174	10.5			174	10.5
175	12			180	13		
		181	14				
		183	16			182	15
184	17						
						191	18
		193	19				
		196	20				
		205	21				
						210	22

$R_1 = 52$	$R_2 = 108.5$	$R_3 = 27$	$R_4 = 65.5$
$\dfrac{R_1}{n_1} = 8.67$	$\dfrac{R_2}{n_2} = 15.50$	$\dfrac{R_3}{n_3} = 5.40$	$\dfrac{R_4}{n_4} = 16.38$

$$H = \left[\frac{12}{N(N+1)} \left(\sum_{i=1}^{k} \frac{R_i^2}{n_i} \right) \right] - 3(N+1) = \left[\frac{12}{22(23)} \left(\frac{52^2}{6} + \frac{108.5^2}{7} + \frac{27^2}{5} + \frac{65.5^2}{4} \right) \right] - 3(23)$$

$$= 10.47 \; (p = .015)$$

may want to examine other neighborhood characteristics that possibly affect home purchase prices. This examination may include such characteristics as age of home, number of square feet, size of lot, and some measure of access to services. Here is another instance where the results (output) from one statistical analysis provide questions (input) that lead effectively into further analysis. In other words, effective geographic problem solving involves this type of cumulative, exploratory process.

KEY TERMS

analysis of variance (ANOVA) or *F*-test, 146
between-group variability, 148
Kruskal-Wallis test, 146
within-group variability, 148

TABLE 10.5

Wilcoxon Rank Sum *W* (Mann-Whitney *U*) Test Results: Home Purchase Prices by Neighborhood

Wilcoxon test number	Neighborhood pair	Z_W	*p*-value
1	1 and 2	−2.000	.046
2	1 and 3	−1.095	.273
3	1 and 4	−1.919	.055
4	2 and 3	−2.517	.012
5	2 and 4	−.284	.776
6	3 and 4	−2.205	.027

MAJOR GOALS AND OBJECTIVES

If you have mastered the material in this chapter, you should now be able to do the following.
1. Recognize those geographic problems or situations for which application of a multiple-sample difference test is appropriate.
2. Explain the rationale and purposes for conducting a difference test for three or more samples, either analysis of variance (ANOVA) or Kruskal-Wallis.
3. Distinguish between the major components of analysis of variance: between-group variability and within-group variability.

CHAPTER 11

Goodness-of-Fit and Categorical Difference Tests

11.1 Goodness-of-Fit Tests
11.2 Contingency Analysis

In the last few chapters, a logical progression of difference tests has been presented. This progression started with *one-sample* difference tests that compare a single sample statistic to a corresponding population parameter (chapter 8). The progression continued through *two-sample* and *dependent-sample* difference tests that analyze two samples for differences or one sample twice, in the case of dependent-sample tests (chapter 9). This sequence finished with *three-or-more-sample* difference tests that determine whether multiple samples have been drawn from multiple populations (chapter 10).

The focus now shifts to special types of difference tests known as **goodness-of-fit tests.** In a goodness-of-fit situation, an **actual** or **observed frequency distribution** is compared with some **expected frequency distribution.** Goodness-of-fit procedures sometimes test the hypothesis that a set of data has a particular frequency distribution. This might be done to confirm or deny the applicability or validity of a geographic model or theory. In these cases, the investigator may expect a certain frequency distribution to exist, based either on theoretical principles or empirical knowledge from past observation or experience. In another context, goodness-of-fit tests determine whether a statistical test assumption is satisfied. For example, many parametric inferential tests make the assumption that samples have been drawn from normally distributed populations; a goodness-of-fit test can check the validity of this assumption.

Contingency analysis is another type of categorical statistical test. The question in contingency analysis concerns the strength of relationship or association between two variables. If data from both variables are assigned to nominal or ordinal categories, then the frequency count of each category from one variable may be cross-classified with the frequency count of each category for the second variable. This cross-tabulation process is summarized by a contingency table, the topic of section 11.2.

11.1 GOODNESS-OF-FIT TESTS

Depending on the circumstances of the problem, a set of data can be tested for "fit" against a variety of expected, sometimes theoretical, frequency distributions. Applications of goodness-of-fit testing are shown for three different expected frequency distributions: (1) uniform or equal, (2) proportional or unequal, and (3) normal.

In some problems, the expected frequency distribution is uniform or equal. For example, suppose an environmental geographer wishes to examine nutrient runoff levels from a series of five adjacent tributaries running into the same bay. Water samples measuring nitrogen and phosphorus levels could be taken on a random selection of days simultaneously from each tributary, and the number of sample days that nutrient levels exceed some critical (threshold) value could be counted for each tributary. Although low levels of such nutrients are acceptable, nutrient levels exceeding some threshold value become pollution problems. It might be expected that the nutrient level is exceeded about the same number of sample days in each tributary, if the nutrient levels are similar from tributary to tributary. This expected frequency count information can be compared with the observed frequency counts of high nutrient level days among the bay tributaries to see whether the two frequency counts are statistically different. A goodness-of-fit procedure can determine whether the actual frequency counts are uniform or equal for this set of tributaries.

Goodness-of-fit procedures can also evaluate whether a set of frequency counts fits some expected distribution that is uneven or proportional. Suppose the geographer expects the distribution of high nutrient level days from tributary to tributary to be something other than uniform. For example, a tributary with a greater proportion of its watershed in farmland might be expected to have more sample days exceeding critical nutrient levels than another tributary with less of its watershed in farmland. One could hypothesize that the expected number of high pollution days is proportional to the percentage of tributary land area under cultivation.

Proportional goodness-of-fit tests can also be applied to evaluate the validity of a geographic model. For example, suppose a recreation planner is studying attendance patterns at a regional park that serves people in several nearby communities. How many park visitors might be expected from each community? To estimate park usage patterns, the potential model of spatial interaction is useful. A simple version of this model states that the volume of spatial interaction is directly related to population size and inversely related to the square of the distance. In this application, the expected number of visitors from a community to a park would be directly proportional to the population of the community and inversely related to the square of the distance separating the community and the park. That is:

$$Pot_{ij} = \frac{Pop_i}{D_{ij}^2} \qquad (11.1)$$

where Pot_{ij} = potential produced at park j by community i
  Pop_i = population of community i
  D_{ij} = distance from community i to park j

The recreation planner can compare the observed number of park visitors from each community with the corresponding number of visitors predicted (expected) by the potential model. The extent to which the model "fits"—that is, the extent to which the observed and expected visitor counts from the set of communities match—can be tested statistically with a goodness-of-fit procedure.

Goodness-of-fit tests are also used to compare an actual frequency distribution with a theoretical probability distribution such as normal or random. A frequency distribution often needs to be tested for normality, an important assumption for the data to be appropriately applied in a parametric test such as t or ANOVA. Suppose a geomorphologist is studying soil erosion rates of several soil orders (alfisols, aridosols, entisols, etc.) in a large regional study area. To test for significant differences in erosion rate by soil order, a spatial random sample is taken from each type of soil. If the samples are drawn from normally distributed populations, a parametric ANOVA could be applied as the difference test. A goodness-of-fit procedure can be used to determine whether this normality requirement has been met, so ANOVA is valid.

In some geographic problems, a frequency distribution should be tested for randomness. In these cases, a Poisson probability distribution would be expected. For example, a goodness-of-fit test for randomness could be applied to a spatial pattern of influenza outbreaks in a metropolitan area. This procedure would compare the observed pattern of influenza cases with the pattern expected if the outbreak follows a random spatial process.

Chi-square (χ^2) Goodness-of-Fit Test

When applied as a goodness-of-fit test, the chi-square statistic compares the observed frequency counts of a single variable (organized into nominal or ordinal categories) with an expected distribution of frequency counts allocated over the same categories. The chi-square test must use absolute frequency counts and cannot be applied if the observations or sampling units are in relative frequency form, such as percentages, proportions, or rates.

Chi-square is a method to determine if a truly significant difference exists between a set of observed frequencies and the corresponding expected frequencies. With this procedure, the focus is on how closely the two frequency counts match, thereby providing a goodness-of-fit measure. The null hypothesis (H_0) states that the population from which the sample has been drawn fits an expected frequency distribution. Thus, H_0 assumes no difference between observed and expected frequency counts. If the magnitude of difference between frequency counts is small across *all* categories, the data are likely to be a random sample drawn from the expected frequency distribution, and H_0 is not rejected. Conversely, the alternate hypothesis (H_A) suggests that the magnitude of difference between frequencies is large in *at least one category*. If H_A is true, the data cannot be considered a random sample from the expected frequency distribution, and H_0 is rejected.

The formula for the chi-square test statistic is

$$\chi^2 = \sum_{i=1}^{k} \frac{(O_i - E_i)^2}{E_i} \qquad (11.2)$$

where O_i = observed or actual frequency count in the ith category
  E_i = expected frequency count in the ith category
  k = number of nominal or ordinal categories

If the observed and expected frequency counts for each category are similar, then all of the differences $(O_i - E_i)$ will be slight, χ^2 will be small, the goodness-of-fit will be strong, and H_0 will not be rejected. However, if at least one difference between frequency counts is large, then the χ^2 statistic will be large, the observed frequency counts will not necessarily come from the population or model theorized by the expected frequency counts, and H_0 will be rejected. In later example problems, the methodology for calculating expected frequencies will be discussed.

Chi-square is a very flexible goodness-of-fit test, in part because the number of nominal or ordinal categories may vary. However, minimum size restrictions apply under certain circumstances. For example, if the nominal variable has only two categories, the expected frequency in both should be at least five. For the test to be valid with more than two categories, no more than one-fifth of the expected frequencies should be less than five, and no expected frequency should be less than two. Sometimes categories need to be combined to increase the expected frequencies to an appropriate size.

Kolmogorov-Smirnov Goodness-of-Fit Test

The Kolmogorov-Smirnov statistic is an alternative to chi-square for testing the similarity between two frequency distributions. Chi-square uses frequencies from either nominal or ordinal classes, but

Kolmogorov-Smirnov (K-S) requires data measured at the ordinal level or interval/ratio level downgraded to ordinal. Technically, K-S requires continuously distributed variables, but only very slight errors result when the technique is applied to discrete data. In the goodness-of-fit application of Kolmogorov-Smirnov, the observed distribution of the sample data is compared to a particular expected distribution, such as normal or Poisson. The null hypothesis states that no significant difference exists between the two frequency distributions.

Geographers most often use Kolmogorov-Smirnov to test a data set for normality. If the observed sample distribution matches a theoretical normal distribution, the investigator can infer that the population from which the random sample was drawn is normally distributed. Then, such parametric statistics as analysis of variance and the difference of means t test can be applied with confidence that the assumption of normality has been met.

More specifically, the Kolmogorov-Smirnov test for normality compares the **cumulative relative frequencies** of the observed sample data with the cumulative frequencies expected for a perfectly normal distribution. Recall that cumulative frequencies can be shown graphically as an ogive (section 2.5). If the two sets of cumulative frequencies closely match, the theoretical and sample distributions can be considered the same and the population considered normal. Any sizable differences between the two cumulative distributions suggest that the sample data were not taken from a normally distributed population.

To calculate an observed cumulative frequency distribution, individual data are ranked from low to high or aggregated into ordinal classes. The ranked data or classes are then converted to Z-scores and cumulative relative frequencies calculated for each value or class. What proportion of the values are *equal to or less than* this Z-value?

The corresponding cumulative relative frequency distribution for an expected (perfectly normal) data set having the same mean and standard deviation must also be determined. That is, if the data were normal, what proportion of the values would lie below (equal to or less than) the given Z-value? **Cumulative normal values** (*CNVs*) of the expected distribution can be found for any Z-value using the following method:

- if $Z > 0$ (above the mean), then CNV
$= .50 + A$ (11.3)
- if $Z = 0$ (equal to mean), then CNV
$= .50$ (11.4)
- if $Z < 0$ (below the mean), then CNV
$= 1.00 - (A + .50)$ (11.5)

where A = probability from normal table (appendix, table A)

Chi-square (χ^2) Goodness-of-Fit Test

Primary Objective: Compare random sample frequency counts of a single variable with expected frequency counts (goodness-of-fit)

Requirements and Assumptions:

1. Single random sample
2. Variables are organized by nominal or ordinal categories; frequency counts by category are input to statistical test
3. If two categories, both expected frequency counts must be at least five; if three or more categories, no more than one-fifth of the expected frequency counts should be less than five, and no expected frequency count should be less than two

Hypotheses:

H_0: Population from which sample has been drawn fits an expected frequency distribution (uniform or equal, proportional or unequal, random, etc.); no difference between observed and expected frequencies

H_A: Population does not fit an expected frequency distribution; there is a significant difference between observed and expected frequencies

Test Statistic:

$$\chi^2 = \sum_{i=1}^{k} \frac{(O_i - E_i)^2}{E_i}$$

Cumulative normal values are graphed for selected Z-values in a perfectly normal expected frequency distribution (figure 11.1). The expected cumulative value corresponding to the mean of a normal distribution is always .50, since half of the values in a normal distribution are equal to or less than mean (equation 11.4). However, to determine cumulative normal proportions for other values of Z, both the probability (A) from the normal table (appendix, table A) and the sign of the Z-value (positive or negative) must be known. For example, if $Z = +1.0$, a value greater than the mean, then equation 11.3 is used. For this Z-value, the probability from the normal table is .3413 and the corresponding CNV is .8413 (.3413 + .5000). This implies that 84.13 percent of the values in a normal distribution are less than or equal to one standard deviation *above* the mean. For negative Z-values, such as $Z = -1.35$, application of equation 11.5 results in a CNV of .0885 (1.0 − (.5000 + .4115) = .0885). According to this calculation, slightly less than 9 percent of the values in a normal distribution lie equal to or less than 1.35 standard deviations *below* the mean (figure 11.1).

In addition to calculating the cumulative relative frequencies for both the observed and expected (normal) distributions, the same information can be plotted as ogives. The K-S procedure allows differences between the two cumulative distributions to be determined in one of two ways. Differences can be calculated directly from the data using the two sets of cumulative values. Alternatively, differences can be determined graphically as vertical deviations for positions along the horizontal axis of the ogive (figure 11.2). The K-S test statistic, D, is the *maximum absolute difference* between the two sets of cumulative values:

$$D = \text{maximum} \left| \text{CRF}_o(X) - \text{CRF}_e(X) \right| \quad (11.6)$$

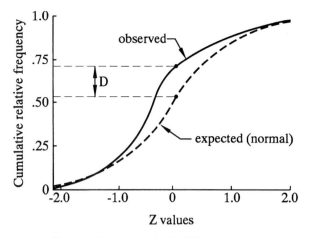

D = maximum vertical difference

FIGURE 11.2
Ogive Showing Observed and Expected (Normal) Cumulative Relative Frequency and Differences (D)

where $\text{CRF}_o(X)$ = cumulative relative frequencies (observed) for variable X
$\text{CRF}_e(X)$ = cumulative relative frequencies (expected) for variable X

When the expected frequency distribution is normal, the cumulative relative frequency ($\text{CRF}_e(X)$) corresponds to the cumulative normal value (CNV).

With continuous data—a common form of data when using the Kolmogorov-Smirnov goodness-of-fit test—the comparison (difference) between observed and expected cumulative relative frequencies must be completed twice, at each end of the data interval in the observed ogive. These so-called *lagged* and *nonlagged* comparisons will be discussed in reference to an actual data set in the example that follows later in this section.

When the deviation between the actual and theoretical (normal) distribution is very large, D is very large and the null hypothesis of no difference between the two distributions could be incorrect. In this instance, the actual distribution is likely to be statistically nonnormal. Conversely, if all the differences between the actual and expected distributions are small, then D will be small. In this case, the null hypothesis is not rejected, and the observed distribution is considered normal.

The significance of the Kolmogorov-Smirnov D can be determined in two ways. In the classical or traditional approach to hypothesis testing, the calculated (or graphically measured) D value can be compared with a value taken from a table of critical D values. The null hypothesis can be rejected if the value of D exceeds the value found in the table. A p-value can be roughly estimated by interpolating within the table of D values.

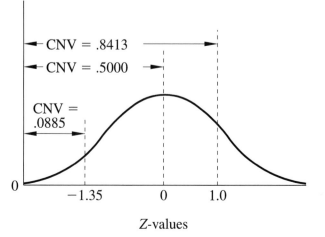

FIGURE 11.1
Cumulative Normal Values (CNV) for Selected Z-Value Positions

Kolmogorov-Smirnov Goodness-of-Fit Test

Primary Objective: Compare random sample frequency counts of a single variable with expected frequency counts (goodness-of-fit)

Requirements and Assumptions:

1. Single random sample
2. Population is continuously distributed (test less valid with discrete distribution)
3. Variable is measured at ordinal scale or downgraded from interval/ratio to ordinal

Hypotheses:

H_0: Population from which sample has been drawn fits an expected frequency distribution (uniform or equal, proportional or unequal, random, etc.); no difference between observed and expected frequencies

H_A: Population does not fit an expected frequency distribution; there is a significant difference between observed and expected frequencies

Test Statistic:

$$D = \text{maximum} \left| \text{CRF}_o(X) - \text{CRF}_e(X) \right|$$

In the *p*-value approach, *D* is converted to *Z*, and the statistical level of significance (*p*-value) is calculated. Most statistical software packages calculate both a *D* value and its corresponding *Z*-value. The computation procedure for converting *D* to *Z* is rather lengthy, statistically complex, and not shown in this text.

Application to Geographic Problem: Chi-square Goodness-of-Fit (Uniform Distribution)

Suppose a newly appointed school district superintendent wants to conduct a geographic study that compares the Scholastic Aptitude Test (SAT) scores among graduating seniors from five comparably sized high schools located in different portions of a metropolitan school district. Are students from all five high schools performing equally well on the SATs? More specifically, do about the same number of students in each school earn test scores above the national median? If the educational experiences are similar at all five high schools, it should be expected that about the same number of students from each school will have above-median SAT scores. If performance levels are unequal, the superintendent has indicated she will begin a five-year improvement program, expending particular effort toward improving the SAT scores of the lower-achieving schools, while continuing to improve the performance levels of the better schools.

Suppose a random sample of 100 students is taken from each school, and the number of sampled students with an SAT score exceeding the national median is recorded (table 11.1, "initial year results"). A chi-square goodness-of-fit test is appropriate since a set of observed frequency counts from nominal categories is being compared with a comparable set of expected frequency counts. Moreover, expected frequency counts are uniform or equal, because each of the five high schools is expected to have an equal number of sampled students with above-median SAT scores.

The chi-square calculation for the initial year is shown in the top portion of table 11.2. A total of 245 sampled students from all five high schools combined scored above the national SAT median. If this group of students were uniformly or evenly distributed among the five high schools, the expected frequency per high school would be 49 (245 divided by 5). The statistical question is whether the actual (observed) frequency counts (42, 45, 51, 47, 60) are significantly different from the expected frequency counts (49, 49, 49, 49, 49). If the frequency counts are different, then the superintendent has statistical verification to begin her five-year SAT improvement program.

The calculated chi-square statistic ($\chi^2 = 3.959$) has an associated *p*-value or significance level of .412. How should this intermediate-sized *p*-value be interpreted? If the *p*-value were very close to 1.0, the actual and expected frequency counts would be very similar, and the logical conclusion would be to not reject the null hypothesis. In this circumstance, it would be clear that each of the five high schools had an equal or uniform number of students exceeding the national median SAT score. Alternatively, if the *p*-value were very close to 0.0, the actual and expected frequency count distributions would be very different, and the decision would be to reject the null

TABLE 11.1

Random Sample of Students with SAT Scores above National Median, by High School

High school	Number of SAT scores above the national median	
	Initial year results	Results after fifth year
John Muir H.S.	42	51
Gifford Pinchot H.S.	45	52
Rachel Carson H.S.	51	56
Garrett Hardin H.S.	47	53
Aldo Leopold H.S.	60	64
Total	245	276

TABLE 11.2

Worktable for Goodness-of-Fit Chi-square Uniform: Number of Students with SAT Scores above National Median, by High School

Initial year results:

Observed number of students with above-median test scores (from table 11.1)

Muir H.S.	Pinchot H.S.	Carson H.S.	Hardin H.S.	Leopold H.S.	Total
42	45	51	47	60	245

If testing goodness-of-fit uniform or equal distribution, the expected frequency of each category is calculated by dividing the total observed frequency count by the number of categories:

$$E_i = \frac{\Sigma O_i}{k} \text{ for all categories } (i)$$

$$E_i = \frac{245}{5} = 49$$

$$\chi^2 = \sum_{i=1}^{k} \frac{(O_i - E_i)^2}{E_i} = \frac{(42-49)^2}{49} + \frac{(45-49)^2}{49} + \frac{(51-49)^2}{49} + \frac{(47-49)^2}{49} + \frac{(60-49)^2}{49}$$

$$= 3.959 \; (p = .412)$$

Results after the fifth year:

Observed number of students with above-median test scores (from table 11.1):

Muir H.S.	Pinchot H.S.	Carson H.S.	Hardin H.S.	Leopold H.S.	Total
51	52	56	53	64	276

$$E_i = \frac{276}{5} = 55.2$$

$$\chi^2 = \sum_{i=1}^{k} \frac{(O_i - E_i)^2}{E_i} = \frac{(51-55.2)^2}{55.2} + \frac{(52-55.2)^2}{55.2} + \frac{(56-55.2)^2}{55.2} + \frac{(53-55.2)^2}{55.2} + \frac{(64-55.2)^2}{55.2}$$

$$= 2.007 \; (p = .734)$$

hypothesis and conclude that the high schools had different SAT test results.

This intermediate-sized p-value is rather inconclusive, but should suggest to the superintendent that there seems to be "room for improvement," particularly if the test scores of Muir (42), Pinchot (45), and Hardin (47) high school students could be substantially improved. The SAT improvement program would be deemed a success (at least statistically) if the magnitude of the chi-square test statistic could be reduced and the resultant p-value shifted significantly toward 1.0. (A p-value of 1.0 will occur only if there is perfect uniformity or equality across all of the frequency count categories—that is, if an equal number of students from each high school were to score above the national median on their SAT test.)

Suppose now the SAT improvement program is implemented. Five years later, the effectiveness of the program is evaluated. The number of students

scoring above the national median increases at each of the five high schools, and the five-school total of students increases from 245 to 276 (table 11.1, "results after fifth year"). Looking at the observed frequency counts, there seems to be a shift toward interschool equality. The magnitude of the chi-square statistic declines from 3.959 to 2.007, confirming this shift toward equity among high schools (table 11.2).

The SAT improvement program has apparently been successful. High schools with the lowest frequency counts during the initial year improved their frequency counts the most by the end of the fifth year. Leopold High School had the highest frequency count during the initial year. Their students managed to improve, but not to the same degree as the other high schools. As a result of these changes in student test performance, the p-value increased considerably from .412 to .734, indicating stronger confidence that the null hypothesis should not be rejected. If the null hypothesis is rejected and a conclusion drawn that high school test scores are different, a 73.4 percent chance exists that this conclusion is incorrect.

Application to Geographic Problem: Chi-Square Goodness-of-Fit (Proportional Distribution)

Chi-square goodness-of-fit is now applied to several simple spatial interaction models to see how well the models fit an actual set of data. Geographers often use spatial interaction models to predict the volume of movement or amount of activity (interaction) between locations. Many of these models are based on the assumption that the amount of movement or interaction from a given origin to a destination is directly related to the population size of the origin and/or inversely related to the square of the distance from the origin to the destination.

To illustrate how these spatial interaction models can be structured as goodness-of-fit problems, consider the 1991 pattern of interprovincial migrants to British Columbia, Canada. British Columbia was Canada's fastest growing province from 1991 to 1996, with 13.15 percent growth. In comparison, Canada's overall growth rate during this period was only 5.7 percent. How well can the observed number of migrants into British Columbia be predicted (fit) using different spatial interaction models to generate expected numbers of migrants?

Three spatial interaction models are tested and evaluated with regard to their relative effectiveness as predictors of actual migration flows into British Columbia. The first model is a "population" model that predicts the volume of in-migration to British Columbia as directly proportional to the population of each origin province. The second model is a "distance" model that predicts the volume of in-migration to British Columbia as inversely proportional to the square of the distance from each origin province to British Columbia. The third model is a "composite" model that predicts the volume of in-migration to British Columbia as directly proportional to the population of each origin province *and* inversely proportional to the square of the distance from each origin province to British Columbia.

During 1991, 80,302 Canadians moved into British Columbia from other provinces. Using the actual number of migrants from each province into British Columbia, the observed frequencies for the models were created by using 1 percent of each province's total in-migration. For example, the actual number of migrants from Newfoundland to British Columbia in 1991 was 1,112. Therefore, the observed frequency allocation for each model is 1 percent of 1,112, which rounds to 11. When similar allocations are made for all other provinces, the total

TABLE 11.3

Observed Frequency Counts: Number of Interprovincial Migrants to British Columbia, Canada, 1991

Province of origin	Number of migrants to British Columbia, 1991
Newfoundland	11
Prince Edward Island	4
Nova Scotia	25
New Brunswick	14
Quebec	50
Ontario	236
Manitoba	73
Saskatchewan	65
Alberta	307
Yukon Territory	10
Northwest Territories	8
Total	803

observed frequency count for each model is 803, approximately 1 percent of the total Canadian movement into British Columbia in 1991 (table 11.3).

How many migrants would be expected to move into British Columbia from each of the other provinces? Each model must meet the general condition that the total number of migrants into British Columbia from all other provinces is 803. This condition applies because the expected total number of in-migrants must exactly equal the observed total. In addition, each model must meet its associated population or distance criteria defining the pattern of expected frequencies.

1. **"Population" model.** This model predicts that the expected number of migrants into British Columbia is directly proportional to the population of each origin province. That is:

$$E_{ij} = Pop_i \qquad (11.7)$$

where E_{ij} = expected number of migrants from province i to British Columbia (j)
Pop_i = population of province i

2. **"Distance" model.** This model predicts that the expected number of migrants into British Columbia is inversely proportional to the square of the distance from each origin province to British Columbia. That is:

$$E_{ij} = \frac{1}{D_{ij}^2} \qquad (11.8)$$

where D_{ij}^2 = distance squared from each province i to British Columbia (j).

3. **"Composite" model.** This model predicts that the expected number of migrants into British Colum-

bia is directly proportional to the population of each origin province *and* inversely proportional to the square of the distance from that origin province to British Columbia. That is:

$$E_{ij} = \frac{Pop_i}{D_{ij}^2} \qquad (11.9)$$

The observed and expected frequency counts from each of the three spatial interaction models are summarized in the upper portion of table 11.4. Each of the columns in this table totals 803, reflecting the fact that each of the spatial interaction models *proportionally allocates* the same total of 803 migrants to the set of provinces of origin. For example, in the ap-

plication of the "population" model, the expected number of migrants from Ontario to British Columbia is 337. How did the model generate this 337 value? Canada's 1991 population (excluding British Columbia) was 24,014,798, and Ontario's 1991 population was 10,084,885, or 41.99 percent of the total. If 803 people move to British Columbia, and 41.99 percent of them are expected to be from Ontario, then the best estimate of migration from Ontario to British Columbia is (803)(.4199) = 337.

In the application of the "distance" and "composite" models, distances from each of the other provinces to British Columbia must be included. To estimate these distances, several alternatives are

TABLE 11.4

Summary Table for Chi-square Goodness-of-Fit Proportional: Interprovincial Migration to British Columbia, Canada, 1991

Province of origin	Observed number of interprovincial migrants	Expected number of interprovincial migrants		
		"Population" model	"Distance" model	"Composite" model
Newfoundland	11	19	8	13
Prince Edward Island	4	4	12	8
Nova Scotia	25	30	12	21
New Brunswick	14	24	14	19
Quebec	50	231	18	125
Ontario	236	337	21	179
Manitoba	73	37	85	61
Saskatchewan	65	33	137	85
Alberta	307	85	340	213
Yukon Territory	10	1	58	29
Northwest Territories	8	2	98	50
Total	803	803	803	803

"Population" model

$E_{ij} = Pop_i$

$$\chi^2 = \sum_{i=1}^{k} \frac{(O_i - E_i)^2}{E_i} = \frac{(11 - 19)^2}{19} + \frac{(4 - 4)^2}{4} + \frac{(25 - 30)^2}{30} + \frac{(14 - 24)^2}{24} + \frac{(50 - 231)^2}{231}$$

$$+ \frac{(236 - 337)^2}{337} + \frac{(73 - 37)^2}{37} + \frac{(65 - 33)^2}{33} + \frac{(307 - 85)^2}{85} + \frac{(10 - 1)^2}{1} + \frac{(8 - 2)^2}{2} = 925.33$$

"Distance" model

$E_{ij} = \dfrac{1}{D_{ij}^2}$

$$\chi^2 = \sum_{i=1}^{k} \frac{(O_i - E_i)^2}{E_i} = \frac{(11 - 8)^2}{8} + \frac{(4 - 12)^2}{12} + \frac{(25 - 12)^2}{12} + \frac{(14 - 14)^2}{14} + \frac{(50 - 18)^2}{18}$$

$$+ \frac{(236 - 21)^2}{21} + \frac{(73 - 85)^2}{85} + \frac{(65 - 137)^2}{137} + \frac{(307 - 340)^2}{340} + \frac{(10 - 58)^2}{58} + \frac{(8 - 98)^2}{98} = 2443.73$$

"Composite" model

$E_{ij} = \dfrac{Pop_i}{D_{ij}^2}$

$$\chi^2 = \sum_{i=1}^{k} \frac{(O_i - E_i)^2}{E_i} = \frac{(11 - 13)^2}{13} + \frac{(4 - 8)^2}{8} + \frac{(25 - 21)^2}{21} + \frac{(14 - 19)^2}{19} + \frac{(50 - 125)^2}{125}$$

$$+ \frac{(236 - 179)^2}{179} + \frac{(73 - 61)^2}{61} + \frac{(65 - 85)^2}{85} + \frac{(307 - 213)^2}{213} + \frac{(10 - 29)^2}{29} + \frac{(8 - 50)^2}{50} = 163.81$$

possible. The method used here takes distances from a mileage guide published by the Canadian Government Travel Bureau. Road distance from Vancouver (the largest city in British Columbia) to the largest city of each of the other provinces is the distance metric in these models. Using algebraic procedures, the expected number of migrants into British Columbia from every other province is generated to meet each model's specifications.

Calculations of the chi-square test statistics are summarized in the lower portion of table 11.4. How can these results be interpreted? In goodness-of-fit models structured like this example, *the lower the magnitude of the predictive test statistic, the better the fit of the model.* The relative predictive power of these three models is estimated from the relative magnitudes of the test statistics. In this application, the smallest test statistic value ($\chi^2 = 163.81$) is associated with the "composite" model, which incorporates both the population and distance squared into prediction. The "distance" model has the largest value ($\chi^2 = 2443.73$), and the "population" model has an intermediate-sized test statistic ($\chi^2 = 925.33$). It makes intuitive sense that the two-variable "composite" model would provide more predictive power than the two single-variable models.

Knowing the overall predictive power of each model is useful, but it is even more informative to identify which specific provinces are good predictors of in-migration and which are poor predictors. If the predictive power of the model is to be improved, it is important to identify those provinces where the observed volume of in-migration differs the most from the number of expected in-migrants. Those large differences identify the large contributors (provinces) of error in the model. Conversely, when the observed

and expected volumes of in-migration are similar, the chi-square model has accurately predicted the volume of migration from that province. Why are predictions from some provinces accurate, while predictions from other provinces are not? Trying to answer that question will lead the geographer to suggest other variables that relate to in-migration.

In the "population" model, Alberta contributes 62.66 percent of the chi-square test statistic magnitude. That is, from a chi-square value of 925.33, nearly 63 percent (579.81) is the result of *more observed migrants than expected* moving from Alberta to British Columbia (table 11.5). This is an expected finding—a migration model based solely on population size of the origin is likely to underestimate the number of migrants from nearby places, and Alberta is the nearest neighboring province of British Columbia. People living in Alberta know a lot about British Columbia and therefore are more likely to move there. Another 15.33 percent of the chi-square test statistic in the "population" model is contributed by Quebec. However, in the case of Quebec, *fewer migrants than expected* are moving to British Columbia. We could hypothesize that the cultural distinctiveness and recent separatist tendencies of Quebec have resulted in less spatial interaction (including migration) between Quebec and other Canadian provinces.

In the "distance" model, Ontario generates the overwhelming percentage of the chi-square test statistic magnitude. More specifically, from a chi-square test statistic of 2443.73, slightly over 90 percent (2201.19) is Ontario's contribution, as *many more migrants than expected* are moving to British Columbia. This is hardly surprising, because in a "distance only" spatial interaction model, the relatively large population of Ontario is not being taken into account.

TABLE 11.5

Summary of Chi-square Differences Not Explained by Spatial Interaction Model: Interprovincial Migration to British Columbia, 1991

Province of origin	Chi-square differences, not explained by model		
	"Population" model	"Distance" model	"Composite" model
Newfoundland	3.37	1.12	0.31
Prince Edward Island	0.00	5.33	2.00
Nova Scotia	0.83	14.08	0.76
New Brunswick	4.17	0.00	1.32
Quebec	141.82	56.89	45.00
Ontario	30.27	2201.19	18.15
Manitoba	35.03	1.69	2.36
Saskatchewan	31.03	37.84	4.71
Alberta	579.81	3.20	41.48
Yukon Territory	81.00	39.72	12.45
Northwest Territories	18.00	82.65	35.28
Total	925.33	2443.73	163.81

In the "composite" model, the goodness-of-fit is much improved over the two previous models. The chi-square test statistic is much lower (163.81), and the dual variables of population and distance, when weighted as equally important and incorporated simultaneously in the model, predict more accurately the magnitude of migration from each province to British Columbia. Quebec, Alberta, and Northwest Territories are the provinces with the largest differences between observed and expected volumes of migration. Quebec's relative isolation from much of Canada is reflected in this example by a lower than expected volume of migration flow to British Columbia. The error in the model associated with Alberta can be explained by citing dramatic underprediction of the population component of the "composite" model. In the case of Northwest Territories, the distance component of the model accounts for most of the error. Just looking at the distance component, many more migrants would be expected from Northwest Territories than were observed.

Summary Evaluation

When using chi-square as a goodness-of-fit model, it is difficult to determine how well the model produces accurate and precise predictions. For example, when using the three spatial interaction models to predict migration into British Columbia, the chi-square procedure generates only a chi-square value, and, if an inferential hypothesis testing approach is used, a *p*-value. The former is an *absolute* measure of how well the observed (actual migration) frequencies match the expected (predicted migration) frequencies, while the latter represents a probability showing the likelihood of making a Type I error. Unfortunately the chi-square procedure does not produce an index measuring the *relative* explanatory ability of each model.

What is known from the chi-square procedure is that the larger the chi-square value, the greater the absolute difference between one or more of the observed and expected frequencies. This large difference implies that one or more of the expected frequencies does not closely match the corresponding observed frequency, therefore suggesting that the model has less predictive power. The extent to which it is less powerful, however, is impossible to fully establish objectively. In problems like the migration example discussed here, the reality of the situation is complex, and numerous factors influence the pattern of migration behavior. Moreover, identifying and operationally modeling these additional factors will prove to be very challenging. In summary, when designing chi-square models in geography, especially complex models involving human behavior, it is very difficult (perhaps impossible) to generate a *p*-

value other than zero. These issues do not invalidate the application of chi-square for modeling because the relative magnitude of chi-square statistics from different models still provides valuable comparative information.

The total frequency count in a chi-square problem creates further uncertainty in the model interpretation. All else being equal, as frequency counts in a chi-square problem increase, so does the absolute magnitude of the chi-square statistic. For example, if the absolute frequency of every category is doubled, the relative frequencies among the categories remain constant, but the larger absolute frequencies cause the magnitude of the chi-square statistic to increase dramatically. Given these facts, it is possible to produce significant results (i.e., large chi-square values and low *p*-values) simply by inflating the frequency counts of the categories.

Application of Kolmogorov-Smirnov Goodness-of-Fit Procedure: Normal Distribution

The application of parametric statistics for geographic problem solving requires one or more samples to be drawn from normally distributed populations. The Kolmogorov-Smirnov goodness-of-fit test is the most common procedure to test for normality before applying more powerful parametric tests to the sample data. If the K-S test shows a close match between the observed distribution of the sample data and the one expected when the data are normally distributed, it can be concluded that the population from which the sample was drawn is, in fact, normally distributed.

In section 9.1, samples of data from less-developed and more-developed countries were used to show whether female participation in the labor force differed statistically for 1980 and 1995. For these data to be applied properly in the two-sample difference of means tests, a researcher needs evidence that the samples are normally distributed. Only then can one have confidence that the normality requirement of parametric difference tests has been satisfied.

Numerical and graphical methods for completing a K-S goodness-of-fit test for normality are illustrated using the 1980 sample data for labor force participation in less-developed countries (see table 9.2). In this sample of 30 countries, female participation in the labor force averages 38.33 percent, with countries ranging from a low of 21 to a high of 56 percent. The standard deviation of 10.42 percent measures the level of variability in the set of data. Can these sample data be considered to have a normal distribution?

Table 11.6 presents the null and alternate hypothesis, the ranked sample data, and the steps necessary to calculate both the observed and expected cumulative relative frequency values used in the K-S test. Results are shown for the first (lowest) data value, a female participation rate of 21 percent. Table 11.7 displays the cumulative relative frequencies both observed and expected for each unique data value in the sample. The cumulative relative frequencies are also shown graphically in an ogive (figure 11.3).

The objective of the K-S goodness-of-fit test for normality is to locate the greatest absolute difference between the observed and expected cumulative relative frequencies along the range of the data. When the observed data is continuously distributed, this observed/expected comparison must occur in two ways. Notice that in the staircase appearance of the observed ogive, each CRFo value can be compared to its corresponding theoretical normal value (CRF_e) at two locations, once at the beginning and again at the end of the interval. First, each observed value (i) is compared with its corresponding expected (normal) value (i) in an equivalent level or *nonlagged* method. Second, each expected value (i) is then compared with the observed CRF value from the next *lower* data level (i − 1). This second comparison can be termed the *lagged* method. The greatest difference between the observed and expected cumulative relative frequencies across all data values for *either* comparison method defines the Kolmogorov-Smirnov D.

For the 30 data values representing female participation in the labor force, the greatest absolute difference between the observed and expected cumulative relative frequencies occurs when $X = 43$ percent and where $D = .140$. At this point in the distribution, the two cumulative relative frequency distributions differ by approximately 14 percent on the cumulative percentage axis (figure 11.3). When this K-S D-value is converted to Z, the resulting test statistic (K-S Z = .765, p = .603) suggests a 60 percent chance that the

---

TABLE 11.6

Worktable for Calculating Observed and Expected Cumulative Relative Frequency Values: Percentage of Females in the Labor Force, Less-Developed Countries, 1980

H_0: population from which sample is drawn is normally distributed
H_A: population from which sample is drawn is not normally distributed

Sample data: percentage of female population in the labor force, less-developed countries, 1980 (see table 9.2)

Step 1: Rank data (X) from lowest to highest:

21	27	34	42	46	49
21	27	35	43	47	50
22	27	39	43	47	50
26	28	39	45	47	51
26	30	39	45	48	56

Step 2: Calculate the mean and standard deviation:

$\bar{X} = 38.33$ $s = 10.42$

Step 3: Calculate the Z-value for each value of X. For the first value ($X = 21$):

$$Z = \frac{X - \bar{X}}{S} = \frac{21 - 38.33}{10.42} = -1.664$$

Step 4: Calculate the *observed* cumulative relative frequency (CRF_o) for each value of X. What proportion of the observed values are equal to or less than each value?

For the first value (21), two values are less than or equal to 21, so

$CRF_o = 2/30 = .067$.

Step 5: Calculate the *expected* cumulative relative frequency (CRF_e) for each value. If the distribution is normal with the given mean and standard deviation, what proportion of the values should be equal to or less than each value?

For the first value (21), where $Z = -1.664$, the value (p) from the normal table = .452

Since $Z < 0$, CRF_e is calculated using equation 11.5 as

$CRF_e = 1.00 - (p + .50) = 1.00 - (.452 + .50) = 1 - .9495 = .048$

Kolmogorov-Smirnov Results for Normality Test: Percentage of Females in the Labor Force, Less-Developed Countries, 1980

X	Z	f	CRF$_o$	CRF$_e$	Difference† (CRF$_o$ − CRF$_e$) Nonlagged	Lagged
21*	−1.664	2	.067	.048	.019	
22	−1.568	1	.100	.058	.042	.009
26	−1.184	2	.167	.118	.049	−.018
27	−1.088	3	.267	.138	.129	.029
28	−.992	1	.300	.161	.139	.106
30	−.800	1	.333	.212	.121	.088
34	−.416	1	.367	.339	.028	−.006
35	−.320	1	.400	.374	.026	−.007
39	.064	3	.500	.526	−.026	−.126
42	.352	1	.533	.638	−.105	−.138
43	.448	2	.600	.673	−.073	−.140**
45	.640	2	.667	.739	−.072	−.139
46	.736	1	.700	.769	−.069	−.102
47	.832	3	.800	.797	.003	−.097
48	.928	1	.833	.823	.010	−.023
49	1.024	1	.867	.847	.020	−.014
50	1.120	2	.933	.869	.064	−.002
51	1.216	1	.967	.888	.079	.045
56	1.696	1	1.000	.955	.045	.012

where X = ranked sample data values
  Z = data in standardized form
  f = frequency
 CRF$_o$ = cumulative relative frequency (observed)
 CRF$_e$ = cumulative relative frequency (expected)

*Calculations for this data value are shown in table 11.6.
**$D_{K\text{-}S}$ = Maximum $|CRF_o - CRF_e|$ = −.140
† Differences between observed and expected values are determined for equivalent data levels (nonlagged) as well as for observed values one level below expected (lagged).
When $D_{K\text{-}S}$ is converted to Kolmogorov-Smirnov Z-value, the test statistic (K-S Z) is K-S Z = .765 (p = .603)

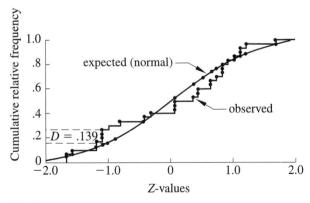

FIGURE 11.3
Observed and Expected (Normal) Cumulative Relative Frequency and Maximum Differences (D) for Less-Developed Country Sample of Female Workforce Data

observed frequency distribution has been drawn from a normally distributed population. Thus, the K-S goodness-of-fit test for normality supports the null hypothesis of no significant difference between the observed data and a normal distribution. To reject

the null hypothesis and support a conclusion that the differences are significant would result in a large, 60 percent chance of committing a Type I error.

K-S test results for all samples used in the difference of means t-test example in section 9.1 are summarized in table 11.8. The resulting p-values range from a low of .468 for the less-developed countries in

Kolmogorov-Smirnov Goodness-of-Fit Results for Four Samples of Female Participation in the Labor Force

Sample*	Mean	Standard deviation	K-S D	K-S Z	p-value**
LDC, 1980	38.33	10.42	.140	.765	.603
LDC, 1995	40.13	7.88	.155	.849	.468
MDC, 1980	30.93	10.56	.212	.794	.554
MDC, 1995	35.93	8.35	.170	.636	.813

*LDC: Less-developed countries
MDC: More-developed countries
**Two-tailed probability

1995—the closest sample to nonnormal—to a high of .813 for the more-developed countries in 1995. Each test for goodness-of-fit confirms that the distribution of the data falls closely enough statistically to a normal pattern. Therefore, parametric tests using these samples can be assumed to have met the requirement of normality.

Summary Evaluation

All four samples in this example show a very strong probability that the observed data are normal. However, the magnitude of each maximum difference (*D*) appears to be very large. How can such a large difference between observed and expected cumulative values still lead to a conclusion that the data are statistically normal? The answer lies in the rigor of hypothesis testing—that is, the unwillingness to commit a Type I error by rejecting a null hypothesis when in fact it is true. In this case, a serious Type I error would be rejection of a distribution as nonnormal, when in fact it is normal. In general, a distribution must be extremely nonnormal to be rejected by the Kolmogorov-Smirnov goodness-of-fit test.

11.2 CONTINGENCY ANALYSIS

The idea of testing an actual frequency distribution against an expected frequency distribution can be expanded to research problems that examine the relationship between two variables, a logical extension of the single-sample or single-variable goodness-of-fit tests. In this instance, however, categorical frequency counts of a single variable are not tested against an expected distribution (such as uniform, proportional, or normal), but rather the frequency distributions of two variables are compared directly with one another.

For example, suppose a city planning agency is developing a five-year capital improvement program (CIP). This program involves the scheduling of public physical improvements based on estimates of future fiscal resources and project costs. As part of the CIP development process, decision makers want citizen input in determining the method for financ-

ing several proposed cultural facilities. The proposals include

- improving the grounds of a local historical site,
- constructing a new wing on the art museum,
- expanding the city aquarium,
- building a new science center with science education facilities, and
- constructing a new branch library.

Several methods of financing are available, including general tax revenue, user fees, administrative fees, and revenue bonds. Suppose a random survey of residents is taken and respondents are asked which of the five projects they prefer. For that preferred project, each respondent is then asked which method of funding he or she prefers. A contingency table can now be constructed showing the number of people who prefer each project/funding source combination (table 11.9). A contingency analysis could then be applied to the frequency counts in this table to determine if the preferred financing methods differ significantly from project to project.

Geographers can also use contingency analysis effectively when comparing maps. Given two spatial pattern (choropleth) maps, the extent of areal relationship between the maps can be examined by selecting *n* identical sets of random point locations from both maps, then allocating each of the sample point combinations to the appropriate cell of a contingency table. For example, suppose a geographer has two maps of the same region. One map shows the suitability of land for general farming (in five ordinal categories: well-suited, moderately suited, somewhat poorly suited, poorly suited, and unsuited), and the other map shows soil association (in five nominal categories: alfisols, entisols, histosols, inceptisols, and spodosols). The spatial relationship between these two maps can be measured through contingency analysis of the frequency counts taken from each cell of the table (table 11.10).

When using contingency analysis, data values from both variables are assigned to nominal or ordinal categories, and the frequency count of each cate-

---

TABLE 11.9

Example Contingency Table Structure: Method of Financing Preferred for Cultural Facilities Projects

Method of financing preferred	Proposed cultural facility projects				
	Improve grounds of historic site	Add new wing on art museum	Expand city aquarium	Build new science center	Build new branch library
General tax revenue	_____	_____	_____	_____	_____
User fees	_____	_____	_____	_____	_____
Administrative fees	_____	_____	_____	_____	_____
Revenue bonds	_____	_____	_____	_____	_____

TABLE 11.10

Contingency Table: Soil Association Cross-Tabulated with Suitability of Land for General Farming

Soil association	Suitability of land for general farming				
	Well suited	Moderately suited	Somewhat poorly suited	Poorly suited	Unsuited
Alfisols	————	————	————	————	————
Entisols	————	————	————	————	————
Histosols	————	————	————	————	————
Inceptisols	————	————	————	————	————
Spodosols	————	————	————	————	————

gory from one variable is cross-tabulated with the frequency count of each category from the second variable. The frequencies are summarized in a two-dimensional contingency, or **cross-tabulation,** table. Each variable must have at least two categories to which data values can be allocated.

The actual frequency count in each cell of the contingency table is compared with the frequency count expected if no relationship exists between the two variables. If at least one difference between actual and expected frequency counts is large, then the variables are not likely to be statistically independent, but rather related in some nonrandom (systematic) fashion. The null hypothesis of no relationship is then rejected. Conversely, if the differences between frequency counts are small, then one would conclude the variables are statistically independent, with only a random association, and the null hypothesis is not rejected.

The chi-square statistic evaluates contingency table frequency counts and is a simple extension of the goodness-of-fit test described in the previous section. *The entire contingency table is evaluated statistically as a single entity.* That is, the analysis determines whether significantly large differences exist between actual and expected frequency counts of the two variables, but does not indicate which cells in the table contain these differences. When both variables have exactly two categories, this analysis poses no problem. With tables having three or more categories for at least one variable, however, further analysis is sometimes needed to learn exactly where the significantly different frequency counts are located.

Certain restrictions apply to the use of chi-square in contingency analysis. Data must be absolute frequency counts rather than relative frequencies such as percentages or proportions. In addition, no more than one-fifth of the expected frequencies should be less than five and none less than two. As discussed with goodness-of-fit procedures, combining categories is sometimes an option if too many expected frequency counts are too low.

The chi-square used in contingency analysis is

$$\chi^2 = \sum_{i=1}^{r} \sum_{j=1}^{k} \frac{(O_{ij} - E_{ij})^2}{E_{ij}} \qquad (11.10)$$

where O_{ij} = observed frequency count in the ith row and j^{th} column
E_{ij} = expected frequency count in the ith row and j^{th} column
r = number of rows in the contingency table
k = number of columns in the contingency table

If the observed and expected frequency counts in each cell of the contingency table are similar, then the differences $(O_{ij} - E_{ij})$ in all of the cells will be small, the statistic will be low, and the null hypothesis will not be rejected. Conversely, if at least one cell in the contingency table has a large difference between observed and expected frequency counts, the test statistic will be large and the null hypothesis will more likely be rejected.

To calculate the expected frequency of cell ij in the contingency table, the sum of all observed frequency counts in row i (R_i) is multiplied by the sum of all observed frequency counts in column j (C_j), then divided by the grand total of all observed frequencies (N):

$$E_{ij} = \frac{R_i C_j}{N} \qquad (11.11)$$

Two items regarding these expected frequencies should be noted. First, the frequency count totals of a row or column are often called **marginal totals.** Thus, R_1 is the marginal total for row 1, C_3 is the marginal total for column 3, and so on. Second, even though all observed frequency counts are integers, the calculated expected frequency values will generally not be integers. Rounding the expected frequency values to integers is not necessary. Doing so would only result in a less precise test.

Contingency Analysis (χ^2)

Primary Objective: Compare random sample frequency counts of two variables for statistical independence

Requirements and Assumptions:

1. Single random sample
2. Variables are organized by nominal or ordinal categories; frequency counts by category are input to statistical test
3. No more than one-fifth of the expected frequency counts should be less than five, and none of the expected frequencies should be less than two

Hypotheses:

H_0: There is no relationship between two variables in the population from which sample has been drawn (variables are statistically independent)

H_A: There is a relationship between two variables in the population from which sample has been drawn (variables are not statistically independent)

Test Statistic:

$$\chi^2 = \sum_{i=1}^{r} \sum_{j=1}^{k} \frac{(O_{ij} - E_{ij})^2}{E_{ij}}$$

Application to Geographic Problem

To illustrate the application of contingency analysis, an example is presented using information from the General Social Survey (GSS). Recall that the two-sample difference of proportions test example in chapter 9 used the GSS to investigate people's attitudes about whether they favor or oppose gun permits.

Another series of questions in the GSS asks each respondent his or her attitude toward eleven different types of music, ranging across the entire musical spectrum—big band, blues or R & B, bluegrass, classical, country western, folk, heavy metal, jazz, Broadway musical, opera, and rap. Respondents were asked to select one of five attitudes toward each type of music: (1) "like it very much"; (2) "like it"; (3) "mixed feelings"; (4) "dislike it"; and (5) "dislike very much."

Nationwide, more than 700 people expressed their attitudes toward these music types. One of the additional pieces of information gathered from these respondents is census division of residence. (See the census division boundaries shown in figure 3.8.) Contingency analysis now examines whether attitudes regarding music type vary significantly from one region of the country to another.

Not surprisingly, country western is the most popular type of music among all GSS respondents. In the top category, "like very much," country western music garnered a higher percentage of responses than any other type of music. Of 739 people surveyed, 200 (27.1 percent) selected this response, sug-

gesting that country western is popular nationwide. Is this popularity consistent across all census divisions, or do people in different regions of the country have varying attitudes toward this music?

Data are now placed in a contingency table, with each of the nine rows identifying a census division and each of the five columns showing the attitude toward country western music (table 11.11). Each cell of this contingency table lists the observed frequency count, followed by the expected frequency count in parentheses. For example, of the 29 total New England respondents surveyed, 5 people indicate they like country western music very much (the observed frequency). The corresponding expected frequency count is 7.8, indicating that this number of New Englanders should like country western very much if no difference in music preference can be found from one census division to another.

The calculations for the contingency analysis of country western music show that significant attitude differences among census divisions exist somewhere (in table 11.12, $\chi^2 = 52.088$ and $p = .014$). Given these results, the null hypothesis should be rejected, and we are 98.6 percent confident that significant attitude differences toward country western music among census divisions exist somewhere in the contingency table.

Where in the contingency table are the largest differences between observed and expected frequency counts? That is, which cells of the table contribute most to the magnitude of the chi-square statistic? A geographer might ask: Can census divisions be identified where country western music is liked much more than expected or liked much less than expected? The three largest differences between observed and expected frequency counts in the contingency table are the following:

1. Middle Atlantic—Mixed Feelings (observed = 39, expected = 24.3, difference = 14.7)
2. East South Central—Like Very Much (observed = 26, expected = 14.3, difference = 11.7)
3. West South Central—Like Very Much (observed = 27, expected = 18.7, difference = 8.3)

In the Middle Atlantic census division, many respondents had "mixed feelings" about country western music. Perhaps this result is due to the eclectic demographics in this region; residents of New York, New Jersey, and Pennsylvania may not have strong positive or negative opinions about country western music.

In the East South Central and West South Central census divisions, more people than expected said they like country western very much. This result is not surprising, as geographers have identified these regions as the cultural hearth of country western

TABLE 11.11

Contingency Table: Census Division of Respondent Cross-Tabulated with Attitude toward Country Western Music, General Social Survey, 1995

Census division	Attitude toward country western music*					Row total
	Like very much	Like it	Mixed feelings	Dislike it	Dislike very much	
New England	5 (7.8)	13 (10.6)	8 (6.8)	3 (2.7)	0 (1.1)	29 (29)
Middle Atlantic	21 (28.1)	30 (37.9)	39 (24.3)	9 (9.9)	5 (3.8)	104 (104)
E. North Central	41 (45.2)	60 (60.8)	40 (39.1)	17 (15.8)	9 (6.1)	167 (167)
W. North Central	8 (13.0)	23 (17.5)	11 (11.2)	4 (4.5)	2 (1.8)	48 (48)
South Atlantic	36 (32.5)	48 (43.7)	22 (28.1)	13 (11.4)	1 (4.4)	120 (120)
E. South Central	26 (14.3)	15 (19.3)	5 (12.3)	5 (5.0)	2 (1.9)	53 (53)
W. South Central	27 (18.7)	24 (25.1)	10 (16.2)	7 (6.5)	1 (2.3)	69 (69)
Mountain	8 (9.5)	16 (12.7)	6 (8.2)	3 (3.3)	2 (1.3)	35 (35)
Pacific	28 (30.9)	40 (41.5)	32 (26.7)	9 (10.8)	5 (4.2)	114 (114)
Column total	200 (200)	269 (269)	173 (173)	70 (70)	27 (27)	739 (739)

All $E_{ij} = \dfrac{(R_i)(C_j)}{N}$

For example: $E_{11} = \dfrac{(R_1)(C_1)}{N} = \dfrac{(29)(200)}{739} = 7.8$

*Each cell of the table contains the observed frequency count, followed by the expected frequency count, in parentheses

TABLE 11.12

Worktable for Chi-square Contingency Analysis: Attitude toward Country Western Music by Census Division

H_0: there is no relationship between two variables (variables are statistically independent with only a random association).

H_A: there is a relationship between two variables (variables are not statisically independent, but related to one another in some nonrandom fashion.

$$\chi^2 = \sum_{i=1}^{r} \sum_{j=1}^{k} \frac{(O_{ij} - E_{ij})^2}{E_{ij}}$$

where O_{ij} = the observed frequency count in the ith row and j^{th} column
E_{ij} = the expected frequency count in the ith row and j^{th} column
r = the number of rows in the contingency table
k = the number of columns in the contingency table

The row variable is census division and the column variable is preference of country western music.

$$E_{ij} = \frac{(R_i)(C_j)}{N}$$

$$E_{11} = \frac{(R_1)(C_1)}{N} = \frac{(29)(200)}{739} = 7.8$$

$$\chi^2 = \sum_{i=1}^{r} \sum_{j=1}^{k} \frac{(O_{ij} - E_{ij})^2}{E_{ij}} = \frac{(5 - 7.8)^2}{7.8} + \frac{(13 - 10.6)^2}{10.6} + \frac{(8 - 6.8)^2}{6.8} + \ldots + \frac{(9 - 10.8)^2}{10.8} + \frac{(5 - 4.2)^2}{4.2}$$

$\chi^2 = 52.088$ p-value = .014

music. To a great degree, this music has derived from the folk ballads of the English and Scotch-Irish who settled the upland South in colonial times. Appalachian folk music, particularly popular in western Kentucky and Tennessee, began to diffuse rapidly about the time of World War I, concurrent with the invention and diffusion of the radio. The nineteenth-century migration of upland southern folk music to states like Arkansas, Texas, and Oklahoma provided natural areas for the expansion of country western music in the mid-twentieth century. In recent decades, Nashville, Tennessee, has emerged as the center of country western music. All of these facts solidly support the popularity of country western music in the East South Central and West South Central census divisions.

Summary Evaluation

Important validity issues are raised in this contingency analysis. Classifying respondents by census division of residence may not be a particularly valid method to distinguish attitude preferences toward various types of music. Categories such as "like very much," like it," and so on may also have validity (definitional) problems. Potential difficulties may arise from the level of spatial aggregation in this example. It is quite likely, for instance, that music preference varies significantly at finer spatial levels. Even within the same county (from city to small town to rural area), it would not be surprising to note a large variability in music preferences. To the extent that such microscale variability in music preferences exists, a census division analysis may be inappropriate.

KEY TERMS

contingency analysis, 154
cross-tabulation, 167
cumulative normal value, 156
cumulative relative frequency, 156
goodness-of-fit tests, 154
marginal totals, 167
observed and expected frequencies, 154

MAJOR GOALS AND OBJECTIVES

If you have mastered the material in this chapter, you should now be able to do the following.
1. Recognize the variety of geographic research problems for which goodness-of-fit tests are appropriate.
2. Explain the difference between an actual or observed frequency distribution and an expected frequency distribution in a goodness-of-fit context.
3. Understand the nature of various frequency distributions (uniform or equal, proportional or unequal, and normal) and identify the corresponding goodness-of-fit test.
4. Define the term cross-tabulation and identify geographic situations that use cross-tabulation.
5. Identify the types of geographic problems for which contingency analysis is appropriate and be able to create and organize an original geographic situation where contingency analysis could be used.

CHAPTER 12

Inferential Spatial Statistics

12.1 Point Pattern Analysis
12.2 Area Pattern Analysis

Geographers often examine spatial patterns on the earth's surface that are produced by physical or cultural processes. These patterns represent the spatial distribution of a variable across a study area. Sometimes geographic variables are displayed as point patterns with dot maps. In chapter 4, descriptive spatial statistics (or geostatistics), such as the mean center and standard distance, were introduced to summarize point patterns. In other instances, explicit spatial patterns representing data summarized for a series of subareas within a larger study region can be displayed effectively using choropleth maps.

Whether data are presented as points or areas, geographers often want to describe and explain an existing pattern. In this chapter, the focus is on inferential spatial statistics that analyze a sample point pattern or sample area pattern to determine if the arrangement is random or nonrandom. When testing a point or area pattern for randomness, the question is whether the population pattern from which the sample has been created was generated by a spatially random process.

Geographers may want to compare an existing spatial pattern to a particular theoretical pattern. Spatial patterns may appear clustered, dispersed, or random (figure 12.1). In case 1, both the point and area patterns have a **clustered** appearance. On the point pattern map, the density of points appears to vary significantly from one part of the study area to another, with many points concentrated in the northwest portion of the area. Perhaps the points represent sites of tertiary economic activity (retail and service functions), which often cluster around a location with high accessibility and high profit potential, such as a highway interchange. On the clustered area pattern map, the shaded subareas could represent political precincts where a majority of registered voters are Democrat, with Republican majority precincts unshaded. Such a clustered pattern would likely occur

if there is a distinctly nonrandom spatial distribution of voters by income, race, or ethnicity in the region.

Other spatial patterns seem evenly **dispersed** or regular. The set of points in case 2 (figure 12.1) appears uniformly distributed across the study area, suggesting that a systematic spatial process produced the locational pattern. The hypothesis in classical central place theory, for example, is that settlements are uniformly distributed across the landscape to best serve the needs of a dispersed rural population. The area pattern in case 2 exhibits a regular or alternating type of spatial arrangement.

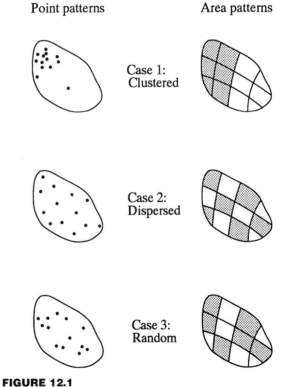

FIGURE 12.1
Types of Point and Area Patterns

This pattern could represent county populations in the same region where the central place distribution of settlements is hypothesized. Shaded counties could have above average populations, while unshaded counties are below average.

The spatial patterns in case 3 (figure 12.1) appear **random** in nature, with no dominant trend toward either clustering or dispersion. A random point, or area, pattern logically suggests that a spatially random (Poisson) process is operating to produce the pattern. The Illinois tornado touchdown locations discussed in section 5.4 illustrate a process that produces a random point pattern. If a choropleth map were designed that showed the number of tornado touchdowns per square mile or kilometer in each Illinois county, a random area pattern would likely result.

In most geographic problems, a point or area pattern will not provide a totally clear indication that the pattern is clustered, dispersed, or random. Rather, many real-world patterns show a combination of these arrangements, with tendencies from "purely" random toward either clustered or dispersed.

If a significant nonrandom arrangement is identified in a point or area pattern, **spatial autocorrelation** is said to exist. The basic property of spatially autocorrelated data is that the values are nonrandomly related or interdependent over space. The clustered patterns in case 1 of figure 12.1 seem to exhibit *positive* spatial autocorrelation, with adjacent or nearby locations having similar values. In case 2, the dispersed patterns have *negative* spatial autocorrelation, with nearby locations having dissimilar values. Random point, or area, patterns (case 3) have *no* spatial autocorrelation.

Section 12.1 presents methods for analyzing point patterns where geographic variables are appropriately displayed as a series of locations or dots on a map. In some problems, the *spacing of individual points* within the overall point pattern is important, and a statistical procedure known as "nearest neighbor analysis" is often applied. For example, an urban geographer might analyze the existing configuration of fire stations in a city to determine whether the pattern is random or more dispersed than random. Suppose one of the goals in the provision of fire service is to provide a locationally equitable or dispersed distribution of fire stations throughout the city. The geographer might suggest the siting of new stations or relocation of existing facilities to meet this goal. The proposed new configuration of fire stations could then be analyzed to determine if it is more dispersed than the existing pattern.

In other instances, the *nature of the overall point pattern* is important, and a statistical test known as "quadrat analysis" is often appropriate. For example, a biogeographer might select a random sample of trees in a national forest to determine which trees are diseased, plot the location of each diseased tree as a point on a map, and analyze the points. The study would determine whether the trees were randomly distributed throughout the study area or clustered in certain portions of the forest. The result of this analysis would provide some guidance concerning the most appropriate type of treatment (e.g., widespread aerial spraying versus concentrated treatment from the ground).

Section 12.2 presents methods for analyzing area patterns. For example, an economic development planner might be interested in evaluating changing spatial patterns of poverty in Appalachia. Through area pattern analysis of a chronological series of county-level choropleth maps showing poverty in the Appalachian region, the spatial distribution of poverty can be evaluated as becoming more concentrated or clustered over time. An urban geographer could analyze a map pattern depicting the number of existing home sales by census tract to determine the degree to which such sales are spatially concentrated in certain portions of the city. If a nonrandom clustering of high turnover rates is found in certain segments of the city, attention could be focused on why the sales exhibit such spatial patterns.

12.1 POINT PATTERN ANALYSIS

The spatial patterns of many geographic variables can be portrayed as dot maps. For example, urban geographers plot the location of settlements as points on a map. Economic geographers study the spatial pattern of retail activities by mapping store locations as dots on the map. Physical geographers show glacial features like drumlins as a series of dots.

For many geographic studies that begin with locations of a variable on a dot map, a primary objective is to determine the form of the pattern of points. The nature of the point pattern can reveal information about the process that produced the geographic result. In addition, a series of point patterns of the same variable recorded at different times can help to determine temporal changes in the locational process. In short, point pattern analysis offers the researcher quantitative tools for examining a spatial arrangement of point locations on the landscape as represented by a conventional dot map.

Nearest Neighbor Analysis

Nearest neighbor analysis is a common procedure for determining the spatial arrangement of a pattern of points within a study area. The distance of each point to its "nearest neighbor" is measured and the average nearest neighbor distance for all points is determined. The spacing within a point pattern can be analyzed by comparing this observed average distance to some expected average distance, such as that for a random (Poisson) distribution.

The nearest neighbor technique was originally developed by biologists who were interested in studying the spacing of plant species within a region. They measured the distance separating each plant from its nearest neighbor of the same species and determined whether this arrangement was organized in some manner or was the result of a random process. Geographers have applied the technique in numerous research problems, including the study of settlements in central place theory, economic functions within an urban region, and the distribution of earthquake epicenters in an active seismic region. In all applications, the objective of nearest neighbor analysis is to describe the pattern of points within a study area and make inferences about the underlying process.

The nearest neighbor methodology is illustrated using the example of seven points shown in figure 12.2. A coordinate system is created and the horizontal (X) and vertical (Y) positions of the points recorded (table 12.1). For each of the points, the nearest neighbor (NN) is determined as the point closest in straight-line (Euclidean) distance. The distances to each nearest neighbor (NND) are then calculated from the coordinates or measured from the map. From the set of nearest neighbor distances, the *average* nearest neighbor distance $(\overline{NND})$ is determined using the basic formula for the mean:

$$\overline{NND} = \frac{\Sigma NND}{n} \qquad (12.1)$$

where n = number of points.

Using the seven nearest neighbor distances in table 12.1, the average nearest neighbor distance is 2.67, indicating that an average distance of 2.67 units separates a point from its nearest neighbor.

The average nearest neighbor distance provides an index of spacing for a set of points. However, the usefulness of this descriptive index comes from comparing the index value for an observed pattern to the results produced from certain distinct point distributions. This objective is analogous to the situation discussed in section 11.1, where an observed frequency distribution was compared to a perfectly normal frequency distribution.

In point pattern analysis, average nearest neighbor distances can be calculated for three distinct point arrangements: random, perfectly dispersed, and perfectly clustered. In each case, the spacing index is determined for an area containing a certain density of points. For example, if points are arranged in a random spatial pattern, the average nearest neighbor distance $(\overline{NND}_R)$ would be determined as follows:

$$\overline{NND}_R = \frac{1}{2\sqrt{Density}} \qquad (12.2)$$

where $\overline{NND}_R$ = average nearest neighbor distance in a random pattern
$Density$ = number of points (n)/area

Since the study area displayed in figure 12.2 has a dimension of 10 by 10 units, the area represented is 100 square units. The corresponding density of points is therefore 7/100, or .07. For an area with this point density, a random arrangement of points within the

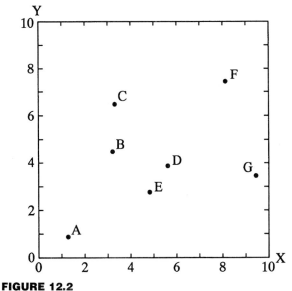

FIGURE 12.2
Location of Points for Nearest Neighbor Problem

TABLE 12.1				

Coordinates and Nearest Neighbor Information for Example*

Point	X	Y	NN	NND
A	1.3	0.9	E	3.94
B	3.2	4.4	C	2.00
C	3.3	6.4	B	2.00
D	5.6	3.8	E	1.36
E	4.8	2.7	D	1.36
F	8.1	7.4	G	4.21
G	9.4	3.4	D	3.82
Sum				18.69

where NN = nearest neighbor
NND = nearest neighbor distance

$$\overline{NND} = \frac{\Sigma NND}{n} = \frac{18.69}{7} = 2.67$$

*See figure 12.2 for graph of point locations.

study area should produce an average nearest neighbor distance:

$$\overline{NND}_R = \frac{1}{2\sqrt{.07}} = 1.89$$

If the arrangement of points shows maximum dispersion or a perfectly uniform pattern, the average nearest neighbor distance ($\overline{NND}_D$) would be determined as

$$\overline{NND}_D = \frac{1.07453}{\sqrt{Density}} \qquad (12.3)$$

Since this pattern represents the most dispersed or separated arrangement of points, it also serves as the maximum value for the nearest neighbor index. The average nearest neighbor distance for a dispersed pattern whose point density matches that in figure 12.2 is

$$\overline{NND}_D = \frac{1.07453}{\sqrt{.07}} = 4.06$$

The pattern of spacing most distinct from dispersion or uniformity is "clustering." When all points lie at the same position, the pattern shows maximum or total clustering of points, and each nearest neighbor distance is zero. Therefore, in a perfectly clustered pattern, the average nearest neighbor distance $\overline{NND}_C$ is also zero, representing the lowest possible value for the index:

$$\overline{NND}_C = 0 \qquad (12.4)$$

Since perfectly clustered and perfectly dispersed patterns provide the extreme spacing arrangements for a set of points, the nearest neighbor index offers a useful method to measure the spacing of locations within an observed point pattern. However, the average nearest neighbor distance is an absolute (as opposed to relative) index, which depends on the units for measuring distance. Therefore, direct comparison of results from different problems or different regions is difficult. Although the minimum value of the nearest neighbor index is always 0 (a perfectly clustered pattern), the maximum value corresponding to a perfectly dispersed pattern is not constant, but a function of the point density. To overcome this, a **standardized nearest neighbor index (R)** is often used. This index is found by dividing the average nearest neighbor distance ($\overline{NND}$) by the corresponding value for a random distribution with the same point density:

$$R = \frac{\overline{NND}}{\overline{NND}_R} \qquad (12.5)$$

With the standardized index, a perfectly clustered pattern produces an R value of 0.0, a random distribution 1.0, and a perfectly dispersed arrangement generates the maximum R value of 2.149 (figure 12.3). Thus, an actual point pattern can be measured for relative spacing along a continuous scale from

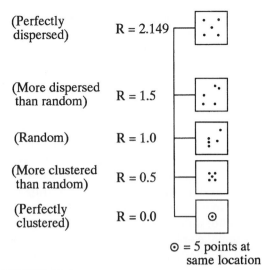

FIGURE 12.3
Continuum of *R* Values in Nearest Neighbor Analysis
(*Source*: Modified from Taylor, P.J. 1977. *Quantitative Methods in Geography*. Boston: Houghton Mifflin)

perfectly clustered to perfectly dispersed. For the set of points from figure 12.2, the standardized nearest neighbor index is

$$R = \frac{2.67}{1.89} = 1.41$$

This spacing pattern is moderately dispersed, since it lies between a perfectly dispersed distribution and a random one.

In addition to its use as a descriptive index of point spacing, the nearest neighbor methodology can also be used to infer results from a sample of points to the population from which the sample was drawn. A difference test can be used to determine if the observed nearest neighbor index ($\overline{NND}$) differs significantly from the theoretical norm ($\overline{NND}_R$), which would occur if the points were randomly arranged. The expectation is that a Poisson process operates over space to produce a random pattern of points. The null hypothesis is that no difference exists between observed and random nearest neighbor values. The test statistic (Z_n) follows a format similar to other difference tests discussed in chapters 8 and 9:

$$Z_n = \frac{\overline{NND} - \overline{NND}_R}{\sigma_{\overline{NND}}} \qquad (12.6)$$

where $\sigma_{\overline{NND}}$ = standard error of the mean nearest neighbor distances

The standard error for the nearest neighbor test can be estimated with the following formula:

$$\sigma_{\overline{NND}} = \frac{.26136}{\sqrt{n(Density)}} \qquad (12.7)$$

where n = number of points

Nearest Neighbor Analysis

Primary Objective: Determine whether a random (Poisson) process has generated a point pattern

Requirements and Assumptions:

1. Random sample of points from a population
2. Sample points are independently selected

Hypotheses:

H_0: $\overline{NND} = \overline{NND}_R$ (point pattern is random)

H_A: $\overline{NND} \neq \overline{NND}_R$ (point pattern is not random)

H_A: $\overline{NND} > \overline{NND}_R$ (point pattern is more dispersed than random)

H_A: $\overline{NND} < \overline{NND}_R$ (point pattern is more clustered than random)

Test Statistic:

$$Z_n = \frac{\overline{NND} - \overline{NND}_R}{\sigma_{\overline{NND}}}$$

Either a directional (one-tailed) or nondirectional (two-tailed) approach can be applied, depending on the form of the alternate hypothesis (H_A). If the problem suggests a clear rationale for the actual point pattern being either dispersed or clustered (as opposed to the null hypothesis of randomness), a one-tailed approach is warranted. However, without an underlying reason or theoretical expectation that the null hypothesis should be rejected on one side as opposed to the other, a two-tailed, nondirectional approach is best.

Application to Geographic Problem

The nearest neighbor methodology is applied to point patterns representing community service sites within a city. Some public services are best located in a highly dispersed pattern to provide relatively equal service distance for all parts of the region. Such a concern for spatial equity is especially important for emergency services, such as police and fire, as well as for neighborhood schools. Other community activities may be sited to offer the region a high degree of efficiency, with less concern for equal spacing of services within the region. Many nonemergency services exhibit more clustered arrangements in their point patterns, possibly reflecting financial rather than spatial constraints.

Four community services in Baltimore, Maryland, are selected for a nearest neighbor analysis of their spacing. Two of the services—police and fire—provide emergency protection from sites of their facilities to various locations within the region. The other two services—elementary education and recreation—provide nonemergency services for persons who

travel to these locations. The facility sites for the four services are shown in figure 12.4.

A nearest neighbor analysis is applied to each pattern to measure the spacing of service sites. The calculation procedure for analyzing the spacing of police facilities is summarized in table 12.2. The nearest neighbor distances (in miles) for each of the 11 sites are shown in figure 12.5. The average nearest neighbor distance for police stations in Baltimore is 1.63 miles. If the stations had been distributed randomly across the city, the average spacing would have been 1.36 miles (table 12.2). To hold the influence of point density constant and allow more useful comparisons of nearest neighbor results, a standardized nearest neighbor index (R) is calculated. This ratio of $\overline{NND}$ to $\overline{NND}_R$ for police stations in Baltimore ($R = 1.20$) suggests that the actual spacing of points is more dispersed than random. Although dispersal of sites is expected for an emergency public service, does the R value of 1.20 differ significantly from 1.0, the result for a random pattern of points?

Equations 12.6 and 12.7 are used to test for a significant difference between an observed nearest neighbor index and the corresponding index value for a random spacing of points. The null hypothesis is that the observed and random nearest neighbor distances are equal. In this problem, a one-tailed (directional) test is logical, since police facilities are hypothesized to have a dispersed locational pattern. Therefore, the alternate hypothesis is that the observed nearest neighbor index is larger than the corresponding random value. The resulting test statistic for police locations in Baltimore ($Z_n = 1.26$; $p = .1038$) indicates that the researcher has less than 90 percent certainty in rejecting the null hypothesis and concluding the spacing of police stations is more dispersed than random.

Inferential interpretations in this problem can be made only with extreme care. It could be argued that the existing pattern of locations is a "natural" sample from the many possible locations where police stations could have been sited. If this argument is not accepted, the inferential assumptions of a random sample of points, independently selected, are not met. The researcher may want to focus on a descriptive comparison of nearest neighbor indices (R-values) for the four community services.

The corresponding nearest neighbor values and test statistic results for each of the four service patterns are shown in table 12.3. The public services are ranked in decreasing size of the R index (from more dispersed to more clustered). The pattern of fire stations in Baltimore is the most dispersed ($R = 1.84$), supporting the assumption of an even distribution of emergency services to meet an equity requirement. The second most dispersed service is elementary

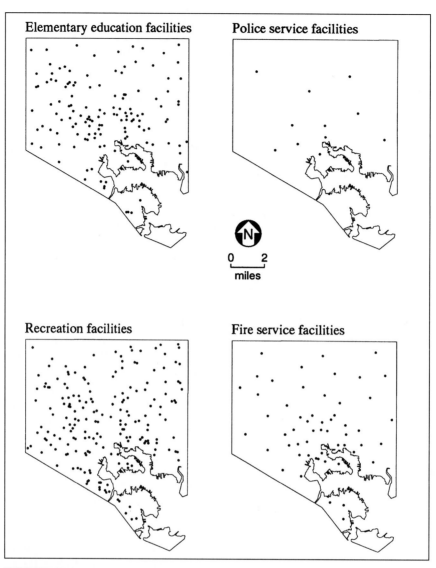

FIGURE 12.4
Location of Selected Public Facilities in Baltimore, Maryland (*Source:* Baltimore City Planning Commission, Dept. of Planning, 1972, 1990.)

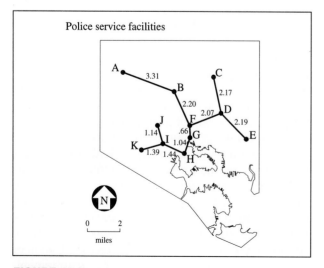

FIGURE 12.5
Nearest Neighbor Distances for Police Service Facilities in Baltimore, Maryland

education facilities ($R = 1.62$). This result also matches the general perception of location strategy, since primary schools are traditionally sited as neighborhood facilities to minimize the travel distance for younger children.

The most clustered spacing pattern among the four services is recreational facilities ($R = 0.85$), the only service where the random nearest neighbor distance exceeds the observed average nearest neighbor distance. Unlike emergency and educational services, recreation does not require an even or dispersed pattern of facilities, which is reflected in its lower R index.

Summary Evaluation

Spacing of facilities for the four selected services in Baltimore confirms the expected arrangements of emergency and nonemergency services (figure 12.6). As hypothesized, the pattern of emergency fire serv-

TABLE 12.2

Worktable for Nearest Neighbor Analysis: Police Service Facilities in Baltimore, Maryland

H_0: $\overline{NND} = \overline{NND}_R$ (point pattern is random)

H_A: $\overline{NND} \neq \overline{NND}_R$ (point pattern is not random)

Calculate mean nearest neighbor distance:

$$\overline{NND} = \frac{\Sigma NND}{n} = \frac{17.97}{11} = 1.63$$

where n = number of points

Calculate random nearest neighbor distance:

$$\overline{NND}_R = \frac{1}{2\sqrt{Density}}$$

where $Density = n/Area$

$$\overline{NND}_R = \frac{1}{2\sqrt{11/80.86}} = \frac{1}{2\sqrt{.136}} = \frac{1}{.737} = 1.36$$

Calculate standardized nearest neighbor distance:

$$R = \frac{\overline{NND}}{\overline{NND}_R} = \frac{1.63}{1.36} = 1.20$$

Calculate test statistic:

$$Z_n = \frac{\overline{NND} - \overline{NND}_R}{\sigma_{\overline{NND}}}$$

where $\sigma_{\overline{NND}} = \dfrac{.26136}{\sqrt{n\,(Density)}}$

$$\sigma_{\overline{NND}} = \frac{.26136}{\sqrt{11(.136)}} = \frac{.26136}{1.223} = 0.214$$

$$Z_n = \frac{1.63 - 1.36}{0.214} = \frac{.27}{.214} = 1.26 \ (p = .1038)$$

TABLE 12.3

Nearest Neighbor Values for Selected Public Facilities in Baltimore, Maryland

Public facility	$\overline{NND}$	Density	$\overline{NND}_R$	R	Z_n	p
Fire	1.12	.680	.61	1.84	11.86	.0000
Elementary education	.63	1.633	.39	1.62	13.33	.0000
Police	1.63	.136	1.36	1.20	1.26	.1038
Recreation	.28	2.325	.33	.85	−3.85	.0001

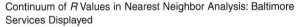

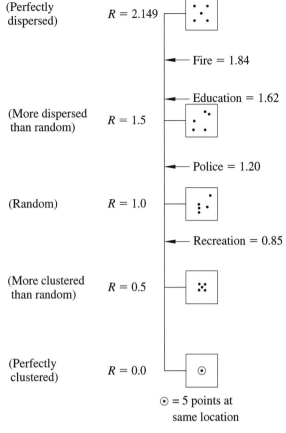

FIGURE 12.6
Continuum of *R* Values in Nearest Neighbor Analysis: Baltimore Services Displayed

ices exhibits the most dispersed spacing, followed closely by the pattern of elementary schools. Also as expected, recreation facilities in Baltimore are more clustered than random. Although police services appear more dispersed than random, statistical evidence suggests that the spacing does not differ significantly from a random pattern.

These results, however, are influenced by the way in which the problem was structured. A critical issue when using the nearest neighbor procedure is the delimitation of the study area boundary. In many research problems (like the analysis of public services in Baltimore), a political boundary defines the study area. In other problems, where no formal or logical boundary exists for enclosing the point locations, the researcher must designate an operational boundary or spatial limit for the region under study. The position of the boundary does not directly affect the distance between nearest neighbors or the average nearest neighbor index. However, boundary position does affect both the area of the study region and the point density, factors that determine the random nearest neighbor distance. Therefore, specification of the study area boundary influences the outcome of a point pattern analysis. Ideally, a functional study area boundary should be defined that is consistent with the type of points being analyzed. Some researchers suggest that the boundary should be defined just beyond the outermost points within the study area.

Another issue related to study area boundary is the specification of nearest neighbors at the edge of a region. In some problems, the nearest neighbor for a point located near the boundary may actually lie outside the study area. Perhaps the nearest fire station or school lies across the border at a shorter distance than the nearest neighbor lying inside the study area. Therefore, how nearest neighbor distances are handled near the study area boundary influences the results of a study. Administrative boundaries do not necessarily offer the best approach to delimiting the study area in point pattern problems.

Quadrat Analysis

An alternate methodology for studying the spatial arrangement of point locations is **quadrat analysis.** Rather than focusing on the spacing of points within a study area, quadrat analysis examines the frequency of points occurring in various parts of the area. A set of quadrats, or square cells, is superimposed on the study area, and the number of points in each cell is determined. By analyzing the distribution of cell frequencies, the point pattern arrangement within the study area can be described.

Whereas nearest neighbor analysis concerns the average spacing of the closest points, quadrat analysis considers the variability in the number of points per cell (figure 12.7). If each of the quadrats contains the same number of points (case 1), the pattern would show no variability in frequencies from cell to cell and would be perfectly dispersed. By contrast, if a wide disparity exists in the number of points per cell for the set of quadrats examined (case 2), the variability of the cell frequencies would be large, and the pattern would display a clustered arrangement. In a third alternative (case 3), the variability of cell frequencies is moderate, and the pattern of points would reflect a random or near random spatial arrangement.

However, the absolute variability of the cell frequencies cannot be used as an effective descriptive measure of the point pattern because it is influenced by the density of points—the mean number of points per cell. This relationship is directly analogous to the influence of the mean on the standard deviation of a variable. Recall from section 3.2 that the coefficient of variation is used for meaningful comparisons of relative variability between distributions. In quadrat analysis, an index known as the **variance-mean ratio (VMR)** standardizes the degree

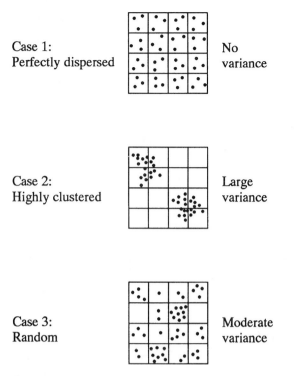

FIGURE 12.7
Quadrat Analysis: The Relationship of Cell Frequency Variability and Point Pattern Form

of variability in cell frequencies relative to the mean cell frequency:

$$VMR = \frac{VAR}{MEAN} \qquad (12.8)$$

where VMR = variance-mean ratio
VAR = variance of the cell frequencies
$MEAN$ = mean cell frequency

The mean and variance are the basic descriptive statistics that summarize the central tendency and variability of a variable. Since the data from quadrat analysis are usually summarized as frequency counts (number of points in each quadrat), formulas for the grouped mean and grouped variance are usually applied (see tables 3.2 and 3.5).

Interpretation of the variance-mean ratio offers insight into the spatial pattern of points within the study area. In a dispersed distribution of points, for example, the cell frequencies will be similar, and variance of cell frequency counts will be very low. In an extreme case (such as case 1 in figure 12.7), if each quadrat contains exactly the same number of points, the variance will be zero, making the variance-mean ratio zero. Conversely, if a point pattern is highly clustered with most cells containing no points and a few cells having many points, the variance of cell frequencies will be large relative to the mean cell frequency. This type of spatial pattern will produce a large value for the variance-mean ratio.

If the set of points is randomly arranged across the cells of the study area, an intermediate value of variance will occur. For a perfectly random point pattern, the variance of the cell frequency is equal to the mean cell frequency. To understand this result, recall from section 5.4 that the Poisson distribution is used to describe the frequency of values for a randomly generated spatial or temporal pattern. In the Poisson distribution, the mean frequency equals the variance of the frequencies. Therefore, when using the variance-mean ratio to describe spatial point patterns, a result close to one (variance equals mean) suggests that the distribution has a random arrangement.

In addition to its use as a descriptive index, the variance-mean ratio can also be applied inferentially to test a distribution for randomness. The test statistic is chi-square, defined as a function of both the *VMR* and number of cells (*m*):

$$\chi^2 = VMR(m - 1) \qquad (12.9)$$

The null hypothesis for this test is expressed as no difference between the observed distribution of points and a distribution of points resulting from a random process (i.e., *VMR* = 1). Either a directional (one-tailed) or nondirectional (two-tailed) approach can be used, depending on the test objective of the alternate hypothesis.

Rejection of the null hypothesis can occur if a point pattern is more clustered than random (*VMR* > 1) or more dispersed than random (*VMR* < 1). The difference test determines whether an observed *VMR* value differs significantly from 1, the theoretical random result. The interpretation of the chi-square test statistic and corresponding *p*-value reflects these two possibilities. A large *VMR* generally produces a larger chi-square value (with a small *p*-value), suggesting greater variability of the cell frequencies and a clustered arrangement of points.

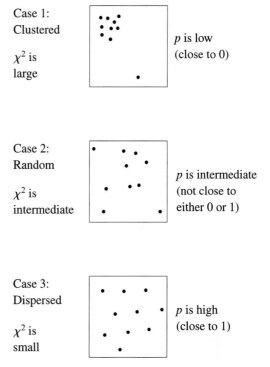

FIGURE 12.8
Quadrat Analysis: The Relationship of Chi-square Value and *p*

Conversely, a small *VMR* produces a smaller chi-square value, indicating lower variability of cell frequencies and a more dispersed distribution of points. In such instances, the corresponding *p*-value is larger. Intermediate chi-square values result from *VMR*s closer to 1, suggesting a lower probability of rejecting the null hypothesis and therefore more likelihood that the distribution of points is actually random. Thus, a continuum of *p*-values from 0 to 1 indicates transition from clustered to random to dispersed (figure 12.8). Either very high or very low *p*-values suggest rejection of the null hypothesis, while intermediate values do not support rejection.

Application to Geographic Problem

The example of Illinois tornado touchdown locations (section 5.4) is reexamined here using quadrat analysis. The objective now is to describe the pattern of points within the study area by analyzing the distribution of cell frequencies. The resulting frequency pattern indicates whether the points have a dispersed, random, or clustered arrangement. The chi-square difference test is then applied to determine whether the observed pattern differs significantly from a theoretical random pattern.

In this analysis, the tornado touchdown points are assumed to be an independent random sample.

Quadrat Analysis

Primary Objective: Determine whether a random (Poisson) process has generated a point pattern

Requirements and Assumptions:

1. Random sample of points from a population
2. Sample points are independently selected

Hypotheses:

H_0: *VMR* = 1 (point pattern is random)
H_A: *VMR* ≠ 1 (point pattern is not random)
H_A: *VMR* > 1 (point pattern is more clustered than random)
H_A: *VMR* < 1 (point pattern is more dispersed than random)

Test Statistic:

$\chi^2 = VMR\ (m{-}1)$

This assumption is justified if the sample is considered a "natural" sample over one time period. This particular set of tornado touchdown sites is one pattern from an infinite number of patterns that could have occurred from 1916 to 1969.

As discussed in section 5.4, 63 of the 85 cells covering the state have at least half of their area in Illinois (figure 12.9). This set of 63 quadrats excludes only 30 of the 480 recorded tornadoes for the state. The mean frequency of 7.14 tornadoes per cell represents the density of points across the study area relative to the quadrat size (table 12.4). The variability in cell frequencies around this mean largely determines the nature of the point pattern. For this set of quadrats, the variance of the tornado frequencies is 12.45 per cell. The two summary statistics produce a *VMR* of 1.74, suggesting a frequency pattern for tornadoes that is on the clustered side of random.

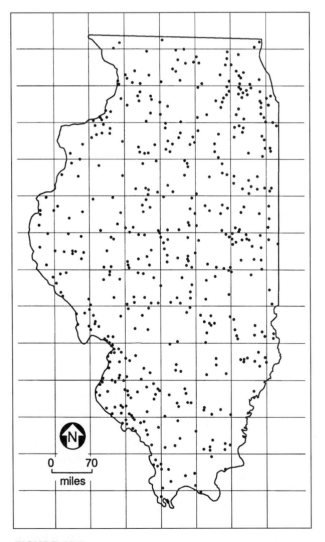

FIGURE 12.9
Illinois Tornado Touchdown Pattern with Quadrats Superimposed

---

TABLE 12.4

Worktable for Quadrat Analysis: Illinois Tornado Example

H_0: *VMR* = 1 (point pattern is random)
H_A: *VMR* ≠ 1 (point pattern is not random)

Calculate mean cell frequency:

$$\text{Mean cell frequency} = \frac{n}{m}$$

where n = number of points
$\quad\quad m$ = number of cells

$$MEAN = \frac{450}{63} = 7.14$$

(See table 5.6 for frequency data.)

Calculate variance of cell frequencies:

$$VAR = \frac{\Sigma f_i X_i^2 - [(\Sigma f_i X_i)^2 / m]}{m - 1}$$

where f_i = frequency of cells with i tornadoes
$\quad\quad X_i$ = number of tornadoes per cell

$$VAR = \frac{3986 - (450^2/63)}{62} = \frac{3986 - 3214.29}{62} = 12.45$$

Calculate variance-mean ratio:

$$VMR = \frac{VAR}{MEAN} = \frac{12.45}{7.14} = 1.74$$

Calculate test statistic:

$$\chi^2 = VMR\,(m - 1) = 1.74(62) = 107.88\ (p = .0003)$$

A difference test is applied to determine whether the *VMR* of 1.74 differs significantly from that of a random pattern (*VMR* = 1). The null hypothesis assumes no difference between the observed and expected *VMR*s. Selection of a directional (one-tailed) or nondirectional (two-tailed) test depends on whether a priori evidence exists to suggest a point pattern that is more clustered than random or more dispersed than random. If no evidence suggests the pattern should be clustered (or dispersed), a nondirectional, two-tailed difference test is appropriate. Conversely, an alternate hypothesis that assumes, a priori, a clustered or dispersed point pattern requires a directional, one-tailed difference test.

Since no climatological evidence suggests that the spatial pattern of tornadoes in Illinois is anything but random, a two-tailed difference test of the null hypothesis seems the logical choice. The resulting chi-square test statistic of 107.88 produces a corresponding *p*-value of .0003 (table 12.4). Thus, the researcher has very strong confidence that the null hypothesis should be rejected and that the resultant point pattern is more clustered than random.

Does the size of quadrats used to cover the study area influence the results from a quadrat analysis? To examine this question, the same pattern of tornadoes in Illinois is analyzed using a cell design with quadrats one-fourth as large. This cell structure produces 263 smaller cells covering 469 of the state's 480 tornadoes (figure 12.10). To be consistent, boundary cells whose area lies more than half outside Illinois are eliminated.

The denser cell network produces more quadrats with low frequencies and a smaller mean cell frequency of 1.78 tornadoes per quadrat. The variance of the cell frequencies about this mean is also lower at 2.31; since the variance remains larger than the mean, the pattern again displays a clustered arrangement.

Standardizing this variance to the mean produces a *VMR* of 1.30, much lower than the value found with the larger quadrat size (*VMR* = 1.74). However, when the ratio is tested for significant difference from *VMR* = 1, the chi-square value of 340.6 produces a very significant result (*p* = .0000). Thus, using the 263 smaller quadrats suggests a very small probability of error when rejecting the null hypothesis of no difference between the observed pattern of points and a random distribution. The researcher can be very confident that the pattern of tornado touchdowns is more clustered than random.

Summary Evaluation

Although quadrat analysis offers a useful approach to studying the spatial arrangement of points, several procedural issues must be considered. Different cell sizes produce varying levels of mean point frequency and variance per cell. Moreover, studies have shown that if the point pattern is held constant but the quadrat size is decreased (i.e., the number of quadrats is increased), the variance of the point frequencies usually declines faster than the mean. This situation suggests that the smaller-sized quadrats often produce smaller *VMR* values. Thus, different cell sizes in a single problem can generally be expected to produce different values of the variance-mean ratio.

Although more research needs to be conducted on optimal quadrat size, two guidelines can be offered. When visual examination of a point pattern reveals distinct clusters, the size of quadrats should match that of the clusters. This cell size allows the results from quadrat analysis to reflect more accurately the visual impression of the pattern. Other researchers have suggested as a general rule of thumb that cell size should be equal to twice the average area per point. In other words, the mean frequency of points should be close to 2.0 per quadrat. This guideline approximates the cell structure used with the 263 cells covering the tornado pattern in Illinois. Taylor (1977) presents a more detailed discussion of cell size in quadrat problems.

Another issue in the use of quadrat analysis also requires some attention. Like nearest neighbor analysis, delimitation of the study area can influence the results of a study. In the use of quadrat analysis, a way to handle cells around the study area boundary needs to be determined. Cells on the boundary will inevitably contain area both inside and outside the study region. In the Illinois tornado problem, cells with more than half of their area outside the study region were eliminated. Alternative decision rules can be used, and these may change the results obtained from the analysis. No matter what decision rule is applied in a particular problem, the geographer needs to be consistent in handling the boundary issue.

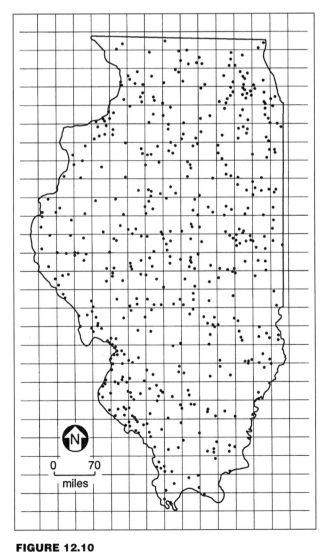

FIGURE 12.10
Illinois Tornado Touchdown Pattern with Smaller Quadrats Superimposed

12.2 AREA PATTERN ANALYSIS

The goals and objectives in **area pattern analysis** are similar to those in point pattern analysis. Just as nearest neighbor and quadrat analysis examine the random or nonrandom nature of point patterns, descriptive and inferential procedures are available to analyze area patterns. A choropleth, or area pattern, map is considered clustered if adjacent contiguous areas tend to have highly similar values or scores. Alternatively, if the values of adjacent areas tend to be dissimilar, then the spatial pattern is considered more dispersed than random (figure 12.1).

Area pattern analysis is appropriate for studying many practical problems in geography. It is particularly valuable to examine the way in which area patterns of a specific variable change over time. For example, suppose a medical geographer is concerned with the diffusion of influenza in a metropolitan area. If the number of cases is reported by census tract for several time periods, the morbidity rate for influenza could be displayed on a series of choropleth maps and the degree of nonrandomness in the patterns analyzed. If the incidence of the disease becomes more dispersed over successive time periods, researchers would be very concerned because the dispersion indicates that the influenza is spreading more evenly throughout the metro area. It could also mean a significant number of cases were appearing in other areas not close to existing areas of high incidence. Conversely, if the morbidity rate patterns were becoming more clustered over successive time periods, then a closer examination of those areas with higher incidence of influenza would be appropriate. Such knowledge could lead to the implementation of effective strategies for disease control.

Suppose a political geographer wants to study the spatial pattern of voter registration rates across a number of precincts within a community. If registration rates have been mapped by precinct before each of the last several elections, a useful comparative analysis of area patterns is possible. If the pattern is becoming more clustered over successive time periods, a different registration strategy may be necessary. An analysis of area pattern trends might also provide insights into demographic or economic variables that seem to be related to the voter registration rate.

Many methods of area pattern analysis are available, depending on the level of data measurement and the way in which the variable under analysis has been organized. Spatial autocorrelation measures are available for variables organized in nominal, ordinal, and interval/ratio form. However, this chapter focuses exclusively on area patterns shown in binary form. Most of the important concepts and theoretical issues dealing with area pattern analysis can be illustrated with binary maps. In addition, variables are frequently converted into binary form for area pattern analysis. Each census tract in the influenza morbidity rate example could be classified as having either an above-average or below-average incidence of influenza. Similarly, each of the community precincts in the voter registration rate example could be classified as above or below average. Those interested in area pattern analysis of spatial data not organized in simple binary form should consult the references cited at the end of the chapter.

The Join Count Statistic

The **join count** is the basic organizational statistic for the analysis of area patterns. A "join" is operationally defined as two areas sharing a common edge or boundary. The procedures involved in calculating a join count statistic are relatively straightforward. The fundamental building block is the number of joins in the pattern and the nature of the join structure in the study area. Figure 12.11 illustrates the simplest situation: a binary classification of data and a two-category choropleth map. Alternative join count procedures are available for groupings of more than two categories. Many variables in geography are binary in nature, or can be downgraded effectively to binary without sacrificing important locational information.

Figure 12.11 shows the join structure for the areas in the study region. For example, area A is joined

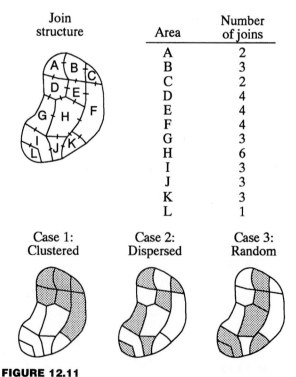

Area	Number of joins
A	2
B	3
C	2
D	4
E	4
F	4
G	3
H	6
I	3
J	3
K	3
L	1

FIGURE 12.11

Join Structure and Examples of Area Patterns Appearing Clustered, Dispersed, and Random to two other areas—*B* and *D*. The number of joins associated with each area is listed adjacent to the join structure map.

To comply with accepted practice, each area can be identified as either "black" or "white." It is then possible to refer to the *observed number* of "black-white," "black-black," and "white-white" joins in the pattern. In case 1 of figure 12.11, the area pattern appears clustered, with relatively few black-white (dissimilar) joins and a rather large number of black-black and white-white (similar) joins. In case 2, a dispersed (alternating checkerboard) pattern has more black-white joins and fewer joins of similar-category areas. With random patterning (case 3), the result is an intermediate number of both similar and dissimilar joins. These situations are all reflected in table 12.5.

To determine if a certain join count distribution has been generated by a random process, the number of black-white joins that would be expected from a theoretical random arrangement of binary observations must be calculated. That is, the *expected number* of black-white joins in a purely random pattern is compared with the one actually observed. If the observed join count is significantly different from the expected join count, the observed pattern can be described as nonrandom.

How is the expected number of black-white joins determined? The locational context in which the area pattern is being evaluated suggests two possible approaches for generating expected join counts. Some area pattern analyses may be conducted under the hypothesis of **free sampling.** Free sampling should be used if the researcher can determine the probability of an area being either black or white based on some theoretical idea or by referring to a larger study area. Once the probability of an area being black or white has been determined, the expected number of black-white joins in a random pattern having those probabilities can also be determined. For example, in the influenza morbidity rate problem, suppose each census tract throughout the metropolitan area is classified as having either above-average incidence of the disease (black) or below-average incidence (white). If the pattern of disease is being analyzed in only one portion of the metropolitan area, logic would suggest that the probability of above- or below-average incidence of influenza should be typical or representative of that found throughout the metro area, and the free sampling hypothesis would be appropriate.

The **nonfree sampling** hypothesis is used when no appropriate reference to a larger study area or general theory is appropriate. In this case, only the study region itself is considered in the analysis, and the expected number of black-white joins is estimated from a random patterning of only those subareas in the study region. For instance, a political geographer studying the spatial pattern of community voter registration rates by precinct may not have any reason to believe this pattern is typical or representative of any larger area (such as the state in which the community is located). Therefore, if it is hypothesized that the area pattern has resulted from the unique characteristics found within the study area, then a nonfree sampling hypothesis is appropriate.

When the researcher cannot refer to a larger study area to determine the expected number of black-white joins, the nonfree sampling context should be used. In many geographic problems, it is difficult to conclude with any degree of confidence that a subarea pattern is typical of a larger area. Therefore, use of the expected number of black-white joins from a nonfree sampling hypothesis frees the researcher from having to make a restrictive assumption. The calculation procedure for area pattern analysis is shown in this text for only the less restrictive, nonfree sampling option.

Nonfree Sampling Test Procedure

To determine if an area pattern has been generated by a random process, the observed number of black-white joins must be compared with the number of such joins expected in a random arrangement. In this framework, the null hypothesis is that the observed number of black-white joins is equal to the expected number of black-white joins in a random area pattern. If reason exists to hypothesize that the observed area pattern is nonrandom in a certain direction, then a one-tailed alternate hypothesis is appropriate. For instance, an area pattern could be hypothesized either

---

TABLE 12.5

Join Counts Associated with Area Patterns

Case	Total joins*	Dissimilar areas joined	Similar areas joined		
		Black-white joins	Black-black joins	White-white joins	Total
1 (Clustered)	19	5	7	7	14
2 (Dispersed)	19	15	0	4	4
3 (Random)	19	12	4	3	7

*See figure 12.11 for map of join structure and area patterns.

as more clustered than random or more dispersed than random. Without an a priori reason to hypothesize a clustered or dispersed pattern, a two-tailed alternate hypothesis is selected, and the conclusion will be that an area pattern is random or nonrandom.

The test statistic (Z_b) for nonfree sampling is

$$Z_b = \frac{O_{BW} - E_{BW}}{\sigma_{BW}} \quad (12.10)$$

where O_{BW} = observed number of black-white joins

E_{BW} = expected number of black-white joins

σ_{BW} = standard error of the expected number of black-white joins

In nonfree sampling, the expected number of black-white joins is calculated by incorporating the observed number of black areas and white areas directly into the equation:

$$E_{BW} = \frac{2JBW}{N(N-1)} \quad (12.11)$$

where E_{BW} = expected number of black-white joins

J = total number of joins

B = number of black areas

W = number of white areas

N = total number of areas (black plus white) = $B + W$

The expected number of black-white joins in the study region is determined by asking how many black-white joins would occur from a theoretical random pattern containing the same number of black and white areas as the observed area pattern. The question becomes how many black-white joins would be generated from a random arrangement of the observed number of black and white areas in the area pattern.

Certain generalizations can be made concerning the expected number of black-white joins. Given a particular total number of areas ($N = B + W$), the value of the denominator in equation 12.11 is constant. If most of the areas in the observed pattern are either black or white, then their product (BW) in the numerator will be a smaller number, producing a smaller expected number of black-white joins. Conversely, if about the same number of black and white areas are found in the observed pattern, then their product will be a larger number, producing a larger expected number of black-white joins. These results seem logical, for a pattern with a large number of black areas and very few white areas (or vice versa) would not be expected to have a large number of black-white joins. The expected number of black-white joins is maximized when an equal number of black and white areas occur in the overall pattern.

Not all randomly generated patterns will contain exactly the same number of black-white joins, so a measure of the amount of variability expected as a result of sampling must be included. The standard error of expected black-white joins is:

$$\sigma_{BW} = \left[E_{BW} + \frac{\Sigma L(L-1)BW}{N(N-1)} \right.$$
$$\left. + \frac{4[J(J-1) - \Sigma L(L-1)]B(B-1)W(W-1)}{N(N-1)(N-2)(N-3)} \right.$$
$$\left. - E_{BW}^2 \right]^{1/2} \quad (12.12)$$

where σ_{BW} = standard error of the expected number of black-white joins

J = total number of joins

ΣL = total number of links = $2J$

B = number of black areas

W = number of white areas

N = total number of areas (black plus white) = $B + W$

Note: $\Sigma L = 2J$: If areas A and B are joined, then A is "linked" to B and B is "linked" to A, making the sum of all links twice the number of joins.

The area pattern may now be analyzed with the nonfree sampling hypothesis (equation 12.10 with supplementary equations 12.11 and 12.12). If the observed number of black-white joins is larger than the expected number of black-white joins, then Z_b will be a positive value, indicating a pattern more dispersed than random. A relatively large number of black-white

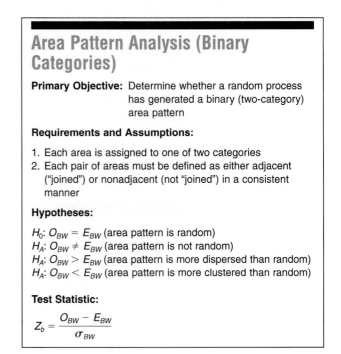

Area Pattern Analysis (Binary Categories)

Primary Objective: Determine whether a random process has generated a binary (two-category) area pattern

Requirements and Assumptions:

1. Each area is assigned to one of two categories
2. Each pair of areas must be defined as either adjacent ("joined") or nonadjacent (not "joined") in a consistent manner

Hypotheses:

H_0: $O_{BW} = E_{BW}$ (area pattern is random)
H_A: $O_{BW} \neq E_{BW}$ (area pattern is not random)
H_A: $O_{BW} > E_{BW}$ (area pattern is more dispersed than random)
H_A: $O_{BW} < E_{BW}$ (area pattern is more clustered than random)

Test Statistic:

$$Z_b = \frac{O_{BW} - E_{BW}}{\sigma_{BW}}$$

(dissimilar) joins will occur in a dispersed pattern (figure 12.11). Conversely, if the observed number of black-white joins is smaller than the expected number of such joins, then Z_b will be negative, and the pattern will appear more clustered than random. The greater the magnitude of Z_b (either positive or negative), the greater the likelihood that the area pattern being analyzed is not random. Just like other Z-score difference tests, this Z_b test statistic for area pattern analysis indicates the number of standard deviations separating the expected random black-white join count from the observed value.

Application to Geographic Problem

Suppose a geographer wants to examine the changing spatial pattern of farm acreage across the conterminous, continental United States over the last century. During this time, has U.S. farm acreage become more clustered or concentrated in certain portions of the country or become more evenly distributed or dispersed throughout the country?

One way this question could be approached is through area pattern analysis. At any particular time it is possible to calculate the average farm acreage per state and subsequently allocate each state to one of two categories—above-average farm acreage or below-average farm acreage. Using this simple binary classification of states, area pattern maps have been constructed for 1890, 1940, and 1987 (figures 12.12, 12.13, and 12.14, respectively).

Before proceeding with the area pattern analysis itself, a little explanatory background information is needed. First, the study area has been designated as the "lower 48" states (excluding Alaska and Hawaii, since these two states are not adjacent or "joined" to any other states). Therefore, when calculating whether a particular state has above-average or below-average farm acreage, the calculation is based on the total farm acreage of only the lower 48 states. Second, the year 1890 is selected as the earliest date in this example problem because that is the first year for which data on farm acreage are available for all 48 states. Strictly speaking, not all 48 states were actually states in 1890; nevertheless, farm acreage data are available for these territories. (Historical geographers will know the following territories earned statehood after 1890: Utah (1896), Oklahoma (1907), Arizona (1912), and New Mexico (1912)). Third, the total farm acreage in the United States increased significantly from 1890 to 1940, but decreased slightly from 1940 to 1987 (table 12.6). Since average farm acreage per state is directly calculated from this total,

FIGURE 12.12
State-level Farm Acreage, 1890

FIGURE 12.13
State-level Farm Acreage, 1940

FIGURE 12.14
State-level Farm Acreage, 1987

TABLE 12.6

Summary of U.S. Farm Acreage Data*

Year	Total farmland (in 1,000 acres)	Average farmland per state (in 1,000 acres)	Number of states having above-average farmland
1890	623,219	12,984	24
1940	1,060,852	22,101	19
1987	961,722	20,036	16

*Data include only the 48 conterminous states.

these values will change concurrently. Finally, note in table 12.6 that the number of states having above-average farm acreage decreases from 24 in 1890, to 19 in 1940, then to just 16 in 1987. This trend suggests that farm acreage is increasingly concentrated in fewer states. Area pattern analysis can now be applied to determine if those few states having large farm acreages are more clustered than random or more dispersed than random.

Across the conterminous United States there are 210 links and 105 joins (recall from equation 12.12 that $\Sigma L = 2J$), and table 12.7 shows the state linkage pattern needed for the area pattern analysis. This general state linkage pattern is appropriate for all

three dates of analysis (1890, 1940, and 1987), although the detailed calculation procedure is now shown only for the 1890 pattern (figure 12.12).

In this example, a "black-white" or dissimilar join occurs when an above-average farm acreage state shares a common boundary with a state with below-average farm acreage. Conversely, a similar join occurs when either two above-average farm acreage states share a common boundary or two below-average farm acreage states share a common boundary. On the 1890 map, 36 of the 105 joins are dissimilar, while the other 69 joins are similar. The research task is to determine if the observed number of dissimilar joins is significantly different from

TABLE 12.7

State Linkage Pattern

State	Number of links (L)	L−1	L (L−1)	State	Number of links (L)	L−1	L (L−1)
Alabama	4	3	12	Nevada	5	4	20
Arizona	4	3	12	New Hampshire	3	2	6
Arkansas	6	5	39	New Jersey	3	2	6
California	3	2	6	New Mexico	4	3	12
Colorado	6	5	30	New York	5	4	20
Connecticut	3	2	6	North Carolina	4	3	12
Delaware	3	2	6	North Dakota	3	2	6
Florida	2	1	2	Ohio	5	4	20
Georgia	5	4	20	Oklahoma	6	5	30
Idaho	6	5	30	Oregon	4	3	56
Illinois	5	4	20	Pennsylvania	6	5	30
Indiana	4	3	12	Rhode Island	2	1	2
Iowa	6	5	30	South Carolina	2	1	2
Kansas	4	3	12	South Dakota	6	5	30
Kentucky	7	6	42	Tennessee	8	7	56
Louisiana	3	2	6	Texas	4	3	12
Maine	1	0	0	Utah	5	4	20
Maryland	4	3	12	Vermont	3	2	6
Massachusetts	5	4	20	Virginia	5	4	20
Michigan	3	2	6	Washington	2	1	2
Minnesota	4	3	12	West Virginia	5	4	20
Mississippi	4	3	12	Wisconsin	4	3	12
Missouri	8	7	56	Wyoming	6	5	30
Montana	4	3	12				
Nebraska	6	5	30	Total	210		822

the number of dissimilar joins expected in a random farm acreage pattern.

The worktable for the 1890 area pattern analysis is presented in table 12.8. The expected number of dissimilar joins in a purely random pattern is calculated by incorporating the observed number of states with above-average and below-average farm acreages directly into the equation. The result indicates that nearly 54 (53.62) dissimilar joins are expected in a random pattern.

To determine whether this contrast between the number of observed dissimilar joins (36) and the number of expected dissimilar joins (53.62) is statistically significant, the standard error of expected dissimilar joins (σ_{BW}) must be incorporated into the test statistic. The standard error of expected dissimilar joins is calculated as 4.89 (table 12.8). The resultant test statistic value (Z_b) is -3.60, with an associated p-value of .0004. These statistics indicate a highly significant tendency toward clustering of states having above-average (and below-average) farm acreages.

Using a similar statistical test procedure, analyses of the 1940 and 1987 patterns (figures 12.13 and 12.14) have also been made. The results of these analyses are summarized in table 12.9, and a definite trend emerges toward increasingly clustered farm acreage in the United States. By 1987, above-average farm acreage is concentrated in only 16 states, and these 16 states are very strongly clustered west of the Mississippi River (figure 12.14). This century-long trend toward clustering of U.S. farm acreage is confirmed by the increasingly large negative Z_b values (reaching -6.11 by 1987) and the associated highly significant p-values (.0000 for both 1940 and 1987).

Many geographers should be able to offer tentative hypotheses related to *why* farm acreage in the United States is becoming increasingly clustered. Agriculture is an evermore competitive enterprise. Those locations having the best natural settings to support food production and best able to take advantage of economies of scale are the locations likely to be most successful. With suburbanization, exurbanization, and low-density sprawl replacing large tracts of agricultural land, only those locations somewhat removed from these regions of sprawl *and* with reasonably high-profit potential are going to be able to survive. Farms located too near areas of metropolitan growth, with nearby competitive land uses that

---

TABLE 12.8

Worktable for Area Pattern Analysis: Farmland Acreage in Conterminous United States

H_0: $O_{BW} = E_{BW}$ (area pattern is random)
H_A: $O_{BW} \neq E_{BW}$ (area pattern is not random)

number of "black" states = 24
number of "white" states = 24

$O_{BW} = 36$

$$E_{BW} = \frac{2JBW}{N(N-1)} = \frac{2(105)(24)(24)}{48(47)} = \frac{120{,}960}{2256} = 53.62$$

$$\sigma_{BW} = \sqrt{E_{BW} + \frac{\Sigma L(L-1)BW}{N(N-1)} + \frac{4[J(J-1) - \Sigma L(L-1)]B(B-1)W(W-1)}{N(N-1)(N-2)(N-3)} - E_{BW}^2}$$

$$= \sqrt{53.62 + \frac{(822)(24)(24)}{48(47)} + \frac{4[105(104) - 822]24(23)24(23)}{48(47)(46)(45)} - (53.62)^2}$$

$$= \sqrt{53.62 + \frac{473472}{2256} + \frac{4[10920 - 822]304704}{4669920} - 2875.1044}$$

$$= \sqrt{53.62 + 209.87 + 2635.51 - 2875.1044}$$

$$= \sqrt{23.9} = 4.89$$

$$Z_b = \frac{O_{BW} - E_{BW}}{\sigma_{BW}} = \frac{36 - 53.62}{4.89} = \frac{-17.62}{4.89} = -3.60 \quad p\text{-value} = .0004$$

TABLE 12.9

Results from Area Pattern Analysis: Trends in U.S. Farm Acreage, 1890, 1940, and 1987

Year	Z_b	p-value
1890	−3.60	.0002
1940	−5.65	.0000
1987	−6.11	.0000

have the potential to generate higher values than agriculture become marginal economic enterprises and have been disappearing from the American landscape at an alarming rate.

These economic trends and changes in land-use patterning have predictable geographic implications. Farm acreage was plentiful in the eastern half of the United States a century ago (look again at the pattern shown in figure 12.12). However, population growth, extensive sprawl, and less than optimal natural settings (e.g., less fertile soils) in this part of the country have sharply reduced the number of large competitive farms. Note on the 1987 map that Illinois is the only state east of the Mississippi River that has above-average farm acreage. Most U.S. farm acreage today is located in the regions of the Rocky Mountains, Great Plains, and Southwest. These areas are generally distant from the most dramatic centers of population growth and sprawl.

Summary Evaluation

As with the point pattern analyses presented earlier in the chapter, caution must be taken when making inferential (p-value) interpretations in this problem. The "natural" sampling argument may again be valid. However, more emphasis should be placed on the descriptive result (clear tendency toward clustering) and the geographic reasons for this trend than on the magnitude of the resulting p-value itself.

A closer examination of the operational definition of "farm" is needed. Although individual state farm acreages have not been provided in this example problem, a rank ordering of the five states with the largest farm acreages in 1987 (Texas, Montana, New Mexico, Nebraska, and South Dakota) reveals a potential validity problem. If a very extensive (large acreage) ranch in semiarid western Texas is classified as a "farm," there may be only a slight relationship between farm acreage and quantity or value of farm production. It is necessary for the geographer to ask: "Just what is being measured and evaluated spatially when looking at the locational pattern of farm acreage?"

Yet another evaluative issue has to do with the fact that a descriptive statistic (mean or average farm acreage) is calculated from spatial data. Some of the issues discussed in chapter 3 (section 3.4) are probably relevant here. For example, there would seem to be little question that boundary delineation has an impact on the farm acreage statistics. States in the United States are really internal subarea boundaries, and it is clear that the total areas of many western states are much larger than the areas of many eastern states. This nonrandom distribution of state size has an immediate and obvious impact on the farm acreage figures that are possible in each state. For example, all other things being equal, one would expect farm acreage in Montana to be much larger than farm acreage in Rhode Island.

KEY TERMS

area pattern analysis, 182
clustered, dispersed, and random patterns, 171
free and nonfree sampling, 183
join count statistic, 182
nearest neighbor analysis, 172
quadrat analysis, 178
spatial autocorrelation (positive and negative), 172
standardized nearest neighbor index (R), 174
variance-mean ratio (VMR), 178

MAJOR GOALS AND OBJECTIVES

If you have mastered the material in this chapter, you should now be able to do the following.
1. Understand the concepts of clustering, dispersion, and randomness in point and area patterns.
2. Recognize situations in which point and area patterns appear clustered, dispersed, or random.
3. Explain the relationships between spatial autocorrelation and types of point and area locational patterns.
4. Identify the types of geographic research problems or situations for which the application of nearest neighbor analysis and quadrat analysis is appropriate.
5. Explain the logic and procedures used in nearest neighbor analysis and distinguish between the logic and procedures used in quadrat analysis.
6. Understand the purposes of area pattern analysis.
7. Define and explain join structure and join count statistic in area pattern analysis.
8. Distinguish between the locational contexts or problem settings in which the hypotheses of free sampling and nonfree sampling are best used.
9. Formulate geographic research problems or situations for which a nonbinary type of area pattern analysis would be needed. Create area pattern examples with various measurement scales.

REFERENCES AND ADDITIONAL READING

Cliff, A. and J. Ord. 1973. *Spatial Autocorrelation*. London: Pion.

Ebdon, D. 1985. *Statistics in Geography. A Practical Approach* (2nd edition). Oxford: Basil Blackwell.

Getis, A. 1964. "Temporal Analysis of Land Use Patterns with Nearest Neighbor and Quadrat Methods." *Annals, Association of American Geographers* 54:391–99.

Goodchild, M. 1988. *Spatial Autocorrelation* (CATMOG series, No. 47). Norwich, England: Geo Books.

Gregory, S. 1978. *Statistical Methods and the Geographer*. London: Longman.

Griffith, D. A. 1988. *Spatial Autocorrelation: A Primer*. Washington D.C.: Assoc. of Amer. Geographers.

Griffith, D. and C. Amrhein. 1991. *Statistical Analysis for Geographers*. Englewood Cliffs, NJ: Prentice-Hall.

Haggett, P., A. Cliff, and A. Frey. 1977. *Locational Models*. London: Edward Arnold.

Odland, J. 1988. *Spatial Autocorrelation*. (Volume 9 in Scientific Geography Series). Beverly Hills, CA: Sage.

Odland, J., R. G. Golledge, and P. A. Rogerson. 1989. "Mathematical and Statistical Analysis in Human Geography," in G. L. Gaile and C. L. Willmott (editors), *Geography in America*. Columbus, OH: Merrill.

Silk, J. 1979. *Statistical Concepts in Geography*. London: G. Allen and Unwin.

Taylor, P. 1977. *Quantitative Methods in Geography: An Introduction to Spatial Analysis*. Boston: Houghton, Mifflin.

Thomas, R. 1977. *An Introduction to Quadrat Analysis* (CATMOG series, No. 12). Norwich, England: Geo Books.

PART FIVE

STATISTICAL
RELATIONSHIP
BETWEEN VARIABLES

Correlation

One of the more important concerns in geographic analysis is the study of relationships between spatial variables. Many geographic studies involve mapping variables and determining the degree of relationship between two or more map patterns. Using visual comparison of maps to measure correspondence or association is subjective because only a general impression of the relationship is gained. Two persons can view the same maps and interpret their association very differently.

Suppose a geographer is investigating urban crime problems within a city. Research and empirical studies on crime patterns from numerous urban areas suggest that a strong underlying association has developed between murders and the presence of drug activity. Although urban researchers from many disciplines are working to understand the complex relationship between these two social problems, the focus of geographic inquiry is often to establish the **spatial association** between the two variables. This general relationship can be seen by comparing a map of homicides to one of known drug sale points in Washington, D.C. (figure 13.1). Although visual examination appears to support the link between murders and drug activity, comparison of the two maps is extremely subjective. An accurate, unbiased estimate of the degree of association between the patterns on the two maps is difficult to obtain.

Correlation analysis provides a more objective, quantitative means to measure the association between a pair of spatial variables. Both the direction and strength of association between the two variables can be determined statistically. In the previous example, the degree of spatial association between

homicides and drug activity could be established more precisely using correlation. Moreover, the technique can be applied either directly to numerical data or to mapped information converted to numerical form.

In section 13.1, the nature of correlation analysis is discussed. The scattergram is highlighted as an analytic tool used to study both the direction and strength of association. Several indices used to measure correlation or association between variables are defined and explained, and each index is illustrated with geographic examples.

The most widely used index of correlation, Pearson's correlation coefficient, is discussed in section 13.2, and examples of association for interval/ratio-level data are included. Other geographic studies involve the use of ordinal or rank-order data. Section 13.3 presents Spearman's correlation index, the most widely used coefficient for measuring association of ordinal data. The use of correlation for measuring map association with examples of dot, choropleth, and isoline maps is explored in section 13.4. The last section of the chapter examines some correlation issues of particular interest to geographers.

13.1 THE NATURE OF CORRELATION

A geographic investigation often begins with graphic display of the data. A common tool for portraying the relationship or association between two variables is a two-dimensional graph called a **scattergram**, or **scatterplot** (see section 2.5). Three examples of scattergrams are shown in figure 13.2. With one variable plotted on each axis, the pattern of

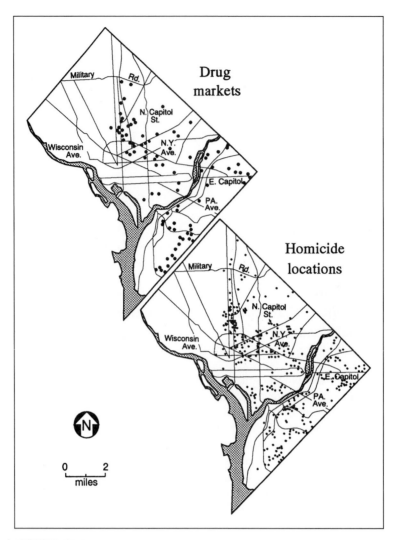

FIGURE 13.1
Homicides and Drug Sale Points in Washington, D.C. (*Source:* Redrawn from maps in *The Washington Post,* January 13, 1989, p. E1, with permission.)

points in a scattergram helps to provide an understanding of the nature of a particular relationship.

Two types of information—direction and strength of association—can be identified from the scattergram. If the general trend of points is from lower left to upper right (figure 13.2, case 1), the **direction of association** is **positive.** In a positive or direct relationship, a larger value in one variable generally corresponds to a larger value in the second variable; alternatively, a smaller value in the first variable usually coincides with a smaller value in the second variable. Such a correspondence will result in a positive correlation.

In geography, a positive correlation is found with many variables related to population size. For example, the association between population and number of retail functions in a sample of settlements usually exhibits a positive or direct relationship. As demonstrated in central place theory, settlements with more people at higher levels of the urban hierarchy generally contain more retail establishments.

The direction of relationship is **negative** if the general trend of points in the scattergram is from upper left to lower right (figure 13.2, case 2). When two variables have a negative or inverse correlation, a larger value in the first variable is generally associated with a smaller value in the second variable; alternatively, a smaller value in the first variable usually corresponds with a larger value in the second variable.

A negative correlation between two variables is clearly illustrated by using the general principle of "distance decay." In such relationships, often studied as examples of "contagious diffusion," some phenomenon or idea declines with increasing distance from a source or origin. Using pollution data from a sample of monitoring sites, the level of air pollution and the distance from the pollution source

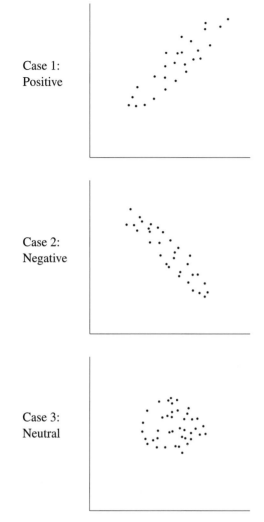

FIGURE 13.2
Generalized Scattergrams Showing Directional Relationships

ables. There may be no logical rationale for expecting a correlation between house size and age of unit.

The **strength of association** between two variables can be determined by examining the amount of point spread in a scattergram. If the point pattern is tightly packed (as in a cigar-shaped pattern), the relationship is said to be strong (figure 13.3, case 2). However, if the points are more widely spread or dispersed (as in a football-shaped pattern), the association is moderate to weak (figure 13.3, case 3). The strongest association, or perfect correlation, occurs when the points are in a straight line or linear pattern (figure 13.3, case 1); the weakest association occurs when the points are distributed with no pattern or association, as in a circular arrangement (figure 13.3, case 4).

Any two variables can be correlated, and the strength and direction of relationship calculated. However, extreme caution must be used when evaluating or interpreting correlations. A relationship or association between variables does not necessarily imply the existence of a cause and effect relationship. For example, a geographer could correlate annual precipitation and pH level of the soil at a sample of locations. Even if a nonzero correlation is derived, it is highly unlikely that any type of causal relationship exists between these variables.

Although the visual examination of a scattergram is a useful way to begin a geographic investigation, the direction and strength of association between

would probably exhibit a negative or inverse relationship. As one moves farther downwind from a site of pollution, the level of exposure to the pollutant declines.

Some variables may not exhibit either a positive or negative relationship. If the pattern of points in the scattergram is **random** (figure 13.2, case 3), no association exists between the two variables. In such examples, the values of one variable are not associated in any nonrandom way with the values of the other. This is sometimes referred to as a **neutral** relationship.

In geographic research, variables may show no relationship or pattern on a scattergram. Suppose a geographer studying housing characteristics in a city wishes to determine if an association exists between house size (square feet of living area) and the number of years since construction. It is unlikely that a significant association exists between these two vari-

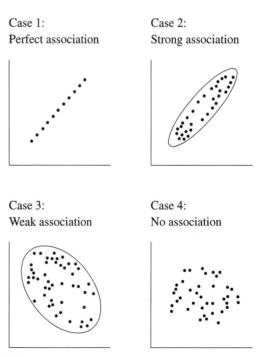

FIGURE 13.3
Generalized Scattergrams Showing Strength of Association

two variables can be determined only in general, often subjective terms. In geographic studies, a more objective, rigorous method is needed for determining the strength of relationship or degree of association between two variables. Statisticians have defined various indices, usually called "correlation coefficients," to measure the strength of relationships. Most of these coefficients are constructed to have a maximum value of 1.0, which indicates perfect positive or direct correlation between variables. A minimum value of −1.0 represents a perfect negative or inverse correlation, and a value of 0.0 is used to denote no correlation or association between variables. Thus, like a scattergram, correlation coefficients can indicate both the direction of the relationship (positive or negative) as well as the strength of association.

As in other areas of inferential statistical analysis, the level of measurement (nominal, ordinal, interval/ratio) largely determines which index of correlation or association is applied to a problem. The most commonly used correlation coefficients are Pearson's product-moment for interval or ratio data and Spearman's rank-order for ordinal or ranked data. Indices to measure association in categorical data (such as measuring the strength of association in a contingency table) are less frequently used in geographic problem solving and will not be discussed here.

13.2 ASSOCIATION OF INTERVAL/RATIO VARIABLES

The most powerful and widely used index to measure the association or correlation between two variables is **Pearson's product-moment correlation coefficient.** In fact, when the term "correlation" appears in geographic literature, it is often assumed that the writer is referring to Pearson's correlation index. To use this measure of association, data must be of interval or ratio scale. It is also assumed that the variables have a linear relationship. In addition, if the index is used in an inferential rather than descriptive manner, both variables should be derived from normally distributed populations.

Pearson's correlation coefficient relates closely to the statistical concept of **covariation**—the degree to which two variables "covary" (vary together or jointly). If the values of the two variables covary in a similar manner, the data contain a large covariation, and the two variables will show strong correlation. Alternatively, if the paired values of the variables show little consistency in how they covary, the correlation will be very weak.

The concept of covariation and its relationship to correlation can be more easily understood by comparing a set of scattergrams, each having four quadrants produced from the mean values of the X (horizontal) and Y (vertical) variables (figure 13.4). In each scattergram, the mean values and total variation in the two variables are held constant. This allows a direct comparison of the relative covariation, which differs in each scattergram.

An understanding of covariation begins first with the deviations of $X(X - \overline{X})$ and $Y(Y - \overline{Y})$ from their respective means. As discussed earlier, these deviations are basic building blocks of standard deviation and standardized scores. The X and Y deviations of each data value (matched pair) are multiplied together and summed for the set of values to produce

$$CV_{XY} = \Sigma(X - \overline{X})(Y - \overline{Y}) \qquad (13.1)$$

where CV_{XY} = covariation between X and Y
$(X - \overline{X})$ = deviation of X from its mean $(\overline{X})$
$(Y - \overline{Y})$ = deviation of Y from its mean $(\overline{Y})$

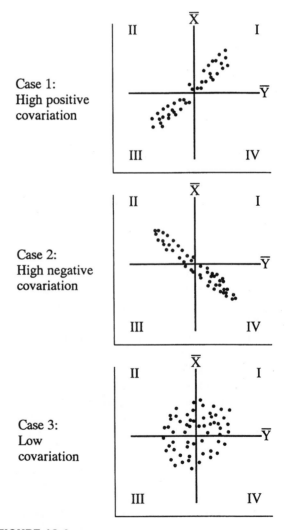

Case 1: High positive covariation

Case 2: High negative covariation

Case 3: Low covariation

FIGURE 13.4
Generalized Scattergrams Showing the Relationship of Covariation to Correlation

Mathematically, covariation is analogous to the important concept of total variation—the sum of the squared deviations from the mean—an integral component of analysis of variance and regression:

$$TV_X = \Sigma(X - \overline{X})(X - \overline{X}) = \Sigma(X - \overline{X})^2 \quad (13.2)$$

$$TV_Y = \Sigma(Y - \overline{Y})(Y - \overline{Y}) = \Sigma(Y - \overline{Y})^2 \quad (13.3)$$

where TV_X = total variation in X
TV_Y = total variation in Y

The point patterns in each scattergram of figure 13.4 represent different examples of covariation. In case 1, virtually all points in the scattergram lie in quadrants I and III. In quadrant I, the deviations of $X(X - \overline{X})$ and $Y(Y - \overline{Y})$ are both positive, since each X and Y value is greater than its respective mean. In quadrant III, where most other points in case 1 are located, the X and Y deviations are both negative, since all X and Y values are less than their respective means. Because the X and Y values for each point in quadrants I and III covary in the same direction from their means, the product of the two deviations will produce a *positive* result for each point. When these individual products are summed according to equation 13.1, the resultant value will be large and positive. Thus, case 1 represents an example of a large positive covariation. As seen earlier, this scattergram also corresponds to a positive or direct correlation between the two variables.

The situation is different for the scattergram in figure 13.4, case 2, where most points lie in quadrants II and IV. In quadrant IV, deviations in X are positive (X values are greater than $\overline{X}$). However, deviations in Y are negative because values of Y are less than $\overline{Y}$. A reversed situation occurs in quadrant II, where deviations in X are negative (X values are less than $\overline{X}$) and deviations in Y are positive. In case 2, the product of the deviations will be *negative* because the values of the X and Y variables in both quadrants II and IV covary in opposite directions from their means. When the products from the X and Y deviations are summed, the resultant covariation will be large and negative. The example of covariation in case 2 represents two variables that have a large, inverse correlation.

In the third scattergram (figure 13.4, case 3), points are scattered in a random pattern, with nearly equal dispersal of points in each quadrant. Some of the X and Y deviations will be positive and others negative. When the deviations are multiplied together for each unit of data, both positive and negative products will occur. When these products are summed for all points, the values generally cancel each other out and produce a covariation close to zero. This low covariation corresponds to a very

small correlation, suggesting little or no relationship between the two variables.

Pearson's correlation coefficient (r) can be expressed mathematically in several different ways:

1. With deviations from the mean and standard deviations (equation 13.4)
2. With X and Y values transformed to Z-scores (equation 13.5)
3. With the original values of the X and Y variables (equation 13.6)

Any of these formulas produces an equivalent result.

Based on the conceptual definition, Pearson's correlation is expressed as the ratio of the covariance in X and Y to the product of the standard deviations of the two variables:

$$r = \frac{[\Sigma(X - \overline{X})(Y - \overline{Y})]/N}{S_X S_Y} \quad (13.4)$$

where r = Pearson's correlation coefficient
N = number of paired data values
S_X, S_Y = standard deviation of X and Y, respectively

If the data are converted to standardized or Z-score form (section 5.5), an alternative formula exists to calculate the Pearson's correlation coefficient:

$$r = \frac{\Sigma[Z_X Z_Y]}{N} \quad (13.5)$$

where Z_X = X variable transformed to Z-score
$= \dfrac{(X - \overline{X})}{S_X}$

Z_Y = Y variable transformed to Z-score
$= \dfrac{(Y - \overline{Y})}{S_Y}$

N = number of paired data values

Using equation 13.5, the correlation between two variables is equal to the sum of the product of Z-scores for each data value divided by the number of paired values. This formula is valid because Z-scores take into account deviations from the mean and the standard deviation.

If one wishes to use the original values of the X and Y variables directly, a computational formula is used to derive Pearson's correlation coefficient:

$$r = \frac{\Sigma XY - ((\Sigma X)(\Sigma Y)/N)}{\sqrt{[\Sigma X^2 - ((\Sigma X)^2/N)]}\sqrt{[\Sigma Y^2 - ((\Sigma Y)^2/N)]}}$$

$$(13.6)$$

Although this formula appears more complex than the previous equations, it uses only the original data, and no prior calculation of means, standard deviations, or deviations from the mean is needed.

In addition to its use as a descriptive index of the strength and direction of association, correlation can also be used to infer results from a sample to a population. The sample correlation coefficient (r) is the best estimator of the population correlation coefficient (ρ). In this application, the null hypothesis is that no correlation exists in the populations of the two variables (H_0: $\rho = 0$). Since the populations of the variables are assumed to have a linear association (see boxed insert), the null hypothesis of no correlation is equivalent to stating that X and Y are independent.

As in most other inferential tests, the alternate hypothesis (H_A) can be stated as directional (one-tailed) or nondirectional (two-tailed). In the one-tailed approach, the researcher has a logical basis for expecting the correlation to be either positive or negative. When no rationale exists for the direction of correlation between two variables, a two-tailed alternate hypothesis should be used.

Although either the t or Z distributions are used to evaluate the significance of a Pearson's correlation coefficient r, the most common approach uses t. Since a Z distribution is appropriate only when sample size exceeds 30, the use of t for testing correlation coefficients offers more flexibility for problems with various sample sizes. The most common test statistic is

$$t = \frac{r}{s_r} \qquad (13.7)$$

where s_r = standard error estimate = $\sqrt{\dfrac{1 - r^2}{n - 2}}$ (13.8)

Pearson Correlation Analysis

Primary Objective: Determine if an association exists between two variables

Requirements and Assumptions:

1. Random sample of paired variables
2. Variables have a linear association
3. Variables are measured at interval or ratio scale
4. Variables are bivariate normally distributed

Hypotheses:

H_0: $\rho = 0$
H_A: $\rho \neq 0$ (two-tailed)
H_A: $\rho > 0$ (one-tailed) or
H_A: $\rho < 0$ (one-tailed)

Test Statistic:

$t = \dfrac{r\sqrt{n - 2}}{\sqrt{1 - r^2}}$

The test statistic for Pearson's r can therefore be rewritten as

$$t = \frac{r\sqrt{n - 2}}{\sqrt{1 - r^2}} \qquad (13.9)$$

Geographic Example of Pearson's Correlation Coefficient

The use of Pearson's product-moment correlation coefficient in geographic analysis is demonstrated with a climatological example. One of the mechanisms influencing precipitation on land areas is the presence of nearby large bodies of water. These water bodies provide an important source of moisture that produces sizable amounts of precipitation when brought over a land area by prevailing winds. However, as distance from the body of water increases, the water's influence decreases, and precipitation levels generally decline. Thus, areas adjacent to water tend to have higher amounts of precipitation than do areas farther away.

An interesting example of this land-water relationship occurs in regions adjacent to the southern and eastern shores of the Great Lakes in New York, Pennsylvania, Ohio, and Indiana. The presence of the Great Lakes and a strong northwesterly wind flow during the winter combine with other influences to create an important regional climatic phenomenon called "lake effect snow." The spatial pattern of snowfall in northeastern Ohio along the southern shore of Lake Erie is examined using Pearson's correlation.

Data from this region are used to examine the influence of distance from Lake Erie on the levels of average annual snowfall. A systematic sample of 38 locations is taken from an isoline map that shows average snowfall amounts for the region (figure 13.5). For each of the sampled points, two variables are recorded: (1) average snowfall, interpolated from the map of isohyets, and (2) straight-line distance from Lake Erie. If Lake Erie influences snowfall within the region, a negative correlation between snowfall and distance is expected. As distance from the lake increases, average annual snowfall amounts should decrease.

A scattergram of the 38 points in the study area shows the relationship between snowfall level (Y) and distance from Lake Erie (X) (figure 13.6). The pattern of points is clearly not simple linear. Contrary to the general relationship just hypothesized, snowfall amounts for locations immediately adjacent to the lake are lower than snowfall levels slightly farther from the lake. A closer look at the climatological processes operating in this lake-effect area suggests the simple hypothesized relationship between dis-

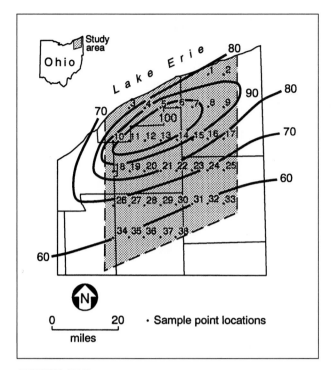

FIGURE 13.5
Northeastern Ohio Study Area: Average Annual Snowfall, in Inches (*Source:* Redrawn from Kent, Robert B. (editor), 1992. *Region in Transition: An Economic and Social Atlas of Northeast Ohio,* (fig. 23.6). Akron: The University of Akron Press.)

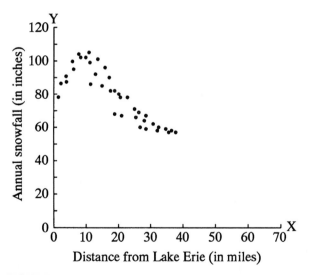

FIGURE 13.6
Scattergram Showing Relationship between Annual Snowfall and Distance from Lake Erie

more moisture. As the moisture-laden air continues moving onto cooler land surfaces south of the lake, it is cooled, the moisture often condenses, and precipitation occurs. Immediately adjacent to the lake, the air temperature is often warm enough for the precipitation to fall as rain. Somewhat farther from the lake (about 7–10 miles), the air is frequently cold enough to create snowfall.

Pearson's correlation cannot be used for this data set because the scattergram shows the relationship between snowfall and distance from Lake Erie to be curvilinear. Several alternate approaches could be used to continue the investigation. One alternative is to apply a nonlinear correlation model matching the curvilinear pattern on the scattergram. While fully valid, this methodology is beyond the introductory level of this text. Another possibility would be to convert the nonlinear data into linear form by using a logarithmic transformation. Pearson's correlation could then be applied to this transformed data. Although this approach has been used effectively in geographic studies, results are often more difficult to interpret.

Yet another alternative is to eliminate that portion of the study area less than about six miles from the lake shore (figure 13.7). The remaining 33 sample points in the revised study region generate a scattergram that seems to show a linear relationship between distance from the lake and amount of snowfall

tance and snowfall is not fully accurate. A more detailed explanation is necessary.

During the winter, the water temperature of Lake Erie is warmer than that of adjacent land areas. As cold air currents from the northwest pass over this warmer water, the air is heated and can hold

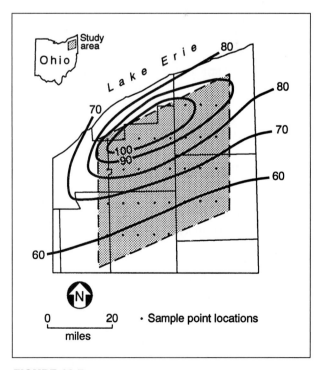

FIGURE 13.7
Revised Northeastern Ohio Study Area: Average Annual Snowfall, in Inches

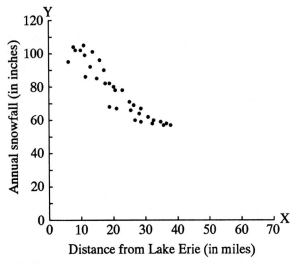

FIGURE 13.8

Scattergram Showing Relationship between Annual Snowfall and Distance from Lake Erie for Revised Study Area

TABLE 13.1

Data for Northeastern Ohio: Average Annual Snowfall and Distance from Lake Erie

Observation	Distance (miles) X	Snowfall (inches) Y
1		
2	(Points 1–5 were eliminated from	
3	study area. See text for explanation.)	
4		
5		
6	7.7	104
7	9.9	102
8	11.2	99
9	12.9	92
10	9.9	95
11	8.3	102
12	10.9	105
13	13.6	101
14	15.8	96
15	17.1	90
16	18.8	82
17	20.6	78
18	11.4	86
19	14.9	85
20	17.5	82
21	20.1	80
22	22.8	78
23	25.0	71
24	26.3	69
25	28.5	67
26	18.8	68
27	21.0	67
28	25.4	66
29	28.0	64
30	30.7	62
31	32.4	60
32	34.6	59
33	36.4	58
34	26.7	60
35	28.5	59
36	32.0	58
37	35.5	57
38	37.7	57

Source: Adapted from Kent, Robert B. (editor). 1992. *Region in Transition: An Economic and Social Atlas of Northeast Ohio,* figure 23.6. Akron: The University of Akron Press.

(figure 13.8). This strategy meets the linearity assumption of the Pearson correlation technique and can be justified on a climatological basis.

The distance (X) and snowfall (Y) values for the revised study area are shown in table 13.1. Since the original X and Y values are directly available, the appropriate computational formula is equation 13.6. The intermediate summation values and the resulting correlation coefficient are shown in table 13.2. The null hypothesis states that no association (correlation) exists between distance and snowfall in the study area. The alternate hypothesis (H_A) states that an inverse association exists between distance and snowfall in the study area. Since the direction of correlation is hypothesized as negative or inverse, a one-tailed test is appropriate.

The resulting r value of -0.93 indicates that a high inverse relationship exists between snowfall and distance from Lake Erie. This correlation is statistically significant with a p-value of 0.000. For locations between about 6 and 40 miles from Lake Erie, the hypothesis of decreased snowfall associated with increasing distance from the lake is confirmed, and the null hypothesis of no correlation can be safely rejected.

Summary Evaluation

The lake effect problem demonstrates the use of correlation analysis in geographic research. Not only was a statistically significant relationship found between distance from the lake and snowfall, but also a solid climatological basis exists to support the correlation.

Another characteristic of this problem is the complexity of delineating a suitable boundary for the study area. At first, it was thought that an analy-

sis would be appropriate from the lake shore to the southern boundary some 40 miles away. The subsequent curvilinear shape of the scattergram indicates that complex climatological processes are operating. Therefore, a simple linear relationship between distance and snowfall does not exist across the entire study area.

Another issue in this problem is the nature of the dependent variable, snowfall. Since the region has very few snowfall monitoring stations, a generalized isoline map represents the only snowfall data available. It would certainly be preferable to

TABLE 13.2

Worktable for Correlation Example: Northeastern Ohio Snowbelt

H_0: no association exists between snowfall and distance
H_A: an inverse association exists between snowfall and distance

Intermediate summation values:

$\Sigma X = 707.2 \quad \Sigma X^2 = 17858.2 \quad \Sigma XY = 50288.7$

$\Sigma Y = 2559 \quad \Sigma Y^2 = 207181 \quad N = 33$

$$r = \frac{\Sigma XY - ((\Sigma X)(\Sigma Y)/N)}{\sqrt{[\Sigma X^2 - ((\Sigma X)^2/N)]}\sqrt{[\Sigma Y^2 - ((\Sigma Y)^2/N)]}}$$

$$= \frac{50288.7 - ((707.2)(2559)/33)}{\sqrt{[17858.2 - ((707.2)^2/33)]}\sqrt{[207181 - ((2559)^2/33)]}}$$

$$= -.93$$

$$t = \frac{r\sqrt{n-2}}{\sqrt{1-r^2}} = \frac{-.93\sqrt{31}}{\sqrt{1-.93^2}} = 14.08 \ (p = .0000, \text{one-tailed})$$

measure snowfall amounts directly at monitoring stations throughout the study region. Because this was not possible, the precision of the data may be questionable.

13.3 ASSOCIATION OF ORDINAL VARIABLES

In geographic problems with data in ranked form, **Spearman's rank correlation coefficient** (r_s) is the most widely used measure of the strength of association between two variables. It is appropriate when (1) variables are measured on an ordinal (ranked) scale, or (2) interval/ratio data are converted to ranks. Spearman's correlation coefficient should be applied when an assumption of Pearson's correlation is not fully met. For example, Spearman's coefficient may be appropriate if samples are drawn from highly skewed or severely nonnormal populations. The statistical power of Spearman's correlation has been shown to be nearly as strong as Pearson's r.

Spearman's rank correlation coefficient is applicable in situations where the X and Y variables have a **monotonic relationship.** Two variables share a monotonically increasing association when X increases as Y increases (or remains constant). Conversely, a monotonically decreasing relationship between variables occurs when X increases, but Y decreases (or remains constant). Because of the nature of ordinal data, Spearman's rank correlation coefficient does not distinguish between a **linear relationship** and a monotonic one.

Similar to Pearson's correlation, r_s ranges from a maximum of 1.0 for perfect positive or direct correlation to a minimum of -1.0 for perfect negative or inverse correlation. When no association exists between variables, $r_s = 0.0$. Spearman's correlation coefficient measures the degree of association between two sets of ranks using the following equation:

$$r_s = 1 - \frac{6(\Sigma d^2)}{N^3 - N} \tag{13.10}$$

where
$d =$ difference in ranks of variables X and Y for each paired data value
$\Sigma d^2 =$ sum of the squared differences in ranks
$N =$ number of paired data values

In a problem where a perfect match occurs in the ranks of all paired values, each difference value (d) equals 0, and the sum of all squared differences also equals 0. In this situation, equation 13.10 generates a perfect positive correlation of 1.0. (table 13.3, case 1). At the other extreme, if the ranks are exactly reversed between the two variables (top rank in variable X matches bottom rank in variable Y, and so on), the Spearman index equals -1.0 (table 13.3, case 2).

TABLE 13.3

Perfect Correlations of Two Ranked Variables

Case 1: Positive correlation

Variable X ranks	Variable Y ranks	Difference in ranks (d)	d²
1	1	0	0
2	2	0	0
3	3	0	0
4	4	0	0
5	5	0	0
			$\Sigma d^2 = 0$

$$r_s = 1 - \frac{6(\Sigma d^2)}{N^3 - N} = 1 - \frac{6(0)}{5^3 - 5} = 1 - \frac{0}{120} = 1.0$$

Case 2: Negative correlation

Variable X ranks	Variable Y ranks	Difference in ranks (d)	d²
1	5	−4	16
2	4	−2	4
3	3	0	0
4	2	2	4
5	1	4	16
			$\Sigma d^2 = 40$

$$r_s = 1 - \frac{6(\Sigma d^2)}{N^3 - N} = 1 - \frac{6(40)}{5^3 - 5} = 1 - \frac{240}{120} = -1.0$$

As in other statistical tests using ordinal data, the presence of tied rankings influences the Spearman's coefficient. However, the effect of ties on the resultant correlation index will be significant only when the proportion of tied rankings to total number of values sampled is very large. In these instances, a correction factor for ties needs to be applied. As a general rule, the correction factor is not necessary if the number of tied rankings is less than 25 percent of the total number of pairs.

Spearman's correlation coefficient is commonly used as a descriptive index of association between two ordinal variables. However, when the paired variables represent a sample drawn at random from a population of bivariate data values, an r_s value can be tested for significant difference from 0. The sample correlation coefficient (r_s) is the best estimator of the population correlation coefficient (ρ_s). In these applications, the null hypothesis is that no relationship exists between the two variables in the population (H_0: $\rho_s = 0$). Confirmation of the null hypothesis is equivalent to affirming independence between the X and Y variables.

Either a one- or two-tailed test is used, depending on the form of the alternate hypothesis. As in the corresponding test of the Pearson coefficient, either the t or Z distributions can be applied to test an r_s value for significance. When the Z distribution is used, the test statistic is determined by the Spearman correlation and the sample size:

$$Z_{r_S} = r_s \sqrt{n - 1} \qquad (13.11)$$

Geographic Example of Spearman's Rank Correlation

The use of Spearman's rank correlation coefficient is demonstrated by examining recent population changes in the United States. Throughout most of the country's history, the percentage of total population living in urban areas has increased, and growth has been particularly strong in densely populated areas. Recently, however, a number of demographers have suggested that this pattern of growth has been altered so that rural areas, small towns, and regions with relatively few people are among the fastest growing locations in the United States. If this hypothesized trend toward nonmetropolitan growth or decentralization is true, then regions with relatively low population densities should have the highest growth rates.

To test this "rural renaissance" or decentralization theory, data on population density and population growth were obtained for U.S. states for five-year intervals from 1960 to 1995. Population per square mile at the beginning of each period is correlated with the percentage of population change dur-

---

Spearman Rank Correlation Analysis

Primary Objective: Determine if an association exists between two variables

Requirements and Assumptions:

1. Random sample of paired variables
2. Variables have a monotonically increasing or decreasing association
3. Variables are measured at the ordinal scale or downgraded from interval/ratio to ordinal

Hypotheses:

H_0: $\rho_s = 0$
H_A: $\rho_s \neq 0$ (two-tailed)
H_A: $\rho_s > 0$ (one-tailed) or
H_A: $\rho_s < 0$ (one-tailed)

Test Statistic:

$$Z_{r_S} = r_s \sqrt{n - 1}$$

---

ing that period. Spearman's correlation values are used here to describe trends in U.S. population growth patterns. If population growth is more pronounced in low-density areas, then Spearman's correlation index will be negative. If a positive correlation coefficient occurs, population growth will be directly related to population density at the state level. In the latter case, the population trend will reflect growth occurring in the high-density areas.

Spearman's correlation coefficient is calculated using the paired rankings of population density and percentage of population change for each of the five time periods. The methodology for the problem is illustrated for the 1990–1995 data, with 12 of the 50 states shown in table 13.4.

The rank correlation analysis for the 50 states shows clear changes in growth trends over the 35-year period (table 13.5). Between 1960 and 1970, population growth was directly related to population density, as indicated by the positive correlation indices for the two earliest time periods. These coefficients reflect a long-standing trend of metropolitanization and continuing growth in previously settled areas of the country. However, because the correlations are relatively close to zero, the tendency for growth to occur in the densest regions was relatively weak in the 1960s.

The occurrence of strong negative rank correlation coefficients between density and population growth during the 1970s provides clear evidence of population movement away from dense areas and toward more sparsely inhabited states. During this decade, the states with the greatest percentage of growth in population had lower population densities. This negative relationship between population

TABLE 13.4

Spearman Correlation Example: State Population Change and Density, 1990–95

State*	Original data population		Ranked data population		Difference (d)	d²
	Percentage change 1990–95	Density** 1990	Percentage change 1990–95	Density 1990		
Alabama	5.47	79.60	28	26	2	4
Alaska	5.80	1.00	32	1	31	961
Arizona	17.54	32.29	49	14	35	1225
Arkansas	5.53	45.14	29	16	13	169
California	6.04	190.40	33	39	−6	36
Colorado	13.60	31.80	46	13	33	1089
Vermont	3.55	60.71	15	21	−6	36
Virginia	6.69	155.83	35	36	−1	1
Washington	11.69	73.18	45	23	22	484
W. Virginia	1.62	74.34	8	25	−17	289
Wisconsin	4.52	89.88	22	27	−5	25
Wyoming	5.29	4.68	26	2	24	576

*Data listed for only 12 states
**Persons per square mile

$$r_s = 1 - \frac{6(\Sigma d^2)}{N^3 - N}$$

$$r_s = 1 - \frac{6(28968)}{50^3 - 50} = 1 - \frac{173808}{124950} = 1 - 1.391 = -.391$$

Source: Bureau of the Census, Dept. of Commerce.

TABLE 13.5

Spearman Correlation Coefficients for State Population Change and Density

Time period	Spearman r_s
1960–1965	+.166
1965–1970	+.176
1970–1975	−.539
1975–1980	−.561
1980–1985	−.406
1985–1990	+.177
1990–1995	−.391

growth and density continued into the first half of the 1980s, but at a slightly weaker level, as seen by the Spearman coefficient of −.406.

The pattern of correlation changes again in the last half of the 1980s as population density and population change show a positive association similar to that recorded in the 1960s. However, this trend is short-lived. The first half of the 1990s again exhibits a return to the earlier trend of moderate, negative correlation between density and population change at the state level. This most recent trend suggests a continuation of the dominant pattern in the 1970s

and 1980s, with greater population change associated with low density, more rural states.

Summary Evaluation

Without further analysis, it is unclear whether growth within any particular state took place in its metropolitan or nonmetropolitan portions. All that is known is the overall statewide growth rate; not known is the extent to which that growth is urban or rural.

In the eastern and midwestern states, which have a longer settlement history, growth since 1970 was probably associated with population moving away from metropolitan areas to rural regions. However, in the more recently settled western states, extensive growth may have occurred in a small number of expanding metropolitan areas, like Phoenix, Arizona, and Las Vegas, Nevada.

The use of Spearman's correlation coefficient in this example documents an interesting transition in population movement patterns within the United States that occurred around 1970. The analysis offers strong evidence of a major change away from a concentration of growth in densely populated states of the country and the emergence of lower-density states as significant areas of population growth. The seemingly anomalous correlation for the 1985–90 period may be attributable to inaccurate state-level population estimates in 1985. Other than this five-year period, the population trends examined here appear remarkably consistent since 1970. However, as just suggested, a more complete understanding of the nature of population change over these periods requires further geographic analysis.

13.4 USE OF CORRELATION INDICES IN MAP COMPARISON

Correlation measures the degree of association between variables. In most cases, data are available in numerical form and can be analyzed directly with correlation techniques. Sometimes, however, spatial information is available only in map form. How can a geographer measure the association between two map patterns when the original data are not readily available?

As discussed in the opening of this chapter, visual comparison of maps may be subjective and lead to biased conclusions. More productive geographic research requires an objective analysis of map patterns. With the use of spatial sampling methods, correlation indices can be applied to numerical data acquired from maps. The methodology is illustrated here with three type of maps: dot maps, isoline maps, and choropleth maps.

Dot Maps

The location of items within a geographic area can often be portrayed effectively with a dot map. A dot on the map represents the location of each item or group of items. The resulting pattern of points defines the geographic distribution of the variable under study. Procedures for analyzing a single point pattern were presented in section 12.1. This section focuses on the relationship between two or more point patterns.

Suppose a geographer wants to investigate the relationship between spatial patterns of two variables portrayed by dot maps covering the same area. An index of association can be calculated using information taken directly from the dot maps. To acquire data for correlating the two dot map patterns, a set of equal-sized quadrats (usually square cells) is placed over the study area. The quadrat size should be adequate to depict the spatial complexity of the pattern and allow the calculation of a representative correlation index. The discussion of cell size in quadrat analysis (section 12.1) also applies to this application of correlating dot maps.

The quadrats should be placed over each map to produce identical cell location and orientation patterns on both maps. If the maps have the same scale, the quadrats used for the first dot map will cover exactly the same areas on the second map. However, if the two maps have different scales, one of two procedures should be followed. The maps could be converted to the same scale, and a set of quadrats used in the manner discussed previously. Alternatively, different-sized quadrats could be used for the two maps if each cell represents the same actual area on the earth's surface.

Each of the quadrats represents an observation, and the frequency of points per quadrat from the two maps are the X and Y values. Using this data set created from the dot maps, either Pearson or Spearman correlation indices can be calculated.

This procedure is demonstrated using the geographic problem relating crime and drug activity discussed at the beginning of the chapter. The two maps in figure 13.1 first appeared in an article in *The Washington Post* illustrating the spatial connection between locations of drug sales and homicides in Washington, D.C. The article used visual correspondence as evidence of a relationship. However, a more objective measure of the degree of association between the two variables can be determined using Pearson's correlation.

The dot maps are constructed at the same scale, which permits the same quadrat pattern to be used for both maps. A set of 16 cells is placed over the dot maps (figure 13.9), and the frequency of drug sale points and homicides is recorded for each cell. Pearson's correlation coefficient is then calculated using the two sets of frequencies collected for the 16 paired data values. The resultant index value ($r = .953$) is significantly different from 0 ($p = .000$) and strongly confirms the hypothesis suggested by the newspaper story that drug activity and murders in Washington, D.C., are spatially related.

Isoline Maps

Isoline maps are useful for presenting data that are distributed continuously across an area, such as precipitation, temperature, barometric pressure, and elevation. By selecting a suitable isoline interval and connecting locations having equal values with an isoline, the pattern of lines on the map represents the geographic distribution of the variable over the area.

The methodology used to investigate the association between two isoline maps is analogous to that used for dot maps. However, instead of placing a set of quadrats over the maps and recording the frequency of dots per cell, a set of sample points is placed systematically on each isoline map. If the maps have equal scales, the same point pattern grid can be used for both maps. If the maps have different scales, they must be converted to the same scale, or the grid system must be altered so that an identical set of point locations occurs on each map.

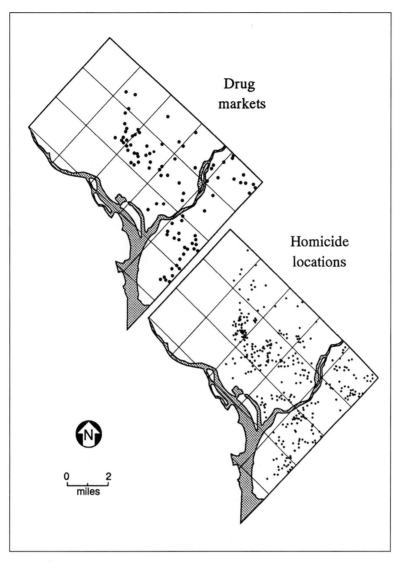

FIGURE 13.9
Dot Maps of Homicides and Drug Sale Points in Washington, D.C., with Quadrats
Superimposed

Once the sample points are properly placed on the maps, the value of the continuously distributed variable is recorded for each matching pair of points. Some interpolation of numerical values is needed where points do not fall directly on an isoline. The set of points represents the values, and the recorded values from the two isoline maps provide the corresponding matched X and Y values. Using the data taken from the maps, a correlation coefficient is calculated that measures the strength of association between the two map variables.

This procedure is illustrated with maps of rural population density and annual precipitation for the People's Republic of China. In a now classic study by Robinson and Bryson (1957), these two variables were used to correlate map patterns in the state of Nebraska.

Geographic studies have shown that various climatic and geomorphological factors profoundly influence the ability of the land to support population. Because they have greater potential for food production, areas with higher precipitation levels tend to have higher population densities. This is especially true in developing areas where a high proportion of the population is involved in primary activities, such as agriculture. People in these areas more directly depend on favorable environmental conditions to survive and lack the resources to modify adverse environments.

A regular grid pattern of 50 points is superimposed on the two isoline maps of China (figure 13.10). For both maps, the value of the variable is interpolated at each point. A Pearson correlation coefficient is then calculated from these data. The resulting value

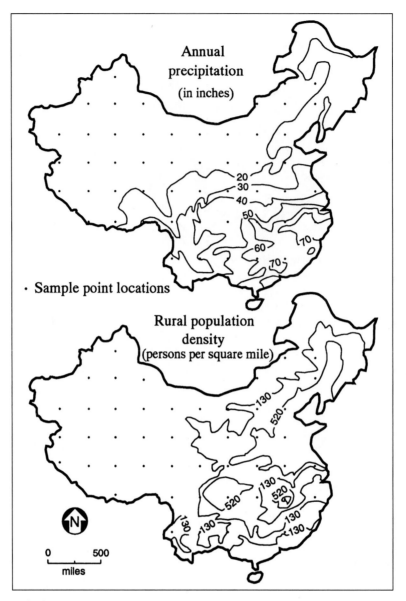

FIGURE 13.10
Isoline Maps of Rural Population Density and Annual Precipitation in China with
Sample Point Locations (*Source:* Central Intelligence Agency. 1971. *People's
Republic of China Atlas.* Washington, D.C.: U.S. Government Printing Office.)

($r = .396, p = .002$) shows a moderately positive and statistically significant relationship between precipitation and population density. Areas in China with higher levels of precipitation (the east and southeast) tend to be associated with higher population densities. Conversely, areas to the west and north with lower levels of precipitation have very low population densities. This methodology for analyzing patterns in isoline maps allows an objective assessment of the strength of association for variables that are continuously distributed.

Choropleth Maps

Choropleth maps are frequently used to display the geographic distribution of data for areas such as states, counties, or census tracts. In creating a choropleth map, the value of each areal unit is allocated to one of several categories, and each category is represented by a particular pattern or color on the map. Many different methods exist to classify areal data for choropleth mapping (see section 2.4).

Sometimes geographers wish to measure the degree of association between two choropleth maps having the same internal subarea boundaries. The maps themselves would have to be used if information is available only in mapped form. For example, an economic geographer investigating spatial patterns of income and unemployment may want to determine whether an association exists between choropleth maps of the two variables. One map may use five categories from highest to lowest income to depict median family income in a set of regions. The other map may show the rate of unemployment in the same areas, classified into a set of four ordinal categories. By assigning a numerical value to the categories (1 for the lowest to n for the highest), the map information can be transformed into numerical data suitable for correlation analysis.

For a problem of this type, Spearman's correlation index is a better choice than Pearson's to show the generalized association between the two variables, because choropleth map patterns are converted to numerical values at the ordinal level. The researcher knows only that the values of a particular map category can be ranked above or below other categories. The interval between map patterns (or between numeric values assigned to the patterns) is not known. Therefore, the proper technique to measure the degree of association between the choropleth patterns is Spearman's correlation index.

This procedure is illustrated with a descriptive example of changes in language use in the Montreal, Quebec, region. Suppose a geographer has a series of choropleth maps that show changes over time in the percentage of population speaking French and English for the 28 municipalities on the island of Montreal. Since the core city of Montreal has considerable language diversity, the maps divide the Montreal municipality into 24 subareas, each containing a generally homogeneous population structure. From this study area of 51 spatial units, the geographer wants to investigate whether the language-use maps are correlated over time.

For those persons speaking French as their primary language at home, two choropleth maps are available. One map shows the change in the percentage of population speaking French during the period from 1971 to 1981 (figure 13.11, case 1). A second map covers change in the same variable for the period from 1981 to 1986 (figure 13.11, case 2). Does the change in French-speaking population for the earlier period (1971–81) correlate positively with the change during the later period (1981–86)? If so, similar spatial processes are operating over time within the Montreal region to affect patterns of language use.

Each of the map patterns is converted into a numerical variable with five categories. The lowest map category, those municipalities showing at least a 7.5 percent *decline* in French-speaking population, is as-

signed a value of 1. The assignment of numerical values to the other map categories continues up to 5, those municipalities having an *increase* in French-speaking population of at least 7.5 percent. Spearman correlation is then calculated for the two variables in rank form. The resulting correlation index ($r_s = .695$) shows a moderately strong positive association between the two time periods. This correlation value indicates a relatively consistent temporal pattern of change for the percentage of French-speaking population within the Montreal region. That is, most of the areas either increased, decreased, or maintained their percentage of French language use during both time periods. If the map patterns had shown greater difference from area to area when the early time period was compared to the later period, the resulting correlation would be weaker. A weaker correlation would suggest that different spatial processes were operating at different times to affect the language-use patterns within the Montreal region.

Using a similar methodology, choropleth maps of the Montreal region showing the change in the percentage of the population speaking English are also correlated over the two time periods. The Spearman's correlation value for the English population ($r_s = .368$) also shows a positive association between the two periods, but at a weaker level. This descriptive analysis suggests that the regional patterns of change in English language use are less consistent over time than those for the French language. Thus, the two choropleth maps showing change in the percentage of English-speaking population differ more widely from the early time period to the later period. Further analysis of these patterns is needed to gain a clearer understanding of the spatial processes that affect the dynamics of language in a multilingual region such as Montreal. Nevertheless, the correlation methodology for relating patterns in choropleth maps is a useful, objective measurement of the strength of association between two variables defined for areas.

13.5 ISSUES REGARDING CORRELATION

When geographers apply statistical analysis to spatial data, the level of aggregation of the observation units may influence the results. This concern is especially important when inferences are drawn from the results of geographic analyses. Significant findings at one level of aggregation may not occur at other levels. For example, correlation results at the individual level are probably not equally significant at higher levels of aggregation. Although it may be observed that the level of income and amount of education are highly correlated for individuals, these same variables may not be similarly associated at the

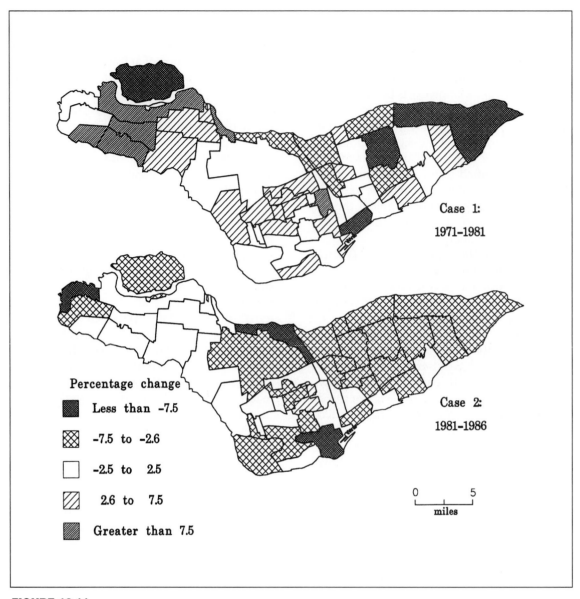

FIGURE 13.11
Choropleth Maps of Percentage Change in Population Speaking French at Home: Montreal, Quebec (*Source:* Modified from Trudel, Daniel, 1992. "A Spatial Analysis of Language-Use Patterns in Montreal, 1971–1986". Master's Thesis, The University of Akron [Figure 4.4])

county or state levels. Thus, different scales of analysis may produce different degrees of correlation.

Another critical geographic concern is the so-called **ecological fallacy** concept, a reversal of the problem of aggregation just discussed. Researchers sometimes use highly aggregated data and then attempt to infer these results to lower levels of aggregation or to the individual level. Such inferences may not be valid. For example, just because crime rates are statistically correlated with the percentage of persons under the poverty level at the state or census tract level, this does not imply that all persons below the poverty level are criminals. The correlation only shows that crime rates are higher in areas with a larger percentage of persons under the poverty level. The data or the statistical analysis cannot support any inferences made beyond this statement, such as to individuals.

KEY TERMS

MAJOR GOALS AND OBJECTIVES

If you have mastered the material in this chapter, you should now be able to do the following.

1. Explain the nature and purposes of correlation analysis.
2. Distinguish the various directional relationships between variables (positive, negative, and neutral) and among different strengths of association between variables (perfect, strong, weak, and none).
3. Explain the concept of covariation, and understand how scattergrams can depict the relationship between covariation and correlation.
4. Create a geographic research problem or situation for which Pearson's correlation coefficient would be appropriate.
5. Create a geographic research problem or situation for which Spearman's rank correlation coefficient would be appropriate.
6. Understand how correlation analysis can be used to measure the degree of association between two map patterns (when the original data are not available).
7. Distinguish between the procedures needed to correlate two dot maps, two isoline maps, and two choropleth maps.
8. Explain how the aggregation problem and ecological fallacy concept might impact on the measured level of correlation.

REFERENCES AND ADDITIONAL READING

Abler, R., J. Adams, and P. Gould. 1971. *Spatial Analysis.* Englewood Cliffs, NJ: Prentice-Hall.

Burt, J. E. and G. M. Barber. 1996. *Elementary Statistics for Geographers.* (2nd edition). New York: Guilford Press.

Ebdon, D. 1985. *Statistics in Geography: A Practical Approach* (2nd edition). Oxford: Basil Blackwell.

Falk, R. and A. D. Well. 1997. "Many Faces of the Correlation Coefficient." *Journal of Statistics Education.* www.stat.ncsu.edu/info/jse/v5n3/falk.html.

Robinson, A. and R. Bryson. 1957. "A Method for Describing Quantitatively the Correspondence of Geographical Distributions." *Annals, Association of American Geographers.* 47: 379–91.

Rogers, J. L. and W. A. Nicewander. 1988. "Thirteen Ways to Look at the Correlation Coefficient." *The American Statistician* 42:59–66.

Taylor, P. 1977. *Quantitative Methods in Geography: An Introduction to Spatial Analysis.* Boston: Houghton, Mifflin.

Regression

In chapter 13 Pearson's correlation coefficient was discussed as the index for computing the degree of association between variables measured on an interval/ratio scale. In correlation analysis, the researcher does *not* have to assume a functional or causal relationship between the two variables. A correlation can be computed for any two variables, as long as the correlation index is consistent with the level of measurement for the data.

Geographers often work in research areas where relationships between variables need to be explored in more detail. The assumption or hypothesis may be that one variable influences or affects another or that a functional relationship ties one variable to another. For these geographic problems, regression analysis is a useful statistical procedure that supplements correlation.

A classic problem studied by geographers is the spatial relationship between level of precipitation in an agricultural region and the population density the area can support. It is hypothesized that the amount of moisture available at locations within a region influences or determines the density pattern of farm population in the region. Data could be collected for the two variables at various sites in the region, and regression used to answer questions about how population density relates to precipitation level. The nature or form of this relationship can be explored and the strength of the relationship determined. Assuming that the relationship is not exact or

perfect, regression allows sources of error in the relationship to be examined and helps uncover additional variables that may influence the geographic pattern. The problem differs from simple correlation analysis because a functional relationship is expected between the variables, and the nature of that relationship needs to be explored more fully. Simply stated, regression should not be used unless a clear rationale or model links one variable to another, as is the case with population density and rainfall.

Regression can be applied successfully in all areas of geography. For example, a medical geographer could examine the relationship between the number of physicians located in the counties of a state and the income level of persons residing in these areas. Regression analysis could be used to predict or estimate the number of county physicians based on a county's income profile. An environmental geographer may wish to examine the form of relationship between the acidity level in various locations in a chain of lakes and distance from a point pollution source. Regression analysis could predict acidity level in a lake based on distance from the pollution source. A political geographer might want to compare the strength of votes for a political party and the educational, financial, or racial composition of voters in the wards of a city. Based on the socioeconomic composition of a city ward, regression could then be applied to predict political party voting strength. A cultural geographer surveying current attitudes at various locations could apply regression to predict the level of support for a controversial issue like abortion as support relates to such socio-economic characteristics as occupation, income, religious belief, or education at those locations.

The discussion in sections 13.1 through 13.4 concerns **bivariate regression,** which examines the influence of one variable on another. The variable creating the influence or effect is called the **independent variable,** and the variable receiving the influence or effect is termed the **dependent variable.** In regression terminology, the dependent variable is affected by (or perhaps caused by) the independent variable. The bivariate regression sections include a detailed examination of a geographic problem that illustrates the technique. This discussion provides an understanding of both the concepts and calculation procedures for simple regression analysis.

The most common application of regression is the identification of linear relationships between variables. In linear form, changes in values of the variables are constant across the range of the data. In these instances, the pattern of points from a scattergram approximates a straight line (figure 14.1, case 1). Real-world relationships, however, sometimes show variables that are related in curvilinear ways where variable changes are not constant across all values (figure 14.1, case 2). For example, the two variables in the original snowbelt study area (chapter 13) depicted a curvilinear relationship when plotted as a scattergram. In this chapter, only regression applied to linear relationships is discussed.

In section 14.5, **multivariate regression** is introduced with discussion limited to a general understanding of the concepts involved. More detailed information, including formulas and computational procedures, can be found by consulting one of the references listed at the end of the chapter.

14.1 FORM OF RELATIONSHIP IN BIVARIATE REGRESSION

Like correlation, bivariate regression attempts to determine how one variable relates to another. The basic question posed by regression is: "What is the *form* or *nature* of the relationship between the variables under study?" As with correlation, the form of relationship between two variables is easily visualized by constructing a scattergram or graph. The point pattern on the graph determines the form of the relationship between variables.

Recall the example in which the average amount of snowfall at locations in northeast Ohio was related to distance from Lake Erie. Data for these interval-level variables were plotted for 33 sites in the study area to produce a scattergram (figure 13.8).

Case 1:
Linear

Case 2:
Curvi-linear

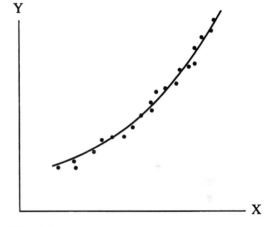

FIGURE 14.1
Linear and Curvi-linear Relationships

The association was negative or inverse because snowfall amounts generally declined at distances farther from the lake. This example is now extended to illustrate the use of regression to analyze geographic data.

Although correlation and regression can both begin with data placed on a scattergram, the rationale for assigning variables to the two axes differs. In correlation, the two variables are assigned arbitrarily to the horizontal and vertical axes, and variables are not identified as being either independent or dependent. Correlation merely determines the degree of association between variables. Thus, when correlation was used to analyze the Snowbelt data for association, independent and dependent variables were not specified, and the assignment of the snowfall and distance variables to axes of the scattergram was arbitrary.

In bivariate regression, however, one variable serves as the dependent variable and the other as the independent variable. Since regression examines the influence of the independent variable on the dependent variable, proper specification of the two variables is necessary, and the assignment of variables to axes of the scattergram is not arbitrary. The independent variable is *always* placed on the horizontal, or X, axis (abscissa) and the dependent variable is always placed on the vertical, or Y, axis (ordinate). If the assignment of variables to axes is reversed, a different regression result will occur.

In the Snowbelt problem, distance from Lake Erie must be the independent variable and snowfall must be the dependent variable. The logical hypothesis is that the level of snowfall is affected or influenced by distance from the lake (and not vice versa); so the snowfall variable must be placed on the vertical axis and distance on the horizontal axis.

The form of association between two variables can be portrayed graphically by plotting the data values on a scattergram. Bivariate regression describes this pattern of points more objectively by placing a line through the scatter of points. This line, called the "least-squares" or "best fitting" line of regression, summarizes the overall trend in the data and essentially defines the form of relationship between the independent and dependent variables.

Although an infinite number of lines could be drawn to summarize the points in a scattergram, the **least-squares regression line** is unique. As the name implies, the line minimizes the sum of squared vertical distances between each data point and the line (figure 14.2):

$$\text{minimize } \Sigma d_i^2$$

where d_i = vertical distance separating point i from the regression line

No other line can be generated where the sum of the squared distances between the points and the line (measured vertically) is a smaller value than that calculated for the least-squares line. This line defines the best estimate of the relationship between the independent and dependent variables. It also serves as a predictive model by generating estimates of the dependent variable using both the values of the independent variable and knowledge of the relationship that connects the two variables.

In a bivariate regression with independent variable (X) and dependent variable (Y), the least-squares regression line is denoted by the following linear (straight-line) equation:

$$Y = a + bX \qquad (14.1)$$

In addition to the two variables, the equation contains two constants or parameters (a and b), which are calculated from the actual set of data. These values uniquely define the equation and establish the position of the least-squares line on the scattergram. The equation of the least-squares line for the Snowbelt example is

$$Y = 113.51 - 1.68X$$

where X = distance from Lake Erie (miles)
Y = annual snowfall (inches)
a = 113.51
b = −1.68

The constant a, called the **Y-intercept,** represents the expected value of Y when the value of X is zero, the point where the regression line crosses the Y axis. It should be noted that a is the predicted best estimate of the Y value when X is zero. In the Snowbelt example, the value of a is 113.51 inches of snow (figure 14.3). This result indicates that more than 113

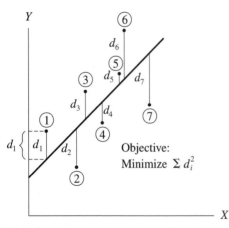

FIGURE 14.2
The Objective of Least-squares Regression

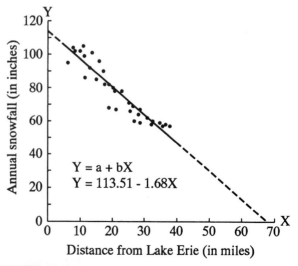

FIGURE 14.3
Interpreting the Regression Line for the Snowbelt Example

inches of snowfall are expected for locations along the lakefront of Lake Erie where distance equals 0.

This interpretation of *a* in the Snowbelt example is invalid, however. Because the relationship between distance and snowfall is curvilinear rather than linear, locations within 6 miles of Lake Erie are excluded from the study area. The regression model uses the 33 locations outside the 6-mile band, and results apply only within the existing bounds of distance (formally termed the **domain of X**). Thus, for locations between 0 and 6 miles from Lake Erie (below the lower bound of distance), interpretation is invalid. In the same way, using the model to estimate snowfall for distances outside the 40-mile limit of the study area would also not be valid.

The other constant in the regression equation, *b*, represents the **slope** of the line. This value, also called the **regression coefficient,** shows the *absolute* change of the line in the *Y* (vertical) direction associated with an increase of 1 in the *X* (horizontal) direction. The slope reveals how responsive the dependent variable is to a unit increase in the independent variable.

Both the sign and magnitude of *b* offer useful information about the bivariate relationship. The sign (+ or −) of the slope determines the direction of relationship between the two variables. If the slope is positive ($b > 0$), the line trends upward from low values of *X* to high values of *X* (figure 14.4, case 1). On the other hand, if the slope is negative ($b < 0$), the line moves downward (figure 14.4, case 2). This interpretation of direction is equivalent to that for the correlation coefficient. When the relationship between the independent and dependent variables is direct, *b* will be positive and the line trends up from left to right. However, when the relationship is inverse, the value of *b* is negative, and the line trends down. When no relationship exists, the value of *b* is zero, and the line parallels the *X* axis (figure 14.4, case 3).

In absolute terms, the magnitude of the parameter *b* indicates the flatness or steepness of the regression line when moving from lower to higher values of *X*. When *b* is large (regardless of sign), the change in *Y* is large relative to a unit increase in *X*. In this situation, the slope of the regression line tends to be steep. When *b* is small, the opposite interpretation occurs; the change in the dependent variable is small when compared to a unit increase in the independent variable, and the line has a flatter slope.

In the Snowbelt example, the calculated slope, or *b* value, is −1.68. Since the regression coefficient is negative, the least-squares line declines from left to right on the scattergram, indicating lower snowfall levels at greater distances from Lake Erie (figure 14.3). Although these results agree with those from

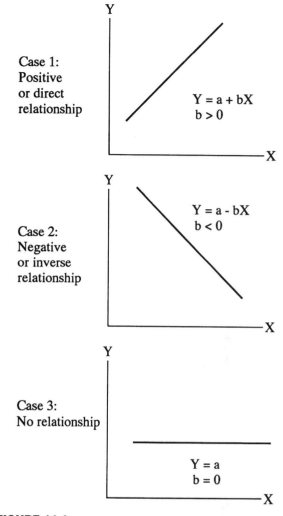

FIGURE 14.4
Interpretation of Slope in Bivariate Regression

the correlation analysis, the slope value measures the inverse relationship in more explicit terms—for each unit increase in *X* (1 mile of distance farther from the lake), the amount of snowfall is reduced by approximately 1.68 inches.

Such interpretations of flatness or steepness, however, can sometimes be misleading, since the value of *b* depends on the units of measurement used for the two variables. If the units of one variable are altered, the value of *b* will also change. For example, note the difference in slope or steepness of the regression line in the Snowbelt example when distance is measured in kilometers (figure 14.5, case 1) rather than miles (figure 14.5, case 2). Therefore, although the slope accurately relates the *absolute* change in the independent and dependent variables, it cannot be used as a valid index of the *relative* relationship between the two variables because its numerical value is tied to the units of measurement.

Case 1:
Distance in kilometers

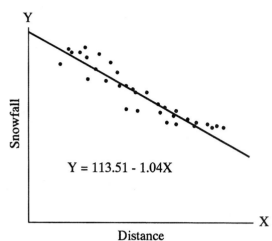

$$Y = 113.51 - 1.04X$$

Case 2:
Distance in miles

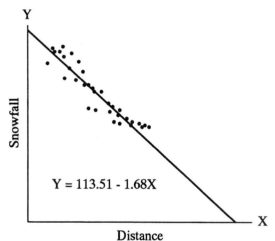

$$Y = 113.51 - 1.68X$$

FIGURE 14.5
Influence of Measurement Units on Magnitude of Slope

The a and b parameters for the least-squares regression line are determined as follows:

$$b = \frac{n\Sigma XY - (\Sigma X)(\Sigma Y)}{n\Sigma X^2 - (X)^2} \qquad (14.2)$$

$$a = \frac{\Sigma Y - b\Sigma X}{n} \qquad (14.3)$$

where
ΣX = sum of the values for variable X
ΣY = sum of the values for variable Y
ΣX^2 = sum of the squared values for variable X
ΣXY = sum of the product of corresponding X and Y values
n = number of observations

Note that the slope or regression coefficient is calculated first, then used to compute the Y intercept or a

value. Table 14.1 shows the intermediate steps used to calculate a and b for the Snowbelt example.

The two parameters uniquely define the least-squares regression line that best summarizes the re-

| TABLE 14.1 |

Bivariate Regression Example for Northeastern Ohio Snowbelt: The Influence of Distance from Lake Erie on Average Annual Snowfall

Site	X	Y	X²	XY
1				
2				
3		(Data points were eliminated from study area)		
4				
5				
6	7.7	104	59.8	804.2
7	9.9	102	97.7	1008.1
8	11.2	99	124.8	1106.0
9	12.9	92	166.2	1186.0
10	9.9	95	36.2	571.5
11		8.3	102	69.2
849.5				
12	10.9	105	119.6	1148.5
13	13.6	101	185.3	1374.7
14	15.8	96	249.2	1515.5
15	17.1	90	292.1	1538.2
16	18.8	82	354.6	1544.2
17	20.6	78	423.2	1604.6
18	11.4	86	129.4	978.1
19	14.9	85	222.5	1267.9
20	17.5	82	307.2	1437.2
21	20.1	80	405.5	1610.9
22	22.8	78	520.3	1779.1
23	25.0	71	624.2	1773.9
24	26.3	69	691.1	1814.0
25	28.5	67	810.2	1907.1
26	18.8	68	354.6	1280.6
27	21.0	67	441.3	1407.5
28	25.4	66	646.1	1677.7
29	28.0	64	785.7	1793.9
30	30.7	62	942.6	1903.5
31	32.4	60	1052.5	1946.5
32	34.6	59	1198.4	2042.4
33	36.4	58	1321.9	2108.8
34	26.7	60	714.2	1603.5
35	28.5	59	810.2	1679.4
36	32.0	58	1024.5	1856.4
37	35.5	57	1259.4	2022.8
38	37.7	57	1418.5	2146.8
SUM	707.1	2559	17858.2	50288.7

$$b = \frac{n\Sigma XY - (\Sigma X)(\Sigma Y)}{n\Sigma X^2 - (X)^2}$$

$$= \frac{(33)(50288.7) - (707.1)(2559)}{(33)(17858.2) - (707.1)^2} = -1.68$$

$$a = \frac{\Sigma Y - b\Sigma X}{n} = \frac{2559 - (-1.68)(707.1)}{33} = 113.51$$

lationship between the independent and dependent variables. This line can be placed on the scattergram by calculating and plotting two points that lie on the line and then connecting them with a straight line. The two points can be determined by selecting any two values of X, substituting them into the regression equation, and calculating the corresponding values of Y.

By selecting certain points to plot, the task of drawing the regression line is usually made easier. For example, the Y-intercept, where $X = 0$ and $Y = a$, usually represents a conveniently graphed point through which the regression line passes. In the Snowbelt example, the Y-intercept occurs at $X = 0$ and $Y = 113.51$ (figure 14.6, point 1). A second convenient point on the regression line is the X-intercept, where $Y = 0$ and $X = -a/b$. The X-intercept for the Snowbelt example is the location where the regression line crosses the X axis, $Y = 0$ and $X = 67.57$ (figure 14.6, point 2). The regression line also passes through the point determined by the means of the X and Y variables. Since the mean values of the two variables are readily available, they represent a useful choice when plotting the regression line. In the Snowbelt example, the means of the two variables are $\overline{X} = 21.4$ miles and $\overline{Y} = 77.5$ inches (figure 14.6, point 3).

14.2 STRENGTH OF RELATIONSHIP IN BIVARIATE REGRESSION

For any realistic application, the scattergram of points will not be depicted perfectly by the least-squares

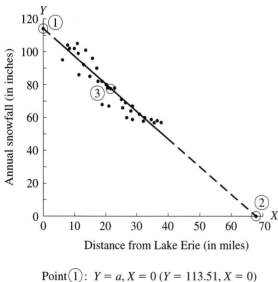

Point ① : $Y = a, X = 0$ ($Y = 113.51, X = 0$)
Point ② : $Y = 0, X = -a/b$ ($Y = 0, X = 67.57$)
Point ③ : $Y = \overline{Y}, X = \overline{X}$ ($\overline{Y} = 77.5, \overline{X} = 21.4$)

FIGURE 14.6
Plotting the Regression Line

line of regression. All points will not lie exactly on the line, which implies that the independent variable cannot fully explain the dependent variable. This "error" in regression analysis can be traced to several sources. Since geographers study patterns on the earth's surface that are frequently produced by complex processes, expecting a single independent variable in a regression model to account fully for the variation in the dependent variable is unreasonable. Even when multiple influences are considered, some portion of most real-world patterns is either attributable to unknown variables or is the result of unpredictable, random occurrences. The inability to measure or operationalize variables in an accurate or valid fashion may be another source of error. As a basis for evaluating the explanatory ability of the regression model, the strength of a relationship must be determined.

The issue of strength in regression analysis can be viewed in both conceptual and practical terms. In bivariate regression, an independent variable (X) is used to explain or account for variation in the dependent variable (Y). The ability of the independent variable to account for the variation in Y provides a measure of strength or level of explanation. To understand this process from another perspective, the strength of relationship in regression is determined by the amount of deviation between the points on the scattergram and the position of the least-squares line. In general, the closer the set of points lies to the regression line, the stronger the linear relationship between the variables. Determining the strength of relationship between two variables in regression is the same as measuring the relative ability of the independent variable to account for variation in the dependent variable.

Abler, Adams, and Gould (1971) created a useful analogy involving a bucket and sponge to illustrate the bivariate regression process (figure 14.7). A bucket full of water represents the total variation in the dependent variable, Y. A sponge denotes the independent variable, X, which will be used to explain variation in Y. When the sponge is dipped into the bucket and removed, some of the water (symbolizing variation in Y) is absorbed. This represents the amount of variation in Y that can be explained by X. Although some water is absorbed by the sponge, some of it remains in the bucket. This residual amount is the portion of the variation in Y that cannot be explained by X.

The sponge analogy can be developed further. The central issue in measuring the strength of relationship is the determination of the relative ability of the sponge to remove water from the bucket. If the sponge is super absorbent, it will remove a large proportion of the water. A less absorbent sponge removes less water.

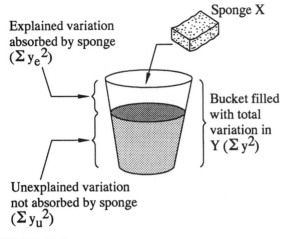

FIGURE 14.7
Bucket and Sponge Analogy in Bivariate Regression (*Source:* modified from Abler, R., J. Adams, P. Gould. 1971. *Spatial Organization: The Geographer's View of the World.* Englewood Cliffs, NJ: Prentice Hall.

Calculating the ratio of volume of water removed by the sponge to the total volume of water originally in the bucket provides a strength index for the sponge.

The bucket and sponge analogy translates directly into regression terminology. The **total variation** in Y, the dependent variable, is represented by the original volume of water in the bucket (figure 14.7). This provides a measure of the variation available for explanation by the independent variable:

$$\Sigma y^2 = \Sigma(Y - \overline{Y})^2 \qquad (14.4)$$

where Σy^2 = total variation in Y

Note that Y ("large y") symbolizes values of the dependent variable, whereas y ("small y") denotes the deviation of each value of the dependent variable from its mean $(Y - \overline{Y})$. Because the calculation of total variation involves summing the squared deviations from the mean, the term is often referred to as the **total sum of squares.** An alternative formula for calculating the total sum of squares is

$$TSS = \Sigma y^2 = \Sigma Y^2 - \frac{(\Sigma Y)^2}{n} \qquad (14.5)$$

where TSS = total sum of squares

As illustrated in the bucket and sponge analogy, the total variation in the dependent variable can be broken into two parts—the **explained variation** and the **unexplained variation**:

$$\Sigma y^2 = \Sigma y_e^2 + \Sigma y_u^2 \qquad (14.6)$$

where Σy_e^2 = explained variation

Σy_u^2 = unexplained variation

The explained variation, also called the "explained sum of squares," is the amount of variation that can

be accounted for by the independent variable X. In the analogy, it is the amount of water absorbed by the sponge. The unexplained variation, also called "residual variation," is the portion of total variation in the dependent variable that cannot be accounted for by the independent variable. It is analogous to the water not removed by the sponge. Because unexplained variation relates directly to analysis of residuals or error in regression, this concept will be discussed later in this section.

The explained variation is calculated by taking the ratio of the square of the covariation between X and Y to the variation in X:

$$\Sigma y_e^2 = \frac{(\Sigma xy)^2}{\Sigma x^2} \qquad (14.7)$$

where Σxy = covariation of X and Y
Σx^2 = total variation of X

As discussed in section 13.2, the covariation between X and Y indicates how the two variables vary together (covary) and is used to help interpret the correlation coefficient. If two variables tend to covary consistently in the same direction, covariation (and the correlation coefficient) is high and positive (see figure 13.4, case 1). If they vary systematically, but in opposite directions, the covariation and resulting correlation is high and negative (see figure 13.4, case 2).

Although the direction of association represented by the sign of the covariation term is important in correlation analysis, it can be ignored when using covariation to determine explained variation. Remember that regression is concerned with the ability of the independent variable to account for variation in the dependent variable. By squaring the covariation in equation 14.7, the influence of a negative covariation is eliminated.

When the squared covariation (numerator of equation 14.7) is large relative to the total variation in X (denominator in equation 14.7), the explained variation is also large. In such cases, the points of the scattergram will tend to lie close to the regression line, and the independent variable X will account for more of the variation in the dependent variable Y. This is equivalent to selecting an absorbent sponge.

On the other hand, if the independent and dependent variables do not covary systematically in a scattergram, the resulting covariation measure (and correlation coefficient) will be low (see figure 13.4, case 3). In this situation, the ratio of the squared covariation to the total variation in X will also be low, producing a smaller amount of explained variation. When the level of covariation of the X and Y variables is weak, the points tend to scatter more widely about the regression line, and the general strength of relationship is weak.

The amount of explained variation is an absolute measure calculated in units of the dependent variable and is comparable to the volume of water removed by the sponge in the analogy. A more useful way to express the strength of a regression relationship is with a relative index that is not tied to the units of measurement. Termed the **coefficient of determination,** or r^2, the relative strength index for regression is simply the ratio of the explained variation to the total variation in Y:

$$r^2 = \frac{\Sigma y_e^2}{\Sigma y^2} \qquad (14.8)$$

The index is often multiplied by 100 for ease of interpretation. In this form, the strength index ranges from 0 to 100 and can be interpreted as the *percentage* of variation in the dependent variable that is explained by the independent variable—or in the bucket and sponge analogy, the percentage of water removed by the sponge.

Information on the strength of relationship for the Snowbelt example is shown in table 14.2. The coefficient of determination ($r^2 = .872$) indicates that 87.2 percent of the variation in snowfall levels across this portion of northeast Ohio can be accounted for by the independent variable, distance from Lake Erie. The remaining 12.8 percent of regional snowfall variation is not explained by the distance variable in this regression model.

Although the coefficient of determination (r^2) is closely related to the correlation coefficient (r), the two indices have different purposes and interpretations. The correlation coefficient shows the direction and level of association between any two variables and does not imply a functional or causal relationship. The coefficient of determination, on the other hand, is used as a regression index to measure the degree of fit of the points to the regression line or the ability of the independent variable to account for variation in the dependent variable. As a result, the use of r^2 requires a logical rationale for the existing relationship and the specification of independent and dependent variables.

14.3 RESIDUAL OR ERROR ANALYSIS IN BIVARIATE REGRESSION

Residual analysis provides additional spatial and nonspatial information about the variation in the dependent variable that cannot be explained by the independent variable. Because geographic relationships seldom allow perfect explanations of the dependent variable, points will only rarely lie on the regression line in the scattergram. The amount of deviation of each point from the regression line is termed the absolute **residual**. It represents the verti-

---

TABLE 14.2

Worktable for Calculating Strength of Relationship for the Snowbelt Example

Total variation:

$$TSS = \Sigma y^2 = \Sigma(Y - \overline{Y})^2 = \Sigma Y^2 - \frac{(\Sigma Y)^2}{N}$$

$$= 207181 - \frac{2559^2}{33} = 207181 - 198438.8$$

$$= 8742.2$$

Explained variation:

$$\Sigma y_e^2 = \frac{(\Sigma xy)^2}{\Sigma x^2}$$

$$\Sigma xy = \Sigma XY - \frac{\Sigma X \Sigma Y}{N} = 50288.7 - \frac{(707.1)(2559)}{33} = -4543.7$$

$$\Sigma x^2 = \Sigma X^2 - \frac{(\Sigma X)^2}{N} = 17858.2 - \frac{(707.1)(707.1)}{33} = 2707.0$$

$$\Sigma y_e^2 = \frac{(\Sigma xy)^2}{\Sigma x^2} = \frac{(-4543.7)(-4543.7)}{2707.0} = 7626.6$$

Coefficient of determination:

$$r^2 = \frac{\Sigma y_e^2}{\Sigma y^2} = \frac{7626.6}{8742.2} = .872$$

---

cal difference between the actual and predicted values of Y:

$$RES = Y - \hat{Y} \qquad (14.9)$$

where RES = residual
$\quad Y$ = actual value of the dependent variable
$\quad \hat{Y}$ = predicted regression line value of Y

The predicted values of Y are generated from the regression line. Each $\hat{Y}$ value represents a best estimate of the dependent variable produced from the a and b parameters that define the regression line. By substituting any value of X, (for example X_1), into the regression equation, the corresponding predicted value ($\hat{Y}_1$) is calculated:

$$\hat{Y}_1 = a + bX_1 \qquad (14.10)$$

By using the parameters of the line and the values of the independent and dependent variables, residuals can also be calculated without having to compute the predicted Y values directly:

$$RES = Y - (a + bX) \qquad (14.11)$$

Why is so much attention given to calculation of the residuals from regression? Geographers gain two general insights when examining residual values associated with a matched pair of data values. First, the size or magnitude of each residual provides the absolute amount of error associated with that data value. For some values, the residual or error is small, and the independent variable accurately predicts the value of the dependent variable. In other cases, the residual is large, indicating poor prediction of the dependent variable. The smaller the magnitude of a residual, the smaller the vertical distance on the scattergram separating the point from the regression line and the less error associated with the data value.

Second, the direction of residuals from the line is important to researchers. For some data values, the actual value of Y exceeds the predicted value, and the residual is positive ($RES > 0$). In these instances, the point lies above the regression line on the graph, and the model underestimates the actual value of the dependent variable. For other values, the predicted value of Y exceeds the corresponding actual value, the residuals are negative ($RES < 0$), and the points lie below the line. In these cases, the regression model has overestimated the actual value of Y.

The insights that can be gained from residual analysis are illustrated by examining two matched pairs of data values from the Snowbelt example—sample locations 14 and 18. Using equation 14.11 and the data values listed in table 14.1, residuals for these values are calculated as follows:

Point 14: $RES_{14} = Y_{14} - (a + bX_{14})$
$= 96 - [113.51 + (-1.68)(15.8)]$
$= 96 - 86.97 = 9.03$

Point 18: $RES_{18} = Y_{18} - (a + bX_{18})$
$= 86 - [113.51 + (-1.68)(11.4)]$
$= 86 - 94.36 = -8.36$

Point 14, located approximately 16 miles from Lake Erie and near the center of the study area, represents a site with a positive residual (figure 14.8, point 14). Positioned above the regression line on the scattergram, with a predicted Y value of 86.97, the model underestimates snowfall at this site by 9 inches.

On the other hand, sample point 18, located about 11 miles from the lake near the northwest corner of the study area, produces a negative residual whose magnitude (absolute error) is quite large (figure 14.8, point 18). With an actual snowfall amount

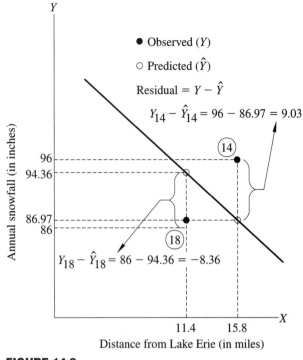

FIGURE 14.8
Interpretation of Residuals from Regression

of 86 inches, but a predicted value of 94.36, the resulting residual (-8.36) lies well below the regression line. The predicted values of Y and the corresponding residuals for each of the 33 study area sites are shown in table 14.3.

Residuals from regression can be used to examine absolute error associated with any data value. Residual analysis can also be used to determine additional information concerning the total error in the model as well as the relative error for any observation. Recall that the unexplained or residual variation is that part of the variation in Y that cannot be explained by the independent variable, X. Also called the "residual sum of squares," this component of variation measures the total error in the regression model and can be derived by summing the squared residuals:

$$RSS = \Sigma y_u^2 = \Sigma(RES)^2 = \Sigma(Y - \hat{Y})^2 \quad (14.12)$$

where RSS = residual sum of squares

Unexplained variation can also be computed by subtracting the explained sum of squares from the total sum of squares:

$$\Sigma y_u^2 = \Sigma y^2 - \Sigma y_e^2 \qquad (14.13)$$

Although total error may be a useful index in some regression problems, a measure of the error associated with a typical or average value is often more

TABLE 14.3

Actual and Predicted Values of the Dependent Variable and Absolute and Standardized Residual Values for Snowbelt Example

Site	Y	X	$\hat{Y}$	RES	SRES
1					
2					
3					
4					
5					
6	104	7.7	100.57	3.43	.57
7	102	9.9	96.88	5.12	.86
8	99	11.2	94.69	4.31	.72
9	92	12.9	91.84	.16	.03
10	95	9.9	103.26	−8.26	−1.38
11	102	8.3	99.57	2.43	.41
12	105	10.9	95.20	9.80	1.64
13	101	13.6	90.66	10.34	1.73
14	96	15.8	87.97	9.03	1.51
15	90	17.1	84.93	5.22	.87
16	82	18.8	81.90	.07	.01
17	78	20.6	78.90	−.90	−.15
18	86	11.4	94.36	−8.36	−1.40
19	85	14.9	88.48	−3.48	−.58
20	82	17.5	84.11	−2.11	−.35
21	80	20.1	79.74	.26	.04
22	78	22.8	75.21	2.79	.47
23	71	25.0	71.51	−.51	−.09
24	69	26.3	69.33	−.33	−.05
25	67	28.5	65.63	1.37	.23
26	68	18.8	81.93	−13.93	−2.32
27	67	21.0	78.23	−11.23	−1.87
28	66	25.4	70.84	−4.84	−.81
29	64	28.0	66.47	−2.47	−.41
30	62	30.7	61.92	.08	.01
31	60	32.4	59.08	.92	.15
32	59	34.6	55.38	3.62	.60
33	58	36.4	52.36	5.64	.94
34	60	26.7	68.65	−8.65	−1.44
35	59	28.5	65.63	−6.63	−1.11
36	58	32.0	59.75	−1.75	−.29
37	57	35.5	53.87	3.13	.52
38	57	37.7	50.17	6.83	1.14

Y = dependent variable (snowfall)

X = independent variable (distance)

$\hat{Y}$ = predicted Y value = $(a + bX)$

RES = residual = $(Y - \hat{Y})$

SE = standard error = $\sqrt{\dfrac{\Sigma RES^2}{n - 2}} = 5.99$

$SRES$ = standardized residual = (RES/SE)

several ways, using the total error, the residuals, or the actual and predicted values of Y:

$$SE = \sqrt{\frac{\Sigma y_u^2}{n - 2}} = \sqrt{\frac{\Sigma RES}{n - 2}} = \sqrt{\frac{\Sigma (Y - \hat{Y})^2}{n - 2}} \quad (14.14)$$

where $\quad SE$ = standard error of the estimate

$\qquad n - 2$ = degrees of freedom

Calculation of total and relative error for the Snowbelt example illustrates the different uses of these regression indices. The unexplained variation or residual sum of squares can be calculated from the residuals or from the actual and predicted Y values:

$$\Sigma y_u^2 = \Sigma (RES)^2 = \Sigma (Y - \hat{Y})^2 = 1115.6$$

It can also be calculated from the total and explained variation:

$$\Sigma y_u^2 = \Sigma y^2 - \Sigma y_e^2 = 8742.2 - 7626.6 = 1115.6$$

These calculations show that of the 8742.2 units of total variation in the dependent variable, 1115.6 units of variation (inches of snowfall, in this case) are left unaccounted for by the single independent variable of distance from Lake Erie. To see how much of this total error is associated with a typical data value or location in northeast Ohio, the standard error of the estimate must be computed:

$$SE = \sqrt{\frac{\Sigma y_u^2}{n - 2}} = \sqrt{\frac{1115.6}{31}} = \sqrt{35.98} = 5.99$$

In this example, the resulting standard error of 5.99 suggests that the typical data value differs from its predicted or regression value by about 6 inches of snowfall. In other words, the independent variable, distance from Lake Erie, can be used to estimate snowfall at locations in this region with a general precision level of 6 inches.

The standard error of the estimate is also used to produce **standardized residual values.** This simple procedure, analogous to generating Z-scores for a distribution, converts absolute residuals into relative residuals:

$$SRES = \frac{RES}{SE} \quad (14.15)$$

where $SRES$ = standardized residual

In this form, standardized residuals relate the magnitude of each residual to the size of the typical residual, represented by the standard error. They can be interpreted as the typical amount of error associated with a value, measured in standard error units.

For example, a value whose residual equals 0 lies on the regression line and has a standardized residual of 0. A residual equal to the standard error value (i.e., a point 1 standard error unit above the regression line) would have a standardized residual

desirable. An index of relative error, called the **standard error of the estimate,** represents the typical distance separating a point from the regression line on the scattergram. Standard error is analogous to standard deviation, which measures the deviation of a typical value from the mean of a distribution. The standard error of the estimate can be calculated in of 1. Standardized residual values have the same sign as their corresponding absolute residual values. Table 14.3 displays the absolute and standardized residual values for the Snowbelt example.

Residual analysis can also be used to identify additional factors that can influence the dependent variable. In bivariate regression, the independent variable explains a portion of the total variation in the dependent variable and leaves the remainder unexplained (residual error.) If this error is interpreted as a new dependent variable, other variables can be identified to explain more of the remaining variation. Thus, residuals from regression should not be viewed as the end of the research process, but rather as an intermediate step in uncovering further influences on the dependent variable.

Geographic use of regression usually involves independent and dependent variables that can be mapped as spatial distributions. In these instances, residual analysis offers a useful method for determining additional influences on the dependent variable: a map of the absolute or standardized residuals. Such maps represent the spatial pattern of error from the regression and show geographically the inaccuracy of the independent variable in estimating the dependent variable. By analyzing spatial trends or nonrandom patterns on residual maps, new variables can often be discovered that serve as additional explanatory influences on the dependent variable.

The use of residual maps in geography is illustrated using the spatial distribution of the absolute residuals from the Snowbelt example (figure 14.9).

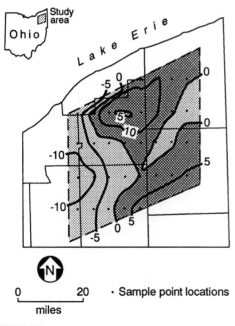

FIGURE 14.9
Residual Map for the Snowbelt Example

The map displays the pattern of error that results when the single variable of distance is used to estimate the level of snowfall across the region. The western portion of the study area appears to have large negative residuals. In this area, the model has greatly overpredicted snowfall levels: the actual levels are much lower than predicted. Locations in the northern and northeastern portions of the study area generally have large positive residuals. In these places, the actual amount of snowfall exceeds the estimated level, indicating substantial underprediction by the regression model.

How does a geographer use a residual map to help identify other variables that may influence the spatial pattern of the dependent variable? To answer this question, two important issues should be considered: (1) Is there another variable with a spatial pattern similar to that of the residuals? (2) Is there a logical, rational reason for this variable to influence the dependent variable? The first factor shows a practical strategy for discovering possible influences in the problem—namely, analyzing the residual map for clues to the new variable. The second factor requires the researcher to have a clear understanding of the geographic patterns and processes being analyzed. Here is another instance where the successful use of statistical analysis in geography requires a strong geographic knowledge or background in the subject under investigation.

14.4 INFERENTIAL USE OF REGRESSION

Sometimes the results from bivariate regression can be tested for significance and results from a sample inferred to the population from which the sample was drawn. This can be done in several ways. For example, inferences can be made concerning the two parameters that define the regression line, the slope, and the Y-intercept. Inferential testing can also be applied to the coefficient of determination, with the measure of strength in regression. However, no matter which inferential procedure is chosen, a stringent set of assumptions applies.

Regression has more assumptions than the other inferential procedures discussed in the text. Variables must be measured on an interval or ratio scale, using one of two modeling schemes. In a **fixed-X model,** the investigator preselects certain values of the independent variable (X), perhaps as part of a controlled experiment. Sample values for the dependent variable (Y) are then derived using a component of randomness. In the **random-X model,** sample values for both the independent and dependent variables are chosen at random. A series of assumptions must be met when using regression in either the fixed-X or random-X models.

1. Since regression analysis places a least-squares line through a scattergram, the variables are assumed to have a linear relationship (figure 14.10, case 1). If the association between variables cannot be represented by a straight line, the regression line will not accurately depict the true relationship between the variables.

2. For every value of the independent variable (X), the distribution of residual or error values ($Y - \hat{Y}$) should be normal, and the mean of the residuals should be zero. If these assumptions are met, a normal distribution of residuals is "centered" on the regression line for any value of X (figure 14.10, case 2). Meeting these requirements makes it virtually certain that variables X and Y are themselves normally distributed. In practice, however, geographers seldom have an adequate number of Y values for every value of X, making it difficult (if not impossible) to validate these assumptions with sample data.

3. For every value of the independent variable (X), the variance of residual error is assumed to be equal. This is known as the "homoscedasticity" or "equal variance" requirement. The funnel-shaped scattergram (figure 14.10, case 3) is almost certainly heteroscedastic, for the variance of residuals at X_1 appears considerably greater than the variance of residuals at X_2.

4. The value of each residual is independent of all other residual values. This requirement assumes that the residuals are randomly arranged or sequenced along the regression line and are not systematically influenced or affected by the magnitude of X. If the data values represent locations, the spatial pattern of residuals is also assumed to be random. If this assumption is violated, autocorrelation is present in the residuals. The scattergram shown in figure 14.10, case 4, appears to be autocorrelated. Note the nonrandom patterning or sequencing of residuals: (a) when the value of X is low, residuals are positive; (b) when the value of X is intermediate, residuals are negative; and (c) when the value of X is high, residuals are positive. The scattergram in case 4 also appears curvilinear, violating assumption 1.

Inferential testing of the regression parameters considers the Y-intercept (α) and slope (β), which define the population regression line relating the independent and dependent variables:

$$Y = \alpha + \beta X \qquad (14.16)$$

The sample Y-intercept (a) and sample slope (b) are the best estimators of their respective population parameters.

Inferential testing of the population intercept is seldom used in geographic problem solving and regression modeling. For many geographic applications, the predicated or estimated value of Y when X is zero has little practical meaning. If the Y-intercept value lies outside the domain of X, testing its significance is not particularly worthwhile. In the Snowbelt example, the relationship between distance from Lake Erie and snowfall is curvi-linear, and as a result, the study area excluded locations within 6 miles of the lake. Thus, the intercept when $X = 0$ has no valid interpretation.

Inferential testing of the population slope or regression coefficient is often very useful in geographic analysis. The slope indicates the responsiveness of the dependent variable to changes in the independent variable, and this relationship may have direct spatial interpretations. In the Snowbelt problem, a geographer can predict the magnitude of decrease in snowfall amount expected at greater distances from Lake Erie. Also, if results from inferential testing show no significant slope (i.e., if $\beta = 0$), then no significant relationship exists between the two variables. In these instances, the regression model has no predictive value.

Inferential significance tests can also be applied to the population coefficient of determination (ρ^2),

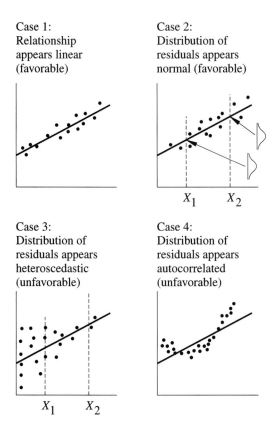

Case 1:
Relationship appears linear (favorable)

Case 2:
Distribution of residuals appears normal (favorable)

Case 3:
Distribution of residuals appears heteroscedastic (unfavorable)

Case 4:
Distribution of residuals appears autocorrelated (unfavorable)

FIGURE 14.10
Scatterplots and Residual Patterns in Regression Analysis

using the sample coefficient of determination (r^2) as the best estimator. Analysis of variance (or F statistic) is used to evaluate the significance of r^2. In this context, the null hypothesis is that the population coefficient of determination is not significantly greater than zero (H_0: $\rho^2 = 0$), and the alternate hypothesis is the converse (H_A: $\rho^2 \neq 0$).

Recall that the total variation in the dependent variable is equal to the sum of two components—the explained variation and the unexplained variation (equation 14.6). It was also shown that the coefficient of determination is the ratio of explained variation to total variation in Y (equation 14.8). Thus, returning to the bucket and sponge analogy, testing the significance of r^2 to determine if it is significantly greater than zero is equivalent to testing whether the sponge (explained variation) removes a significant amount of water (total variation) from the bucket.

These components of variation in Y may be expressed in terms of the "sum of squares," illustrating the direct equivalence between regression and analysis of variance:

$\Sigma y^2 = \Sigma y_e^2 + \Sigma y_u^2$ (from equation 14.6)

$\Sigma y^2 = TSS$ (total sum of squares)

$\Sigma y_e^2 = RSS$ (regression or explained sum of squares)

$\Sigma y_u^2 = ESS$ (error, residual, or unexplained sum of squares)

Therefore: $TSS = RSS + ESS$ and

$$r^2 = \frac{\Sigma y_e^2}{\Sigma y^2} = \frac{RSS}{TSS}$$

The F statistic from analysis of variance is expressed in terms of the coefficient of determination:

$$F = \frac{r^2(n - 2)}{1 - r^2} \qquad (14.17)$$

This F statistic for regression is the square of the t statistic used in the previous chapter to test the significance of r, the simple correlation coefficient (equation 13.9). In the Snowbelt example, the significance testing results in the following:

$$F = \frac{r^2(n - 2)}{1 - r^2} = \frac{.873(33 - 2)}{1 - .873} = 213.09$$

$$p = .0000$$

From this result, it can be concluded that the independent variable (distance from Lake Erie) accounts for a significant amount of the total variation in the dependent variable (snowfall).

Bivariate Regression Analysis

Primary Objective: Determine if an independent variable (X) accounts for a significant portion of the total variation in a dependent variable (Y)

Requirements and Assumptions:

1. Variables are measured on interval or ratio scale
2. Fixed-X Model: Values of independent variable (X) chosen by the investigator, and values of dependent variable (Y) randomly selected for each X
   Random-X Model: Values of both X and Y randomly selected
3. Variables have a linear association
4. For every value of X_i, the distribution of residuals ($Y - \hat{Y}$) should be normal, and the mean of the residuals should equal zero
5. For every value of X, the variance of residual error is equal (homoscedastic)
6. The value of each residual is independent of all other residual values (no autocorrelation)

Hypotheses:

H_0: $\rho^2 = 0$
H_A: $\rho^2 \neq 0$

Test Statistic:

$$F = \frac{r^2(n - 2)}{1 - r^2}$$

14.5 BASIC CONCEPTS OF MULTIVARIATE REGRESSION

Although bivariate regression provides an effective way to understand the use and interpretation of regression in geographic research, most real-world problems require a multivariate approach. **Multiple regression,** where a set of independent variables is used to explain a single dependent variable, offers a logical extension to the simple, two-variable regression procedure. In the bucket and sponge analogy, this approach is equivalent to using additional sponges to absorb a greater volume of water from the bucket.

As in the bivariate case, the selection of independent variables for multiple regression is an important step in applying this statistical method properly. In geographic studies, the focus of investigation is on explaining the spatial pattern of the dependent variable. Independent variables must be obtained that relate to or explain this dependent variable. Since use of regression suggests a functional relationship between the variables, the independent variables need to be evaluated carefully to ensure that they all show a logical relationship to the dependent variable.

Multivariate regression proceeds in a way similar to that used in bivariate analysis. The statistical technique determines the functional relationships linking the independent variables to the dependent variable. The strength of these relationships is measured and the remaining error analyzed. Because multiple regression uses more than one independent variable, the procedure provides information on the absolute and relative ability of each variable to explain the dependent variable.

Multiple regression first examines the functional relationship between the independent variables and the single dependent variable. In bivariate regression, this objective could be explored graphically by observing the pattern of points plotted in a two-dimensional scattergram. However, one dimension (axis on a graph) is required for each variable. Except for problems using only two independent variables, graphical representation is not practical in multivariate regression. Nevertheless, the form of the relationship connecting the independent variables to the dependent variable can be presented and interpreted algebraically.

Using the least-squares objective discussed earlier in this chapter, a regression line is generated to represent the prevailing trend of points in the two-variable scattergram:

$$Y = a + bX_1 \qquad (14.18)$$

In those multivariate problems having two independent variables, data values are located in a three-dimensional space, and a best-fitting plane (figure 14.11) can be derived:

$$Y = a + b_1 X_1 + b_2 X_2 \qquad (14.19)$$

For problems having three or more independent variables, the same procedure is applied algebraically:

$$Y = a + b_1 X_1 + b_2 X_2 + \ldots + b_n X_n \quad (14.20)$$

where

$$
\begin{aligned}
Y &= \text{dependent variable} \\
X_1 \ldots X_n &= \text{independent variables} \\
a &= Y\text{-intercept or constant} \\
b_1 \ldots b_n &= \text{regression coefficients}
\end{aligned}
$$

In both the bivariate and multivariate equations, the constant, or a-value, shows the value of Y when the values of all X variables are zero. The major difference between the equations lies in the regression coefficients. In the two-variable problem, one regression coefficient or b-value is produced to show the influence of a single independent variable on the dependent variable. In the multivariate case, however, one regression coefficient (b_i) is calculated for each independent variable (X_i). Each coefficient indicates the absolute influence of an independent variable on the dependent variable. Like the bivariate case, each regression coefficient shows the

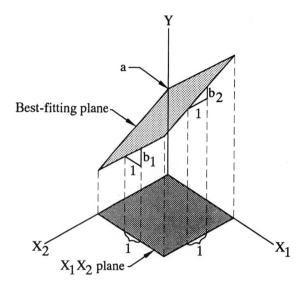

FIGURE 14.11
Three-dimensional Graph of Multiple Regression Plane

change in Y associated with a unit change in the given X variable.

Because multiple regression uses more than one independent variable to explain a dependent variable, the relative importance of each variable is a major concern. At first, it may appear that the regression coefficients could be used directly to determine the importance of independent variables. However, because the measurement units influence these parameters, they cannot be used directly.

This situation is analogous to the problems encountered when using standard deviation to compare relative levels of variation between two or more variables. As discussed in chapter 3, when variables are measured on different scales, the coefficient of variation provides a better measure for comparing the relative variation between variables. In multiple regression, a valid comparison of the *relative* ability of each independent variable to explain the variation in the dependent variable is accomplished by computing **standardized regression coefficients,** also known as "beta values." These coefficients serve as relative indices of strength, allowing a direct comparison of the influence of each independent variable in accounting for the variation in the dependent variable. Among the set of independent variables, the one with the largest beta value produces the strongest relationship to the dependent variable. Like the interpretation of regression coefficients, the sign of each beta value (positive or negative) determines whether this relationship is direct or inverse.

In problems having more than one independent variable, measuring the total amount of variation explained by the set of independent variables is also important. In earlier discussions on bivariate regression, two indices provided absolute and relative measures of strength. Explained variation measured the total level of variation in the dependent variable that could be explained or accounted for by the single independent variable. The explained variation was then converted to a relative index, called the "coefficient of determination," or r^2.

Similarly, in multivariate regression, a **coefficient of multiple determination** (R^2) is calculated as

$$R^2 = \frac{\Sigma y_e^2}{\Sigma y^2}$$

(14.21)

where R^2 = coefficient of multiple determination

      Σy_e^2 = explained variation

      Σy^2 = total variation in Y

This index measures the ratio of variation explained by the set of independent variables to the total variation in the dependent variable. When the index is multiplied by 100, the result is interpreted as the percentage of variation explained.

KEY TERMS

bivariate and multivariate regression, 211
coefficient of determination, 217
coefficient of multiple determination, 224
dependent and independent variables, 211
domain of X, 213
explained and unexplained variation, 216
fixed-X and random-X models, 220
least-squares regression line, 212
multiple regression, 222
residuals, 217
slope or regression coefficient, 213
standard error of the estimate, 219
standardized regression coefficients
   (beta values), 223
standardized residual values, 219
total sum of squares, 216
total variation, 216
Y-intercept, 212

MAJOR GOALS AND OBJECTIVES

If you have mastered the material in this chapter, you should now be able to do the following.

1. Explain the nature and purposes of regression analysis.

2. Recognize geographic problems for which regression analysis is a more useful statistical procedure than correlation.

3. Understand how the form or nature of relationship between variables is measured with a least-squares regression line.

4. Distinguish between positive (direct), negative (inverse), and no relationship between variables and identify the corresponding slope of regression line.

5. Explain how strength of relationship between variables is measured, including an understanding of total variation, explained and unexplained variation, covariation, and the coefficient of determination.

6. Explain the importance of residual analysis in regression for geographic research (including an explanation of the value of residual maps).

7. List (and briefly explain) the assumptions that must be met when using regression for inferential problem solving.

8. Explain the basic concepts and purposes of multivariate regression.

REFERENCES AND ADDITIONAL READING

Abler, R. F., J. S. Adams, and P. R. Gould. 1971. *Spatial Organization: The Geographer's View of the World*. Englewood Cliffs, NJ: Prentice-Hall.

Clark, W. A. V. and P. L. Hosking. 1986. *Statistical Methods for Geographers*. New York: John Wiley and Sons.

Draper, N. R. and H. Smith. 1998. *Applied Regression Analysis* (3rd edition). New York: Wiley.

Ebdon, D. 1985. *Statistics in Geography: A Practical Approach*. Oxford: Basil Blackwell.

Falk, R. and A. D. Well. 1997. "Many Faces of the Correlation Coefficient." *Journal of Statistics Education*. www.stat.ncsu.edu/info/jse/v5n3/falk.html.

Griffith, D. A. and C. G. Amrhein. 1991. *Statistical Analysis for Geographers*. Englewood Cliffs, NJ: Prentice-Hall.

Griffith, D. A. and C. G. Amrhein (contributor). 1997. *Multivariate Statistical Analysis for Geographers*. Englewood Cliffs, NJ: Prentice-Hall.

Johnston, R. J. 1987. *Multivariate Statistical Analysis in Geography: A Primer on the General Linear Model*. London: Longman.

Neter, J., W. Wasserman, and M. Kutner. 1983. *Applied Linear Regression Models*. Homewood, IL: Irwin.

Rogers, J. L. and W. A. Nicewander. 1988. "Thirteen Ways to Look at the Correlation Coefficient." *The American Statistician* 42:59–66.

CHAPTER 15

Epilogue: Geographic Problem Solving in Practical Situations

One of the primary goals of this textbook has been to provide you with an introduction to the broad spectrum of statistical techniques that geographers use to approach and solve spatial problems. It is neither possible nor desirable to discuss all quantitative procedures available to geographers in a single introductory textbook. For example, this text does not discuss such important techniques available to geographers as multivariate statistical analysis, surface modeling techniques, input/output analysis, or Lorenz curve/Gini coefficient. If you are interested in learning more about the techniques covered in the textbook, we encourage you to consult more advanced references. We hope this book has provided you with a solid foundation in statistical problem solving and that this learning experience will serve as a springboard for further study.

Another of our goals in writing this textbook has been to provide the foundation to make you an effective scientific problem solver. A litmus test for mastery of this material is your ability to select the appropriate statistical technique or procedure in a given geographic situation. In many real-world settings, geographers are called on to solve problems and to select methods for finding appropriate answers. After successfully completing the material in this book, you should be able to decide which method will solve a geographic problem most effectively.

Consider the following list of geographic scenarios that illustrate the types of real-world issues and tasks typically faced by professional geographers:

1. Suppose you are working for an environmental planning agency or a consulting firm that is conducting natural hazards research. You need to distribute questionnaires to a representative sample of residents along the floodplain of a river to learn their attitudes toward alternative flood management policies.
2. Suppose you are working for the World Health Organization in eastern Africa to determine the current spatial distribution of the population infected with HIV. You need to gather information that will allow you to develop the best strategy to reduce the spread of AIDS.
3. Suppose you are a spatial analyst working for a metropolitan police department. The department is implementing a Geographic Information System referenced to a criminal records file that contains a variety of socioeconomic data. The goal of this system is to monitor spatial patterns related to crime, allocate police personnel to districts, and reduce crime rates.

Having completed the material in this book, you should now know the best approaches to problem solving in each of the above scenarios. The following list of questions are important points that often need to be resolved:

1. What are the issues that need to be analyzed?
2. What are target and sample populations?
3. What subgroups must be identified within the population?
4. How can a usable spatial sampling frame be generated?
5. What sampling procedure should be used?
6. What type of data collection should be used?
7. What questions should be included in the survey?
8. How should each question be worded?
9. What logistical problems must be overcome?
10. What sample size should be used?
11. What summary descriptive statistics should be calculated?
12. What inferential tests should be used to identify differences between subgroups?

TABLE 15.1

Organizational Structure for Selection of Appropriate Inferential Test

Level of measurement	Type of question Questions about differences			
	One sample	Dependent sample (matched-pairs)	Two samples	Three or more (k) samples
Nominal scale (categorical variables)	Chi-square goodness-of-fit (χ^2)			
Ordinal Scale	Chi-square goodness-of-fit (χ^2)	Wilcoxon matched-pairs signed-ranks (Z_W)	Wilcoxon (Mann-Whitney) rank sum (Z_W)	Kruskal-Wallis analysis of variance by ranks (H)
	Kolmogorov-Smirnov goodness-of-fit (D)			
Interval/ratio scale	One-sample difference of means (Z or t)	Matched-pairs (t_{mp})	Two-sample difference of means (Z or t)	Analysis of variance (ANOVA)(F)
	One-sample difference of proportions (Z)		Two-sample difference of proportions (Z_p)	
	Nearest neighbor analysis (Z_n)			
	Quadrat analysis (χ^2)			
	Area pattern analysis (Z_b)			

Most of these questions are open-ended, dealing mainly with issues of sample design and implementation. They involve collecting and preparing data as well as other early steps in the geographic research process discussed in chapter 1. However, the final two questions *refer directly to the statistical methods* geographers use to solve problems, which is the primary focus of the textbook. Question 11 concerns the selection of an appropriate descriptive method, and question 12 concerns inferential problem solving.

The remainder of this chapter contains a series of geographic situations or scenarios. With each scenario, you will be given enough information to make an informed decision regarding which statistical method(s) to use. Perhaps the most difficult decision involves selection of the inferential test. Recall from chapter 8 (table 8.7) that the selection of an inferential method may require considering such dimensions as

1. type of question asked,
2. level of measurement used,
3. parameter(s) considered,
4. number of samples, and
5. relationship of samples (independent or dependent).

To help choose the most appropriate inferential technique, the body of table 8.7 is completed here (table 15.1). This updated framework includes every inferential technique presented in chapters 8–14. It can be used as a decision-making guide for most of the geographic situations that follow.

As you read each of the following geographic situations or scenarios, carefully consider the information presented. This information is designed to help direct you toward a certain statistical technique or procedure. Keep in mind that not all of these geographic situations refer to inferential procedures. For a number of items, the best choice is a descriptive statistic (e.g., coefficient of variation, mean center) or a probability method (e.g., binomial, probability map).

Following the list of geographic situations, we have included a corresponding list of reasonable answers. With many of these scenarios, enough information is provided to direct you to a single best answer, but other questions in the list may have more than one reasonable solution. With some questions, you may be able to identify a single technique that offers the only reasonable approach to the problem. In other cases, you may be able to identify a group of procedures that would apply to that problem. Do not view the list of solutions as definitive. If you are innovative, you may identify valid alternatives that we have not listed that could apply to the situation.

TABLE 15.1 *(CONTINUED)*

Level of measurement	Type of question	
	Questions about strength of relationship between two variables	Questions about form or nature of relationship between two variables
Nominal scale (categorical variables)	Contingency analysis (χ^2)	
Ordinal scale	Contingency analysis (χ^2) Spearman rank correlation coefficient (Z_{r_s})	
Interval/ratio scale	Pearson correlation coefficient (t_r)	Bivariate regression analysis (F)

GEOGRAPHIC PROBLEM-SOLVING SITUATIONS OR SCENARIOS

1. Wanda Wolf, the head of scientific research at Redrock National Park, wants to determine the attitudes of park visitors regarding various park activities and services. She decides to give a survey questionnaire to every tenth car entering the park. What type of sample is Wanda planning to take?

2. A cartographer is studying the map-interpretation abilities of different people. A random sample of 12 seniors and 16 freshmen (28 total people) answer a set of interpretive questions from the map, and the 28 scores are placed in a single, overall rank order. How would the cartographer test the hypothesis that higher map interpretation scores occur for the seniors?

3. The executive body overseeing establishment of the new European Union has 200 members appointed in proportion to the *population* of the 15 participating countries. How would a researcher test to see if the distribution of representatives by country is also proportional to the *financial assets* of the states?

4. The U.S. Department of Housing and Urban Development is attempting to reduce crime rates in various public housing projects. Data are collected from a sample of projects for a number of variables thought to be associated with this crime, including resident age, income, unemployment, and existence of support programs. What test could be used to determine the degree of influence of any of these factors on crime?

5. A farmer in Manitoba is concerned about the threat of hailstorms and wants to know whether he should purchase insurance. He has a 35-year record of hailstorm frequency and knows that anywhere from 0 to 4 storms have hit his farm in the past, with an average frequency of 1.4 hailstorms per year. How can he calculate the likelihood of a hailstorm hitting his farm next year?

6. An economic geographer wants to examine the relationship between population size and the number of retail establishments for a set of Australian cities. What technique should be applied?

7. An oceanographer wants to analyze the pattern of manganese nodules in the North Atlantic to determine if the pattern is more clustered than random. What technique should be used?

8. A small village in southeastern England has a 78-year record showing the number of days in May with measurable precipitation (greater than 0.10 centimeters). Some years are totally dry, with no measurable precipitation, but during one year, rainfall occurred 16 days of the month. The average for the 78-year record is 3.2 days with measurable precipitation. What method should be used to calculate the likelihood of 10 or more days of rainfall for this upcoming May?

9. As part of a recreation planning survey at a park, questions are asked of all persons visiting the park on a particular day. What type of sample is built into this sample design?

10. A regional planner obtains random samples of home costs from four Michigan counties. Suppose these home costs are only slightly skewed in each county and that home prices within each county vary about the same amount. What test should the planner use to learn if the differences in home cost by county are significant?

11. A geographer wants to see if monthly rents for student housing vary according to the distance of the rental property from campus. He suspects the closer an apartment unit to campus, the higher the rent. How can he test whether this distance decay effect is present?

12. Officials in Somerset, England, know from a recent sample survey that 3.2 percent of their residents are "senior elderly," that is, age 80 or older. What will they be calculating if they try to estimate the likelihood that the true population proportion of senior elderly is between 3.0 and 3.4 percent?

13. A work-study student assigned to the administrative office at the university is trying to find out what activities the students want in the soon-to-be-constructed student union. The only

specification the office has is that each student must have an equal chance of being selected. What is the simplest type of sampling that the student could recommend?

14. Many European politicians suspect a strong relationship between the age of an individual and his or her willingness to "give up" a national currency for the Euro. How could a geographer test this hypothesis using one or more samples totaling at least 1,000 persons from the countries participating in the European Union?

15. The Barranca Board of Education wants some information about local residents' attitudes toward spending more money on public education. Plans are to take a random sample of 285 community voters and ask them (among other questions in the survey) their political party affiliation (Democrat, Republican) and their attitude about increasing the level of public education expenditures (strongly agree, agree, neither agree nor disagree, disagree, or strongly disagree). What technique could be used to determine if there is a relationship between political party affiliation and attitude toward funding for public education?

16. A social services planner in Chicago, Illinois, is concerned about people's access to a set of service centers scattered around the metropolitan area and does not want anyone living too far away from any of the centers. How could she test whether the pattern of these centers is actually more dispersed than random?

17. From a recent nationwide study it is known that the typical American watches 25 hours of television per week (with a population standard deviation of 5.6 hours). Suppose 50 New Orleans residents are randomly selected and their viewing hours are calculated (both average and standard deviation). What technique should be used to determine whether New Orleans television viewing habits differ from the nationwide viewing habits?

18. A geographer is looking at two maps of Nashville, Tennessee. One map has a point indicating the location of every pawnshop, and this point pattern seems clustered near the downtown area. The other map shows the location of grocery stores, a pattern that is rather evenly dispersed throughout the metro region. What descriptive spatial statistic can measure these patterns?

19. A demographer has recorded the number of children per family in 140 households in the Indian state of Kerala. She suspects from a quickly sketched histogram that this distribution is very

positively skewed. What inferential statistic could she use to test this suspicion?

20. Geographic consultant Pierre Portage is studying visitor activity patterns in the provincial park system of Quebec. Administrators of the park system want to know if park attendance levels differ from May to August, using the same set of parks in each case. What technique should Pierre advise them to apply?

21. A geographer has two isoline maps of Pakistan: one showing mean annual precipitation and the other showing population density. Suppose the same random sample of points is selected from each map (i.e., if a particular location is chosen on the first map, that same location on the second map is also included in the analysis). Assume that both variables are from normally distributed spatial populations. What technique will determine the degree to which the two map surfaces are associated?

22. A physical geographer is preparing to take soil samples from an area that is 20 percent marshland. She suspects that environmental degradation is most severe in the marshland portion of the study area; so she wants to be sure it is proportionally represented in the sample. What type of spatial point sampling should she apply?

23. Environmentalists are investigating the effects of thermal plume (higher temperatures) in Chesapeake Bay around a water discharge pipe at Calvert Cliffs nuclear power plant. Suppose power plant personnel wish to estimate surface water temperatures at various distances from the discharge point. What statistical procedure would be most appropriate?

24. Geoscientist Gary Gabbro is studying soil erosion rates of several soil orders (alfisols, aridosols, etc.). He eventually wants to use the ANOVA test to evaluate differences in erosion rate by soil order. He has taken a systematic spatial sample of observations from each soil type, but is not sure if any of the underlying populations from which these samples have been taken is normally distributed, an ANOVA assumption. What should he do before running the ANOVA?

25. Chinese agricultural officials are conducting an extensive study of crop damage resulting from insect infestations in the lower Liao River Valley. It is hypothesized that one possible factor associated with the extent of insect damage is the amount of fertilizer used by different farmers. What test would be used to determine the importance of fertilizer use levels on the amount of insect damage?

26. An urban geographer wants to analyze a map pattern depicting the number of existing home sales by census tract in Detroit, Michigan. He suspects some blockbusting is occurring and wants to determine the extent of home sales clustering in certain neighborhoods. How should he proceed?

27. A random sample exit poll of Jefferson County voters resulted in the following opinions regarding a commercial-retail blue law that would prohibit businesses from being open on Sunday. In Drumlintown, 48 percent were in favor of such a law, while outside Drumlintown, 54 percent were in favor. How would a geographer test the hypothesis that support for the blue law is higher outside Drumlintown than in the city?

28. A geography student has a table listing all of the states, and after the state name is a 0 or 1: 0 if the state has not passed legislation making English the official language and 1 if it has. If the student makes a map using this information, what technique can she use to determine if the spatial pattern of states making English the official language is random?

29. The people of Great Britain seem less interested in joining the European Union when compared to many other European countries. By contrast, a recent referendum in Germany showed 71 percent of their voters favored joining the Union. Using recent information from a sample of British voters, how would a British economist discover if the percentage of British favoring the Union is significantly lower than the percentage of Germans who share this attitude? Is this a one-tailed or two-tailed test?

30. A medical geographer wants to predict the number of physicians living in a city based on the median family income of the city. Suppose this relationship has been tested for another set of cities in the state, but not for this particular place. What statistical technique is appropriate?

31. A seismologist in New Zealand has been monitoring earthquakes in the region for years. Many years no "noticeable" quakes (above 3 on the Richter scale) occurred, but one year there were six such quakes. How will the seismologist calculate the probability of more than one earthquake of this magnitude next year?

32. Suppose an urban planner in San Francisco, California, wants to determine if residents in rental units differ from residents in owner-occupied units with regard to the percentage favoring city rent control legislation. If random samples of residents are taken from both types of residents, what statistical test should be used?

33. Transportation planners in Minneapolis, Minnesota, are concerned about the effect of weather on the ridership levels of public transit. The feeling is that days with "bad" weather conditions (cold and snow) tend to attract more riders than "good" weather days (warm and sunny). Given daily ridership data from local bus routes, and knowing the overall population variability in these rates, how could the difference in ridership levels be statistically validated?

34. A researcher is working with a 50-year climatological record of Atlanta, Georgia, and knows the number of days that have temperatures above 90°F for each of those years. When she constructs a histogram and frequency polygon showing the distribution of 50 values, it doesn't look at all symmetric. A few years had many more hot days than the other years. What descriptive statistic can be used to measure this lack of symmetry?

35. Experts who handle snow-making equipment for Tamarac Lodge and Ski Resort suspect a relationship between the direction of the overnight wind and the need to make snow during the late winter/early spring season. Using a sample(s) of data on these conditions during this period for the last five seasons, how could they test this relationship?

36. In a list of 50 test scores, the average score is 74.1 out of 100. What statistic could be calculated to determine how much the typical score varies from this mean?

37. From a random sample of trees in a national forest, a biogeographer wants to identify which trees are diseased, plot their locations on a map, and analyze the subsequent point pattern to determine whether the trees are randomly distributed throughout the national forest or clustered in certain portions of the forest. This will provide some guidance concerning the most appropriate type of treatment (widespread aerial spraying vs. concentrated treatment of specific areas from the ground). What test should be used?

38. Jill Doakes, currently with the U.S. Bureau of the Census, has been asked to determine the location of the geographic center of population at the end of 1999. This is the point where a rigid map of the country would balance if equal weights (each representing the location of one person) were positioned on it. What statistical technique should she use to make this calculation?

39. The cost of a "market basket" of groceries is obtained from a random sample of grocery stores in Boston, Atlanta, Minneapolis, and San Diego. The total cost of the same list of food items (same

national brands and sizes) is recorded for each store in every city surveyed. What test should be used to learn if the market basket costs between cities are different?

40. A region contains three different types of rock: limestone, calcareous marl, and sandstone. Suppose a similar-sized area is sampled in each rock type, and the number of springs in each rock type is tabulated. What test can be used to determine if the density of springs differs by rock type?

41. A Japanese city has five major shopping centers. A random sample of shoppers at each center is asked how much money they spent shopping in the center that day. What statistical test could be used to see if the typical shopper at the different centers spends differing amounts of money? Assume the pattern of expenditures at each center is not normal, but very positively skewed.

42. Elmwood, Oklahoma, has two restaurants. Suppose a random sample of patrons at each restaurant is asked the distance they have traveled to eat (the distribution of distances is severely nonnormal). What test is appropriate to determine whether patrons travel different distances to eat?

43. A geographer plans to make a choropleth map of Texas counties, showing median family income. However, she wants to use a method of classification that will highlight the unusual counties (outliers) in the data. That way, those looking at the map will quickly spot the "extreme" counties. What method of classification should she use?

44. A climatologist is conducting a study of "lake effect snow" along the shores of Lake Ontario. He has collected data from a number of monitoring sites at different distances from the lake and recorded the total amount of snowfall at each of these monitoring sites through last winter. What statistical technique should he use to determine if there is any relationship between snowfall amount and distance from the lake?

45. A choropleth map of states in Mexico used the following legend categories: 11.37 to 16.26, 16.27 to 21.36, 21.37 to 26.36, 26.37 to 30.14. What method of classification has been used?

46. The housing department at Everbrown University has just completed a sample of 86 students. Each student surveyed was asked the distance of his/her rental unit from campus and the monthly rental cost of that unit. Given this information, what technique could housing personnel use to predict the monthly rent for a student who selects an apartment 1.5 miles from campus?

47. A recent earthquake in eastern Turkey resulted in the following number of deaths and injuries in 11 villages located at increasing distances from the earthquake epicenter: 520, 410, 320, 310, 50, 170, 210, 140, 100, 50, 20. That is, 520 deaths and injuries occurred in the village closest to the epicenter, 410 deaths and injuries occurred in the village second closest to the epicenter, and so on. If no further information is available, how can these data be used to decide if proximity to the epicenter is related to the number of casualties and injuries?

48. A random sample of residents whose homes are on the floodplain of the Mississippi River are surveyed regarding their attitudes toward various flood management policies. After a flood has occurred, the same set of residents is surveyed again about these issues. What test should be used to determine if attitudes regarding flood management policies have changed?

49. Steve ("Smokey") Jones is a ranger in Plywood National Forest, and he has access to forest fire records over a 65-year period. Most years are fire-free, but one dry year 16 fires burned ten acres or more. He has discovered the average is 1.8 fires per year. How can he calculate the odds of a fire in Plywood next year?

50. A 1997 survey of winter guests at Tamarac Lodge showed that 28 percent of the respondents did not ski during their visit. A consultant expects that a significantly higher percentage of visitors to Tamarac Lodge this upcoming winter will not ski. How can she test this hypothesis? Is this a one-tailed or two-tailed test?

51. In previous years, the overall average travel time for a boat trip down the Colorado River through Grand Canyon National Park was six days. Because of river congestion and overuse, park personnel think the trip might take longer now. How can they determine if boat trips now take longer?

52. A geographic researcher has been studying the relationship of distance from Lake Ontario and the amount of "lake effect snow." Some friends of hers live in this study area and have a cabin two miles from the lake. What technique should she apply to estimate how much snowfall they received last winter?

53. A glaciologist studying temperature variation in Antarctica has proposed that temperatures are significantly warmer today than 300 years ago. Using 100 test sites around the continent, he has (1) measured the average temperature over the last year, and (2) estimated the temperature at those sites 300 years ago from the chemical composition of ice bores. How would he test for warming over this time period?

54. A geographer believes that tertiary economic activity in a metropolitan area is clustered near locations having high accessibility (such as major highway interchanges). If each sampled retail establishment in the analysis is plotted as a point on an area map, what technique could be used to analyze this pattern statistically?

55. An economic geographer is studying the spatial patterns of income in Melbourne, Australia, and it is known that the average family income is $38,346. A sample survey of 15 Melbourne families reveals an average of $40,480. How can the geographer test to see if the sample of 15 families is representative of Melbourne?

56. A cartographer is studying the relative effectiveness of different types of maps. The same map is produced by computer in two ways: color and black-and-white. A random sample of 15 people is selected to answer a set of interpretive questions from the color map, and another random sample of 15 people is selected to answer the same set of questions from the black-and-white map. The score for each participant is recorded. How would the cartographer test the hypothesis that higher map interpretation scores occur when the color map is used? Is this a one-tailed or two-tailed test?

57. Environmentalists are investigating the effects of thermal plume (higher temperatures) in Chesapeake Bay around a water discharge pipe at Calvert Cliffs nuclear power plant. What technique could they use to see if the temperature of surface water is related to distance from the pipe?

58. An environmental planner has been collecting sulfur dioxide level readings from a sample of monitoring sites located at different distances downwind from a coal-burning utility power plant in the Ohio River Valley. What statistical technique can she use to predict the sulfur dioxide level at an unmonitored site 12 miles downwind from the power plant?

59. Texas is known for taking capital punishment seriously; more executions occur here than in any other state. However, a geographer is curious about whether attitudes vary across the state. As one part of the study, people from El Paso (west Texas) and Beaumont (east Texas) are sampled. In each city, the geographer has estimated what percentage of population is in favor of the death penalty. What technique should be used to see if these attitudes differ statistically?

60 A hurricane has recently moved up the Atlantic coast of the United States, causing damage in major coastal cities. Suppose the amount of damage (in millions of dollars) is as follows—with the cities listed by location from south to north (i.e., the southernmost city is listed first; the northernmost city last): 786, 451, 507, 410, 202, 51, 171, 35, 27, 58. How can these data be evaluated to learn if location northward or southward is significantly related to the amount of damage?

61. The agricultural extension agent in an Oregon county is concerned about the adverse effects of hailstorms on farms in her region. Hailstorm activity has been recorded over the last 55 years. Some years, no hailstorms occur anywhere in the county. However, in one extreme year, nine hailstorms occurred. How can she calculate the probability of the county experiencing exactly two hailstorms next year?

62. The Vermont Tourist Bureau wants to target its television advertising for the upcoming winter season. The bureau director feels that visitors to the state come equally from the Boston, New York, and metropolitan regions in southern Canada. However, many resort owners and bureau members believe that Boston is the clear leader for tourists. Using a sample(s) of data from the previous season on the origins of visiting tourists to each of 50 resorts in the state, how could this question be resolved?

63. The chair of a local environmental action group feels that a strong relationship exists between the volume of recycled material left along the curb for pickup and the age of the homeowner. What method would be used to confirm this observation statistically?

64. Suppose a demographer for the Ontario provincial government is looking at both the number of migrants moving into economic development areas and the number of migrants leaving the same sample set of areas. How can she test the number of in-migrants and the number of out-migrants for statistically significant differences?

65. A climatologist is looking at modified climographs from ten weather stations and is focusing on the plots of average monthly precipitation for each of the stations. Some graphs have a very high spike (showing that almost all the annual precipitation falls in just a couple of months), whereas other graphs seem to indicate precipitation falls rather evenly throughout the year. What descriptive statistic can be used to formally quantify these visual observations?

66. A climatologist wants to analyze a tornado "touchdown" point pattern map for the state of Arkansas. The overall pattern appears to be somewhat clustered, with the highest density of touchdowns in the northwestern portion.

What inferential technique can be used to determine whether the pattern is more clustered than random?

67. Recent financial difficulties have affected the economies of many Asian countries. A researcher has both the 1995 and current annual volume of exports from the same set of randomly selected Asian countries. He notices the distributions of these export volumes is very positively skewed, with a few countries (e.g., Japan) having much larger volumes of exports than other countries. How can he test to learn whether the volume of Asian exports has declined?

68. Personnel at an Ontario provincial park want to locate a new campground on a riverfront site that will not be flooded too frequently. Budget managers advise a location be chosen that will be flooded no more frequently than once a decade. Historical records show that flooding has occurred nine different years on that site in the last century. What procedure should be used to determine the probability that this site will not be flooded too frequently in the next 25 years?

69. An urban geographer is part of an interdisciplinary team collecting data from city residents. One action that has been suggested is to survey every home in selected city blocks to save time and money. What sampling strategy would this be?

70. An hydrologist is studying the volume of material carried by two rivers (expressed in grams of solid material per liter of water). One river drains an agricultural area, while the other drains a forested area. To test for differences in volume of material carried, 12 random water samples are taken from both rivers (one sample a week through the summer months). What test should be used? Is this a one-tailed or two-tailed test?

71. An economic development planner is interested in evaluating changing spatial patterns of poverty in Appalachia. She has a choropleth map series at the county level that shows each county as either above regional median family income (shaded) or below this income figure (unshaded). How can she test whether poverty is becoming more spatially concentrated over time?

72. A fluvial geomorphologist is studying surface runoff levels in two different environments: (1) a natural setting with no modification by human activity, and (2) an urban setting, highly modified through development. To test for significant differences in runoff, spatial random samples are taken from both settings. Suppose the runoff data are ordered from high to low within the two sampled areas. What test should be used to see whether the mean runoff from the urban area exceeds that in the rural setting?

73. The community of Sliding Stone, Pennsylvania, has five clearly identifiable neighborhoods. A geography major at the local university thinks lot sizes differ by neighborhood, so she selects a random sample of ten lots from each neighborhood. She assumes that lot sizes within each neighborhood vary by differing amounts. What statistical technique should she use to verify her hypothesis?

74. London, England, has a long, normally distributed record of annual precipitation. Someone wants to know what statistical procedure should be used to estimate the likelihood that more than 1,000 millimeters of precipitation will fall next year. As a step in this process, what statistic needs to be calculated?

75. A British political geographer wants to determine if the Labour party candidate for Parliament has a similar level of support among the voting population in two neighboring towns. The proportion of a random sample of voters in each town favoring the candidate is recorded. How may one test if the difference in support for the candidate between the two towns is significant?

76. A geographer is helping store management refine their "typical customer" demographic profile. She has an alphabetical list of all customers who have a store credit card and wants to send a detailed survey to every 20th cardholder. This 5 percent sample, with the possibility of careful follow-up on nonrespondents, will save the store considerable money and time. What type of sampling is she suggesting?

77. Suppose there are two choropleth maps of Texas counties—one showing median family income in one of five ordinal categories, and another showing the percentage of labor force in professional occupations classified into four ordinal categories. What technique should be used to learn if the two choropleth map patterns are related?

78. A private high school is situated in the center of a five-neighborhood region. Each neighborhood has about the same high school age population. A random selection of 100 students at the high school resulted in the following counts (22, 24, 13, 24, and 17). That is, 22 students selected are from neighborhood 1, 24 from neighborhood 2, and so on. What technique can be used to help determine if student enrollments vary by neighborhood?

79. An urban geographer is examining citizens' attitudes toward growth in a rapidly expanding suburban area. The research design is set up to survey a random sample of 230 people at two different times. People's attitudes toward growth

will first be measured before construction of a major new regional shopping mall; then the same people will be resurveyed after the mall has been open about six months. How can the geographer test whether respondents' attitudes toward growth have changed significantly?

80. A cultural geographer is interested in looking for a possible relationship between ethnic/racial type and eating preferences (type of restaurant food preferred). How would he test to see if significant differences exist in a sample population(s) in a city like Atlanta, Georgia?

81. The U.S. Department of Housing and Urban Development continues to try to reduce crime rates in public housing projects. Data are available for many variables, including age, income, unemployment, and so on. What test could be used to estimate the crime rate in a housing project not included in the sample?

82. A researcher wants to depict graphically the average annual snowfall amounts that are 75 percent likely to be exceeded for 85 cities distributed across western and central Canada (British Columbia, Alberta, Saskatchewan, and Manitoba). What should be done?

83. Suppose an urban geographer is studying historical changes in housing characteristics in London, England. She wishes to determine if house size (square feet of living area) and the number of years since home construction are associated. What method could she use to learn if such a relationship exists?

84. In recent years, it seems that there has often been a surplus of oil on the international market. Has this affected the volume of oil exported by the Organization of Petroleum Exporting Countries (OPEC)? With both the 1990 and current annual volume of oil exported from each OPEC country, how could a geographer test whether the volume of oil exports has decreased significantly?

85. In studies concerning crop damage caused by insect infestations in the lower Liao River Valley of China, the role of fertilizer is examined in detail. What test would be appropriate if officials want to predict the amount of insect damage on a farm not included in the study sample, solely based on the amount of fertilizer used?

86. A recreation geographer is testing the effectiveness of an attendance prediction model from a set of cities to a system of regional state parks. It is expected that attendance will be directly related to city population size and inversely related to distance from the city to the state park. What statistical technique could be used to test the validity of this model?

87. Political geographer Gerry Mander has a metropolitan area map showing the 78 election districts. For each district, Gerry knows whether the majority of registered voters is Republican or Democrat. What technique should he use to analyze this map to learn if political party affiliation is more clustered than random? Is this a one-tailed or two-tailed test?

88. A medical geographer has data from the Centers for Disease Control listing the number of AIDS cases per 100,000 population by county for all 3,143 counties in the United States. His task is to construct a choropleth map based on a classification of these AIDS figures. What method of classification would be most appropriate if the goal is to have roughly an equal number of areas allocated to each of five categories?

89. From a representative sample of British voters, how would a political geographer determine if the percentage of voters supporting conversion to the Euro differs significantly by gender?

90. Rates of tuberculosis infection appear higher in those neighborhoods of Lagos, Nigeria, that have lower levels of industrial pollution. The World Health Organization (WHO) planner in the area, Georgia Peach, argues this finding appears contrary to expectations. If she takes samples from a set of "more polluted" and "less polluted" neighborhoods, how can she test to see if tuberculosis infection rates differ?

91. A cartography student has set up a research paper project. She wants to contrast the map interpretation skills of male versus female university students. Taking map interpretation test scores from 18 male and 20 female students, she places them in overall rank order, from highest score to lowest score. What technique should she use to determine if male and female map interpretation skills are similar or different?

92. A sample of residents is surveyed to determine the number of miles traveled per week for reasons not related to work. A year later, following large increases in gasoline prices, the same individuals are surveyed a second time to monitor changes in their discretionary travel behavior. How could researchers test whether the change in gasoline price led to decreased automobile use?

93. A regional planner is studying income inequalities in a set of counties in northern Appalachia. Median family income by county is known for both 1990 and 1995. How can the planner descriptively measure the relative dispersion of income among counties in 1990, then compare that directly with relative dispersion of 1995 income?

94. On the wall of the fire department headquarters in Chicago, Illinois, a large map shows the location of the fire stations and recent fires that have occurred. Someone is using a ruler to measure the straight-line distance from the fire station to a fire. A visiting geographer says, "No, that is not the true distance!" What measure of distance should be used?

95. An Ontario demographer is examining the number of migrants moving into economic development areas and the number of migrants leaving the same sample of areas. Information is only available from a ranked list: which economic development area had the largest number of in-migrants, second largest number of in-migrants, and so on. How can she test for significant differences between the in-migrants and out-migrants?

96. Joe Doakes is working on a new "Atlas of Australia" and has been asked to construct an isoline map showing the likelihood that more than 20 inches of precipitation will occur at various weather stations across the country. What should Joe create to complete this task?

97. Suppose the average age of New Zealand residents is 29.6 years. Raymond Yachtclub (the new planning director of Wellington) wants to determine if the city has a typical age profile, with a mean age similar to the national average. If information is collected from a sample of 80 Wellington residents, what test should he use?

98. According to a recent nationwide study, only 54 percent of all Americans can locate Mexico on a world map (a true fact). From a sample of 120 students in a geography course, 63 percent were able to locate Mexico. What test should be used to determine if the sample of students is a typical nationwide group? Is this a one-tailed or two-tailed test?

99. A community in Texas has six neighborhood parks. The park and recreation planner takes surveys of a random sample of visitors to each of these parks, asking each visitor the distance traveled from home to the park (such data are assumed to be from nonnormal populations). What test could be used to determine if park visitors travel different distances to the six neighborhood parks?

100. A geographer in Birmingham, England, is conducting a housing survey, including questions about a proposed rent control initiative. He knows that about 38 percent of the people of Birmingham live in apartments, and he wants about 38 percent of the surveys given to apartment dwellers. What type of sampling should he use?

101. A medical geographer is concerned with the recent diffusion of influenza in Moscow, Russia. If the number of cases is known for each local administrative unit for several consecutive weeks, the morbidity rate for influenza could be displayed on a series of choropleth maps (one for each week), and the degree of randomness, clustering, or dispersal in each pattern analyzed. What technique would be used to do this?

102. A geographer working for the Soil Conservation Service wants to know if average soybean yields per acre from three soil types are identical. A number of soybean yields is collected from a sample of farms in each of the soil types. The populations are assumed to be normally distributed. What statistical test should be used?

103. The proportion of total "low-weight" live births appears to be rising in rural villages of Nigeria, according to Georgia Geographicus, a health care planner with the World Health Organization (WHO). She thinks the proportion of low-weight births is now significantly higher than that reported last year by WHO for the entire African continent. How can she test whether this is correct?

104. The efficiency and equity of the Los Angeles, California, fire protection system are being examined by geographic consultant Joe "Sparky" Doakes. He has been asked to look at the spacing of fire stations throughout the city. Among other issues, fire officials have expressed concern that some homes may be too far from the nearest fire station and that a more dispersed pattern of stations might be needed. What technique should Sparky use to help address these concerns?

105. From a sample of less-developed countries and more-developed countries, data are collected on the percentage of deaths in each country caused by heart disease and other heart-related problems. What technique should be used to learn if there is a difference between LDC and MDC death rates as a result of these causes?

106. Winter tourism levels seem to be increasing steadily in the Canadian Rockies. A survey in 1995 calculated the seasonal revenue from tourism for 45 communities, each having fewer than 3,000 year-round residents. How could a recreational planner learn if seasonal revenue from last year has increased since 1995 for these same communities?

107. An environmental geographer wants to predict lake acidity level based on its distance from a pollution source. An inverse relationship is sus-

pected between acidity level and distance from the source. What technique should be used?

108. The SAT scores of 50 Middlebury State students, 41 Salisbury State students, and 83 Boysenbury State students are randomly selected, and an overall ranking of scores is created. What statistical test should be used to determine whether the SAT scores between schools are different?

109. An urban planner is examining the locational variation of housing quality in a metro area. The study region contains 87 local election districts, and for each election district, the number of residential units failing to meet at least one housing code standard is listed. Next, each of the 87 districts is classified as either having more substandard units than average or fewer substandard units than average, and these results are mapped. The planner suspects that the substandard units are mostly concentrated in election districts near the city center. What inferential procedure should be used to test this hypothesis?

110. Marcia Monadnock is the resource manager and recreation activities planner at a ski resort in the Catskill Mountains. She thinks that weather conditions in New York City on the Fridays before a weekend have a psychological and behavioral impact on winter skiers traveling to the resort. Specifically, she thinks if it is a "warm" Friday, there will be fewer skiers, and if it is a "cold" Friday, there will be more skiers, regardless of the actual conditions on the slopes. How could she see if her perception is correct, using data from the "twelve critical weekends" last winter?

111. Coralie Ollis is a geographer doing some consulting for a major car insurance company. She has pulled a random sample of 375 accidents from company files, and her hypothesis is that the average dollar value of insurance claims varies significantly by season (spring, summer, autumn, winter). How should she test this hypothesis statistically?

112. From a random sample of small villages in the Scottish Highlands, a researcher has collected information on average male life expectancy and average female life expectancy for each village. What technique can be used to see if male and female life expectancies are different?

113. A climatologist stationed at Thule, Greenland, wants to learn if December daily low temperatures at a weather station differ under two types of conditions: (1) at least one inch of snow on the ground, and (2) less than one inch of snow on the ground. Data are collected for a number of sample days under both types of surface conditions, and average low temperatures recorded. What statistical test should be used?

114. After completing a local pollution abatement program, Florence Floss wants to know if average pollution levels from a sample of ten sites are significantly lower than national Environmental Protection Agency standards. The national average EPA standard and the amount of variability around that mean are known from existing reports. What statistical procedure should Ms. Floss follow?

115. An economic geographer has two ranked lists of European countries: ranks of inflation in 1950 and ranks of current inflation for those same countries. How can she test whether the current inflation pattern is different from the 1950 pattern?

116. Religious affiliation (Protestant, Catholic, Jewish, other) and voting preference (Democrat, Republican, Independent) data were collected from each of 250 people in a New York City precinct and the results tabulated. How could the idea be tested that religious affiliation is associated with voting preference?

117. The historical record at a site along the lower Brahmaputra River in Bangladesh indicates that flooding occurs at least once during the rainy season four years out of ten (40 percent of the time). How would the probability be estimated that flooding will occur at least once in six of the next ten years?

118. A medical geographer is concerned about attendance at a regional health care clinic that serves several neighborhoods. She wants to develop and test a model that predicts clinic use as a function of neighborhood population and inversely related to distance from the clinic. What technique might she use?

119. The mayor of Concord received 60.8 percent of the vote in his first successful election to that office. He is running for reelection, and a poll of 110 expected voters has been taken. What technique will tell his campaign manager if the current level of support differs significantly from the first election?

120. Legislators in Aspen, Colorado, want to determine if year-round residents differ from seasonal residents with regard to the percentage favoring a law to strengthen the existing noise control ordinance. If random samples are collected from both types of residents what test should be used?

121. How would a geomorphologist test the relationship between type of bedrock material and

the level of soil acidity for a sample of sites in France? That is, do different types of bedrock support soil having varying levels of acidity?

122. In Banff National Park, Alberta, some observers think the number of bear sightings has increased dramatically. The park headquarters has data listing the number of sightings at each campground from 20 years ago. How could a resource manager test this historical data against a newly collected set of sightings from the same campgrounds to determine if any increase has occurred?

123. The Chinook City Council feels that the turnover rate of land parcels is not equivalent across all types of land use. The council has set up a scheme to tabulate type of land use (residential, commercial-retail, industrial) against the timing of the last sale of that land parcel (sold in the last year, sold one to five years ago, sold more than five years ago). If they randomly select 300 land sales from the city records, how can they determine if a relationship exists between date of sale and type of land use?

124. Transportation planners in the metropolitan area of Eskerville know that the average commuter travels 7.3 miles to work (each way). A detailed study is to be made of 20 commuters (who have a sample average commuting distance of 7.8 miles). How can planners test whether the commuting distances of sampled commuters are typical of all Eskerville commuters?

125. A cultural geographer surveying attitudes regarding abortion in Chicago, Illinois, wants to predict the level of support for legal abortion in a ward of the city, based on the known percentage of ward population that is Catholic. Such a relationship has already been examined for a sample of other Chicago wards. What technique will the geographer use?

126. A climatologist has 85 years of annual snowfall data for Denver, Colorado. What inferential statistical procedure can be applied to learn if the distribution of annual snowfalls is normal?

127. Economic development planners for Kentucky have two choropleth maps of Kentucky counties—one showing categories of median family income, another showing categories of percentage of labor force in professional occupations. They want to learn if these two choropleth map patterns are related. The only current information is the maps themselves; the data from which the maps were made are not readily available. What technique should they use?

128. An environmental planner has been collecting sulfur dioxide level readings from a sample of monitoring sites located at different distances downwind from a coal-burning utility power plant in the Ohio River Valley. She suspects that as one moves further downwind from the power plant, the level of sulfur dioxide exposure declines. How can she test this relationship statistically?

129. City planners want to determine the attitudes of residents concerning the need for constructing a public swimming pool, and a 5 percent sample of families will be randomly selected. The average family size in the entire community is known, and the planners want to test that the average family size in the 5 percent sample group is representative of the family size communitywide. What test should be used?

130. Experts studying the rate of inflation for European states suspect a pattern. They feel that countries occupied by Germany during World War II have a higher rate of inflation than those not occupied by the Third Reich. How could they test this hypothesis? Is this a one-tailed or two-tailed test?

131. A research worker in Kenya is studying the distribution of several medical problems: (1) an eye infection known as "river blindness"; (2) those testing HIV positive; and (3) an infectious form of "sleeping sickness." Data are collected on the number of people infected with each medical problem from a random sample of 100 people on a river floodplain and a second random sample of 125 people living in an adjacent highlands area. How can one test the hypothesis that the incidence of types of infection is related to the physical location (river floodplain, highland)?

132. A corn farmer in Nebraska has discovered that at least 3 inches of precipitation are needed during the growing season to avoid irrigation. Historical precipitation data indicate that in 15 of the last 18 years precipitation was sufficient. If the farmer can afford to irrigate only one year in five, how can she determine whether it is profitable to plant corn at this location?

133. Suppose an economic geographer is given a choropleth map of Pennsylvania counties, with each county classified in a binary way as either above or below the statewide average regarding the percentage of county population below the poverty level. She hypothesizes that the Appalachian portion of the commonwealth is poorer, and she needs to select a technique that can test whether the map pattern is more clustered than random. What technique should she use? Is this a one-tailed or two-tailed test?

134. Economic observers suspect that a strong relationship can be found between a country's pop-

ular vote (by number) for the European Union and a country's total volume of stock activity recorded annually in their national market. How could this relationship be tested?

135. An environmental planner is conducting a study of hazard perception along a major river. A questionnaire lists three different management strategies: strengthen the levee, rechannel a portion of the water upstream, and relocate particularly hazardous structures. The study contains three distinct groups (strata): those living on the 25-year floodplain, those living on the 100-year floodplain, and those living on higher ground above the 100-year floodplain. How can the environmental planner test whether the preferred management strategy is related to the respondent's location?

136. In 1995, the Ontario Bureau of Tourism compiled a "top ten" list from a provincewide sample survey, ranking cities in the province by "livability." The Bureau wants to take these ten places from the 1995 list and ask Ontario residents to rank them in terms of "livability" at the present time to see if the rankings change. What statistical test should the Bureau use to measure the degree to which the two top ten lists are related?

137. The residential subdivision where Fred Foehn lives is less than half a mile from an interstate highway. Fred has been measuring noise levels at noon each day in his front yard and noting the corresponding wind direction at the same time. Fred thinks noise level is related to wind direction. What technique can he use to test this hypothesis?

138. A new vocational-technical school may be built in a community. A recent citywide survey indicated that 53 percent were in favor of constructing such a facility. If a sample is taken, how can it be determined if the sample respondents' views are typical of city residents at large?

139. A cartographer is designing a world map showing birthrates by country. According to the study director, the following categories of birthrate are to be used: less than 10, 10 to 19.9, 20 to 29.9, 30 to 39.9, and 40 or greater. What method of classification is being used in this map design?

140. Urban geographers have long thought that home lot size increases with distance from the central business district (CBD). How would a geographer provide a statistical test of this hypothesized relationship to estimate home lot size six miles from the CBD?

141. A demographer is studying population growth rates within Nigeria to see if this variable is re-

lated to religious affiliation. The researcher has collected data for a sample of families that show both religious affiliation and number of children. How could this relationship be examined?

142. A community planner in Middlesex, England, has assisted in the recent development of a new town (planned community) project. Among other attractions, the development has on-site health care facilities; so a logical hypothesis is that residents in this planned town are older than the typical resident in Middlesex. If a sample of 50 residents is taken, and these data are compared with the statistics for all of Middlesex that are already known, what statistical technique should be applied?

143. Complete historical climate records for Fairbanks, Alaska, indicate that 18.6 percent of the days have clear weather, with no appreciable cloudiness. Suppose a sample of 50 days in the last year is selected at random, and 26 percent of these days had clear weather. What test should be used to determine if the sample days are typical or representative of the overall climate record?

144. In a study of human impacts on soil chemistry along a segment of bayfront beach, an environmental researcher hypothesizes that the impacts are most severe in the tidal wetland portion of the study area, which encompasses about 15 percent of the total study area. Therefore, he wants to "oversample" in that area, taking about 50 percent of the soil samples from the tidal wetlands. What type of sampling is he advocating?

145. Information from Europe at the onset of World War II showed that European nations on average had an inflation rate of 7.9 percent. How would someone test whether the overall current rate of inflation for these countries differs significantly from the level before World War II?

146. A planner records the level of noise 200 yards behind a noise barrier running along an interstate highway at 12:00 noon each day for 4 months. The noise barrier is 10 yards off the highway. She also records the level of noise over the same period of time 210 yards away from the same highway, but in an adjacent area *not protected* by the noise barrier. Using a sample or samples from these two data sets, how would she test whether there is a significant difference in the percentage of days that exceed an acceptable noise level?

147. The U.S. Geological Survey wants to map the likelihood that a magnitude 5.0 or greater earthquake (Richter scale) will occur across San Fernando Valley in southern California. What should the survey do?

148. A spatial analyst for the Center for Climatological Research is currently examining five different point pattern maps, with each map showing the spatial distribution of severe hailstorms in a five-county area in Colorado for a particular year. How could he analyze and directly compare the amount of relative dispersion or variability on the five maps?

149. A fluvial geomorphologist is studying surface runoff levels in two different environments: (1) a natural setting with no modification by human activity, and (2) an urban setting, highly modified through development. To test for significant differences in runoff, spatial random samples are taken from both settings. The runoff data are measured at ratio scale, but the population variances are not known. What test should be used to see whether the mean runoff from the urban area exceeds that in the rural setting?

150. A geographer is studying total fertility rates in African countries and wants to correlate these data with other variables using Pearson's product moment correlation coefficient. How would she test this total fertility rate variable to see if it is normally distributed for input into the Pearson's correlation test?

151. A geographer with the World Bank is studying global economic development trends and has collected country-level data on the percentage of labor force in "telecommunications" jobs. He thinks that countries classified as "high," "upper-middle," "middle," "lower-middle," and "lower" income have different percentages of their labor force in this job category. From some previous testing of the data, he has discovered that some distributions are very positively skewed and distinctly nonnormal. What statistical test should he use to explore his hypothesis?

152. A climatologist has measured sulfur dioxide levels in the lower atmosphere repeatedly at seven different weather stations along a line of latitude across Northwest Territories, Canada, for an entire year. How can he test for differences in sulfur dioxide levels between stations?

153. In Pennsylvania counties, percentage of unemployment data are available for 1990 and 1995. How would Terrance Tundra (the current state economic development planner working with labor statistics) measure the relative dispersion or variability of unemployment rates in 2000, to compare with the two earlier dates?

154. The United Nations Population Fund is evaluating the success of a family planning program recently begun in rural areas of Somalia. Previous studies have shown that a 38 percent rate of effective contraceptive use after two years is a suitable international standard for success. How would a demographer test whether the Somali program is effective?

ANSWERS TO GEOGRAPHIC PROBLEM-SOLVING SITUATIONS OR SCENARIOS:

1. systematic sample
2. two-sample difference test—Wilcoxon/Mann-Whitney
3. chi-square goodness-of-fit—proportional
4. correlation analysis
5. Poisson probability
6. Pearson's correlation coefficient
7. point pattern analysis (quadrat)
8. Poisson probability
9. cluster sample
10. analysis of variance (ANOVA)
11. Pearson's correlation coefficient
12. confidence interval
13. random sample
14. contingency analysis
15. contingency analysis
16. point pattern analysis (nearest neighbor)
17. one-sample difference of means—Z
18. standard distance
19. Kolmogorov-Smirnov goodness-of-fit—normal
20. matched-pairs t
21. Pearson's correlation coefficient
22. stratified proportional or systematic
23. regression analysis
24. Kolmogorov-Smirnov goodness-of-fit—normal
25. Pearson's correlation coefficient
26. area pattern analysis
27. two-sample difference of proportions
28. area pattern analysis
29. one-sample difference of proportions, one-tailed
30. regression
31. Poisson probability
32. two-sample difference of proportions
33. two-sample difference of means—Z
34. skewness
35. chi-square goodness-of-fit—equal
36. standard deviation
37. point pattern analysis (quadrat)
38. mean center (weighted mean center)
39. analysis of variance (ANOVA)
40. chi-square goodness-of-fit—equal
41. Kruskal-Wallis analysis of variance
42. two-sample difference test—Wilcoxon/Mann-Whitney
43. natural breaks
44. Pearson's correlation coefficient
45. equal intervals based on range
46. regression analysis

47. Spearman's rank correlation coefficient
48. matched-pairs *t*
49. Poisson probability
50. one-sample difference of proportions, one-tailed
51. one-sample difference of means—*Z* or *t*
52. regression analysis
53. matched-pairs *t*
54. point pattern analysis (quadrat)
55. one-sample difference of means—*t*
56. two-sample difference of means *t*, one-tailed
57. Pearson's correlation coefficient
58. regression analysis
59. two-sample difference of proportions
60. Spearman's rank correlation coefficient
61. Poisson probability
62. contingency analysis
63. Pearson's correlation coefficient
64. matched-pairs *t*
65. kurtosis
66. point pattern analysis—quadrat
67. Wilcoxon matched-pairs signed-ranks
68. binomial probability
69. cluster sample
70. two-sample difference of means *t*, two-tailed
71. area pattern analysis
72. two-sample difference test—Wilcoxon/Mann-Whitney
73. Kruskal-Wallis analysis of variance
74. standard score
75. two-sample difference of proportions
76. systematic sample
77. Spearman's rank correlation coefficient
78. chi-square goodness-of-fit—equal
79. matched-pairs
80. contingency analysis
81. regression analysis
82. probability map
83. correlation
84. matched-pairs *t*
85. regression analysis
86. chi-square goodness-of-fit—proportional
87. area pattern analysis, one-tailed
88. quantile breaks
89. two-sample difference of proportions
90. two-sample difference of proportions
91. two-sample difference test—Wilcoxon/Mann Whitney
92. matched-pairs *t*
93. coefficient of variation
94. Manhattan distance
95. Wilcoxon matched-pairs signed-ranks
96. probability map
97. one-sample difference of means—*Z*
98. one-sample difference of proportions, two-tailed
99. Kruskal-Wallis analysis of variance
100. stratified sample

101. area pattern analysis
102. analysis of variance (ANOVA)
103. one-sample difference of proportions
104. point pattern analysis (nearest neighbor)
105. two-sample difference of proportions
106. matched-pairs *t*
107. regression
108. Kruskal-Wallis analysis of variance
109. area pattern analysis
110. two-sample difference of means—*t*
111. analysis of variance (ANOVA)
112. matched-pairs *t*
113. two-sample difference of means—*Z* or *t*
114. one-sample difference of means—*Z* or *t*
115. Wilcoxon matched-pairs signed-ranks
116. chi-square goodness-of-fit—equal
117. binomial probability
118. chi-square goodness-of-fit—proportional
119. one-sample difference of proportions
120. two-sample difference of proportions
121. chi-square goodness-of-fit—equal
122. matched-pairs *t*
123. contingency analysis
124. one-sample difference of means—*t*
125. regression
126. Kolmogorov-Smirnov goodness-of-fit—normal
127. Spearman's rank correlation coefficient
128. Pearson's correlation coefficient
129. one-sample difference of means—*Z* or *t*
130. two-sample difference of means—*Z* or *t*, one-tailed
131. contingency analysis
132. binomial probability
133. area pattern analysis, one-tailed
134. Pearson's correlation coefficient
135. contingency analysis
136. Spearman's rank correlation coefficient
137. contingency analysis
138. one-sample difference of proportions
139. equal intervals not based on range
140. regression analysis
141. contingency analysis
142. one-sample difference of means—*Z*
143. one-sample difference of proportions
144. disproportionate stratified point sample
145. matched-pairs *t*
146. two-sample difference of proportions
147. probability map
148. coefficient of variation
149. two-sample difference of means—*t*
150. Kolmogorov-Smirnov goodness-of-fit—normal
151. Kruskal-Wallis analysis of variance
152. analysis of variance (ANOVA)
153. coefficient of variation
154. one-sample difference of proportions

Statistical Tables

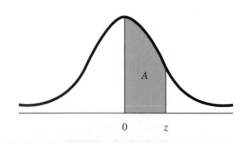

The Normal Table

Note: To get *A* for a given value of *Z*, insert a decimal point before the four digits. For example, $Z = 1.43$ gives $A = 0.4236$.

z	0.00	0.01	0.02	0.03	0.04	0.05	0.06	0.07	0.08	0.09
0.0	0000	0040	0080	0120	0160	0199	0239	0279	0319	0359
0.1	0398	0438	0478	0517	0557	0596	0636	0675	0714	0753
0.2	0793	0832	0871	0910	0948	0987	1026	1064	1103	1141
0.3	1179	1217	1255	1293	1331	1368	1406	1443	1480	1517
0.4	1554	1591	1628	1664	1700	1736	1772	1808	1844	1879
0.5	1915	1950	1985	2019	2054	2088	2123	2157	2190	2224
0.6	2257	2291	2324	2357	2389	2422	2454	2486	2517	2549
0.7	2580	2611	2642	2673	2704	2734	2764	2794	2823	2852
0.8	2881	2910	2939	2967	2995	3023	3051	3078	3106	3133
0.9	3159	3186	3212	3238	3264	3289	3315	3340	3365	3389
1.0	3413	3438	3461	3485	3508	3531	3554	3577	3599	3621
1.1	3643	3665	3686	3708	3729	3749	3770	3790	3810	3830
1.2	3849	3869	3888	3907	3925	3944	3962	3980	3997	4015
1.3	4032	4049	4066	4082	4099	4115	4131	4147	4162	4177
1.4	4192	4207	4222	4236	4251	4265	4279	4292	4306	4319
1.5	4332	4345	4357	4370	4382	4394	4406	4418	4429	4441
1.6	4452	4463	4474	4484	4495	4505	4515	4525	4535	4545
1.7	4554	4564	4573	4582	4591	4599	4608	4616	4625	4633
1.8	4641	4649	4656	4664	4671	4678	4686	4692	4699	4706
1.9	4713	4719	4726	4732	4738	4744	4750	4756	4761	4767
2.0	4772	4778	4783	4788	4793	4798	4803	4808	4812	4817
2.1	4821	4826	4830	4834	4838	4842	4846	4850	4854	4857
2.2	4861	4864	4868	4871	4875	4878	4881	4884	4887	4890
2.3	4893	4896	4898	4901	4904	4906	4909	4911	4913	4916
2.4	4918	4920	4922	4925	4927	4929	4931	4932	4934	4936
2.5	4938	4940	4941	4943	4945	4946	4948	4949	4951	4952
2.6	4953	4955	4956	4957	4959	4960	4961	4962	4963	4964
2.7	4965	4966	4967	4968	4969	4970	4971	4972	4973	4974
2.8	4974	4975	4976	4977	4977	4978	4979	4979	4980	4981
2.9	4981	4982	4982	4983	4984	4984	4985	4985	4986	4986
3.0	4987	4987	4987	4988	4988	4989	4989	4989	4990	4990
3.1	4990	4991	4991	4991	4992	4992	4992	4992	4993	4993
3.2	4993	4993	4994	4994	4994	4994	4994	4995	4995	4995
3.3	4995	4995	4996	4996	4996	4996	4996	4996	4996	4997
3.4	4997	4997	4997	4997	4997	4997	4997	4997	4998	4998
3.5	4998	4998	4998	4998	4998	4998	4998	4998	4998	4998

The area, *A*, stays at 0.4998 until $Z = 3.62$. From $Z = 3.63$ to 3.90 $A = 0.4999$. For $Z > 3.90$, $A = 0.5000$, to four decimal places.
From Leon F. Marzillier, *Elementary Statistics* ©1990 by Wm. C. Brown Publishers.

TABLE B

Table of Random Numbers

31871	60770	59235	41702	89372	28600	30013	18266	65044	61045
87134	32839	17850	37359	27221	92409	94778	17902	09467	86757
06728	16314	81076	42172	46446	09226	96262	77674	70205	98137
95646	67486	05167	07819	79918	83949	45605	18915	79458	54009
44085	87246	47378	98338	40368	02240	72593	52823	79002	88190
83967	84810	51612	81501	10440	48553	67919	73678	83149	47096
49990	02051	64575	70323	07863	59220	01746	94213	82977	42384
65332	16488	04433	37990	93517	18395	72848	97025	38894	46611
42309	04063	55291	72165	96921	53350	34173	39908	11634	87145
84715	41808	12085	72525	91171	09779	07223	75577	20934	92047
63919	83977	72416	55450	47642	01013	17560	54189	73523	33681
97595	78300	93502	25847	19520	16896	69282	16917	04194	25797
17116	42649	89252	61052	78332	15102	47707	28369	60400	15908
34037	84573	49914	59688	18584	53498	94905	14914	23261	58133
08813	14453	70437	49093	69880	99944	40482	04254	62842	68089
67115	41050	65453	04510	35518	88843	15801	86163	49913	46849
14596	62802	33009	74095	34549	76634	64270	67491	83941	55154
70258	26948	60863	47666	58512	91404	97357	85710	03414	56591
83369	81179	32429	34781	00006	65951	40254	71102	60416	43296
83811	49358	75171	34768	70070	76550	14252	97378	79500	97123
14924	71607	74638	01939	77044	18277	68229	09310	63258	85064
60102	56587	29842	12031	00794	90638	21862	72154	19880	80895
33393	30109	42005	47977	26453	15333	45390	89862	70351	36953
92592	78232	19328	29645	69836	91169	95180	15046	45679	94500
27421	73356	53897	26916	52015	26854	42833	64257	49423	39440
26528	22550	36692	25262	61419	53986	73898	80237	71387	32532
07664	10752	95021	17030	76784	86861	12780	44379	31261	18424
37954	72029	29624	09119	13444	22645	78345	79876	37582	75549
66495	11333	81101	69328	84838	76395	35997	07259	66254	47451
72506	28524	39595	49356	92733	42951	47774	75462	64409	69116
09713	70270	28077	15634	36525	91204	48443	50561	92394	60636
51852	70782	93498	44669	79647	06321	04020	00111	24737	05521
31460	22222	18801	00675	57562	97923	45974	75158	94918	40144
14328	05024	04333	04135	53143	79207	85863	04962	89549	63308
84002	98073	52998	05749	45538	26164	68672	97486	32341	99419
89541	28345	22887	79269	55620	68269	88765	72464	11586	52211
50502	39890	81465	00449	09931	12667	30278	63963	84192	25266
30862	61996	73216	12554	01200	63234	41277	20477	71899	05347
36735	58841	35287	51112	47322	81354	51080	72771	53653	42108
11561	81204	68175	93037	47967	74085	05905	86471	47671	18456

From Leon F. Marzillier, *Elementary Statistics* ©1990 by Wm. C. Brown Publishers.

TABLE C

Student's *t* Distribution

							Degrees of freedom									
t	1	2	3	4	5	6	7	8	9	10	11	12	13	14	15	16
0.1	0317	0353	0367	0374	0379	0382	0384	0386	0387	0388	0389	0390	0391	0391	0392	0392
0.2	0628	0700	0729	0744	0753	0760	0764	0768	0770	0773	0774	0776	0777	0778	0779	0780
0.3	0928	1038	1081	1104	1119	1129	1136	1141	1145	1148	1151	1153	1155	1157	1159	1160
0.4	1211	1361	1420	1452	1472	1485	1495	1502	1508	1512	1516	1519	1522	1524	1526	1528
0.5	1476	1667	1743	1783	1809	1826	1838	1847	1855	1861	1865	1869	1873	1876	1878	1881
0.6	1720	1953	2046	2096	2127	2148	2163	2174	2183	2191	2197	2202	2206	2210	2213	2215
0.7	1944	2218	2328	2387	2424	2449	2467	2481	2492	2501	2508	2514	2519	2523	2527	2530
0.8	2148	2462	2589	2657	2700	2729	2750	2766	2778	2788	2797	2804	2810	2815	2819	2823
0.9	2333	2684	2828	2905	2953	2986	3010	3028	3042	3054	3063	3071	3078	3083	3088	3093
1.0	2500	2887	3045	3130	3184	3220	3247	3267	3283	3296	3306	3315	3322	3329	3334	3339
1.1	2651	3070	3242	3335	3393	3433	3461	3483	3501	3514	3526	3535	3544	3551	3557	3562
1.2	2789	3235	3419	3518	3581	3623	3654	3678	3696	3711	3723	3734	3742	3750	3756	3762
1.3	2913	3384	3578	3683	3748	3793	3826	3851	3870	3886	3899	3910	3919	3927	3934	3940
1.4	3026	3518	3720	3829	3898	3945	3979	4005	4025	4041	4055	4066	4075	4084	4091	4097
1.5	3128	3638	3847	3960	4030	4079	4114	4140	4161	4177	4191	4203	4212	4221	4228	4235
1.6	3222	3746	3960	4075	4148	4196	4232	4259	4280	4297	4310	4322	4332	4340	4348	4354
1.7	3307	3844	4062	4178	4251	4300	4335	4362	4383	4400	4414	4426	4435	4444	4451	4458
1.8	3386	3932	4152	4269	4341	4390	4426	4452	4473	4490	4503	4515	4525	4533	4540	4546
1.9	3458	4026	4232	4349	4421	4469	4504	4530	4551	4567	4580	4591	4601	4609	4616	4622
2.0	3524	4082	4303	4419	4490	4538	4572	4597	4617	4633	4646	4657	4666	4674	4680	4686
2.1	3585	4147	4367	4482	4551	4598	4631	4655	4674	4690	4702	4712	4721	4728	4735	4740
2.2	3642	4206	4424	4537	4605	4649	4681	4705	4723	4738	4750	4759	4768	4774	4781	4786
2.3	3695	4259	4475	4585	4651	4694	4725	4748	4765	4779	4790	4799	4807	4813	4819	4824
2.4	3743	4308	4521	4628	4692	4734	4763	4784	4801	4813	4824	4832	4840	4846	4851	4855
2.5	3789	4352	4561	4666	4728	4767	4795	4815	4831	4843	4852	4860	4867	4873	4877	4882
2.6	3831	4392	4598	4700	4759	4797	4823	4842	4856	4868	4877	4884	4890	4895	4900	4903
2.7	3871	4429	4631	4730	4786	4822	4847	4865	4878	4888	4897	4903	4909	4914	4918	4921
2.8	3908	4463	4661	4756	4810	4844	4867	4884	4896	4906	4914	4920	4925	4929	4933	4936
2.9	3943	4494	4687	4779	4831	4863	4885	4901	4912	4921	4928	4933	4938	4942	4945	4948
3.0	3976	4523	4712	4800	4850	4880	4900	4915	4925	4933	4940	4945	4949	4952	4955	4958
3.1	4007	4549	4734	4819	4866	4894	4913	4927	4936	4944	4949	4954	4958			
3.2	4036	4573	4753	4835	4880	4907	4925	4937	4946	4953	4958					
3.3	4063	4596	4771	4850	4893	4918	4934	4946	4954							
3.4	4089	4617	4788	4864	4904	4928	4943	4953								
3.5	4114	4636	4803	4876	4914	4936	4950									
3.6	4138	4654	4816	4886	4922	4943										
3.7	4160	4670	4829	4896	4930	4950										
3.8	4181	4686	4840	4904	4937											
3.9	4201	4701	4850	4912	4943											
4.0	4220	4714	4860	4919	4948											
4.1	4239	4727	4869	4926	4953											
4.2	4256	4739	4877	4932												
4.3	4273	4750	4884	4937												
4.4	4289	4760	4891	4942												
4.5	4304	4770	4898	4946												
4.6	4319	4779	4903	4950												
4.7	4333	4788	4909													
4.8	4346	4796	4914													
4.9	4359	4804	4919													
5.0	4372	4811	4923													
5.8	4456	4858	4949													
6.4	4507	4882														
7.0	4548	4901														
10.0	4683	4951														
12.8	4752															
31.9	4900															
63.7	4950															

TABLE C

(continued)

Degrees of freedom

t	17	18	19	20	21	22	23	24	25	26	27	28	29	30	∞
0.1	0392	0393	0393	0393	0394	0394	0394	0394	0394	0394	0395	0395	0395	0395	0398
0.2	0781	0781	0782	0782	0783	0783	0784	0784	0785	0785	0785	0785	0786	0786	0793
0.3	0928	1162	1163	1164	1164	1165	1166	1166	1167	1167	1168	1168	1168	1169	1179
0.4	1529	1531	1532	1533	1534	1535	1536	1537	1537	1538	1538	1539	1540	1540	1554
0.5	1883	1884	1886	1887	1889	1890	1891	1892	1893	1894	1894	1895	1896	1896	1915
0.6	2218	2220	2222	2224	2225	2227	2228	2229	2230	2231	2232	2233	2234	2235	2257
0.7	2533	2536	2538	2540	2542	2544	2545	2547	2548	2549	2550	2551	2552	2553	2580
0.8	2826	2829	2832	2834	2837	2839	2841	2842	2844	2845	2847	2848	2849	2850	2881
0.9	3097	3100	3103	3106	3108	3111	3113	3115	3116	3118	3120	3121	3122	3124	3159
1.0	3343	3347	3351	3354	3357	3359	3361	3364	3366	3367	3369	3371	3372	3373	3413
1.1	3567	3571	3575	3578	3581	3584	3586	3589	3591	3593	3595	3597	3598	3600	3643
1.2	3767	3772	3776	3779	3782	3785	3788	3791	3793	3795	3797	3799	3801	3802	3849
1.3	3945	3950	3954	3958	3962	3965	3968	3970	3973	3975	3977	3979	3981	3982	4032
1.4	4103	4107	4112	4116	4119	4123	4126	4128	4131	4133	4136	4138	4139	4141	4192
1.5	4240	4245	4250	4254	4258	4261	4264	4267	4269	4272	4274	4276	4278	4280	4332
1.6	4360	4365	4370	4374	4377	4381	4384	4387	4389	4392	4394	4396	4398	4400	4452
1.7	4463	4468	4473	4477	4481	4484	4487	4490	4492	4495	4497	4499	4501	4503	4554
1.8	4552	4557	4561	4565	4569	4572	4575	4578	4580	4583	4585	4587	4589	4590	4641
1.9	4627	4632	4636	4640	4644	4647	4650	4652	4655	4657	4659	4661	4663	4665	4713
2.0	4691	4696	4700	4704	4707	4710	4713	4715	4718	4720	4722	4724	4725	4727	4772
2.1	4745	4750	4753	4757	4760	4763	4766	4768	4770	4772	4774	4776	4777	4779	4821
2.2	4790	4794	4798	4801	4804	4807	4809	4812	4814	4816	4817	4819	4820	4822	4861
2.3	4828	4832	4835	4838	4841	4843	4846	4848	4850	4851	4853	4854	4856	4857	4893
2.4	4859	4863	4866	4869	4871	4874	4876	4877	4879	4881	4882	4884	4885	4886	4918
2.5	4885	4888	4891	4894	4896	4898	4900	4902	4903	4905	4906	4907	4908	4909	4938
2.6	4906	4910	4912	4914	4916	4918	4920	4921	4923	4924	4925	4926	4927	4928	4953
2.7	4924	4927	4929	4931	4933	4935	4936	4937	4939	4940	4941	4942	4943	4944	
2.8	4938	4941	4943	4945	4946	4948	4949	4950	4951	4952	4953	4954	4955	4956	
2.9	4950	4952	4954	4956	4957	4958	4960								

Note: The column headed by ∞ is for use when *df* > 30. It coincides with column 1 of table A.

For *t* > 2.9 and *df* > 16, *A* > 0.4950.

All missing entries in the table have *A* > 0.4950. The reason this value of *A* was chosen is that for *A* > 0.495, *p*-value < 0.005. Therefore, if the value you want for *A* is missing, you may assume the significance level is 0.005 or higher.

For *t* > 5.0, selected values for *t* are given for *df* = 1, 2, 3, to show when significance is reached at various significance levels.

For example, *t* = 7.0, *df* = 2 gives *A* = 0.4901, *p*-value = 0.0099 (< 0.01). Therefore, there is significance at the .01 level. Any value of *t* less than 7.0 would not be significant at the .01 level.

TABLE D

Values of _t_ for Selected Confidence Levels

					Level of confidence					
_df_	0.68	0.80	0.85	0.90	0.925	0.95	0.975	0.98	0.99	0.995
1	1.82	3.08	4.17	6.31	8.45	12.7	25.5	31.8	63.7	12.7
2	1.31	1.89	2.28	2.92	3.44	4.30	6.21	6.97	9.92	14.1
3	1.19	1.64	1.92	2.35	2.68	3.18	4.18	4.54	5.84	7.45
4	1.13	1.53	1.78	2.13	2.39	2.78	3.50	3.75	4.60	5.60
5	1.10	1.48	1.70	2.02	2.24	2.57	3.16	3.36	4.03	4.77
6	1.08	1.44	1.65	1.94	2.15	2.45	2.97	3.14	3.71	4.35
7	1.07	1.41	1.62	1.89	2.09	2.36	2.84	3.00	3.50	4.03
8	1.06	1.40	1.59	1.86	2.05	2.31	2.75	2.90	3.36	3.83
9	1.05	1.38	1.57	1.83	2.01	2.26	2.69	2.82	3.25	3.69
10	1.05	1.37	1.55	1.81	1.99	2.23	2.63	2.76	3.17	3.58
11	1.04	1.36	1.55	1.80	1.97	2.20	2.59	2.72	3.11	3.50
12	1.04	1.36	1.54	1.78	1.95	2.18	2.56	2.68	3.05	3.43
13	1.03	1.35	1.53	1.77	1.94	2.16	2.53	2.65	3.01	3.37
14	1.03	1.35	1.52	1.76	1.92	2.14	2.51	2.62	2.98	3.33
15	1.03	1.34	1.52	1.75	1.91	2.13	2.49	2.60	2.95	3.29
16	1.03	1.34	1.51	1.75	1.90	2.12	2.47	2.58	2.92	3.25
17	1.02	1.33	1.51	1.74	1.90	2.11	2.46	2.57	2.90	3.22
18	1.02	1.33	1.50	1.73	1.89	2.10	2.45	2.55	2.88	3.20
19	1.02	1.33	1.50	1.73	1.88	2.09	2.43	2.54	2.86	3.17
20	1.02	1.33	1.50	1.72	1.88	2.09	2.42	2.53	2.85	3.15
21	1.02	1.32	1.50	1.72	1.87	2.08	2.41	2.52	2.83	3.14
22	1.02	1.32	1.49	1.72	1.87	2.07	2.41	2.51	2.82	3.12
23	1.02	1.32	1.49	1.71	1.86	2.07	2.40	2.50	2.81	3.10
24	1.02	1.32	1.49	1.71	1.86	2.06	2.39	2.49	2.80	3.09
25	1.01	1.32	1.49	1.71	1.86	2.06	2.38	2.49	2.79	3.08
26	1.01	1.31	1.48	1.71	1.85	2.06	2.38	2.48	2.78	3.07
27	1.01	1.31	1.48	1.70	1.85	2.05	2.37	2.47	2.77	3.06
28	1.01	1.31	1.48	1.70	1.85	2.05	2.37	2.47	2.76	3.05
29	1.01	1.31	1.48	1.70	1.85	2.05	2.36	2.46	2.76	3.04
30	1.01	1.31	1.48	1.70	1.84	2.04	2.36	2.46	2.75	3.03
∞	1.00	1.28	1.44	1.64	1.78	1.96	2.24	2.33	2.58	2.81

If you want to construct a 0.95 confidence interval estimate with _df_ = 15, use _t_ = 2.13. Use the last line when _df_ > 30. The values in it coincide with the normal table values.

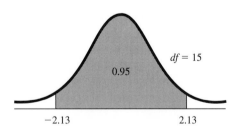

$df = 15$

0.95

−2.13　　2.13

Critical Values of the *F* Distribution

$\alpha = .05$

df_1

df_2	1	2	3	4	5	6	7	8	9	10	11	12
1	161	200	216	225	230	234	237	239	241	242	243	244
2	18.5	19.0	19.2	19.2	19.3	19.3	19.4	19.4	19.4	19.4	19.4	19.4
3	10.1	9.6	9.3	9.1	9.0	8.9	8.9	8.8	8.8	8.8	8.8	8.7
4	7.7	6.9	6.6	6.4	6.3	6.2	6.1	6.0	6.0	6.0	5.9	5.9
5	6.6	5.8	5.4	5.2	5.1	5.0	4.9	4.8	4.8	4.7	4.7	4.7
6	6.0	5.1	4.8	4.5	4.4	4.3	4.2	4.2	4.1	4.1	4.0	4.0
7	5.6	4.7	4.4	4.1	4.0	3.9	3.8	3.7	3.7	3.6	3.6	3.6
8	5.3	4.5	4.1	3.8	3.7	3.6	3.5	3.4	3.4	3.4	3.3	3.3
9	5.1	4.3	3.9	3.6	3.5	3.4	3.3	3.2	3.2	3.1	3.1	3.1
10	5.0	4.1	3.7	3.5	3.3	3.2	3.1	3.1	3.0	3.0	2.9	2.9
11	4.8	4.0	3.6	3.4	3.2	3.1	3.0	3.0	2.9	2.8	2.8	2.8
12	4.8	3.9	3.5	3.3	3.1	3.0	2.9	2.8	2.8	2.8	2.7	2.7
13	4.7	3.8	3.4	3.2	3.0	2.9	2.8	2.8	2.7	2.7	2.6	2.6
14	4.6	3.7	3.3	3.1	3.0	2.8	2.8	2.7	2.6	2.6	2.6	2.5
15	4.5	3.7	3.3	3.1	2.9	2.8	2.7	2.6	2.6	2.6	2.5	2.5
16	4.5	3.6	3.2	3.0	2.8	2.7	2.7	2.6	2.5	2.5	2.4	2.4
17	4.4	3.6	3.2	3.0	2.8	2.7	2.6	2.6	2.5	2.4	2.4	2.4
18	4.4	3.6	3.2	2.9	2.8	2.7	2.6	2.5	2.5	2.4	2.4	2.3
19	4.4	3.5	3.1	2.9	2.7	2.6	2.5	2.5	2.4	2.4	2.3	2.3
20	4.4	3.5	3.1	2.9	2.7	2.6	2.5	2.4	2.4	2.4	2.3	2.3
21	4.3	3.5	3.1	2.8	2.7	2.6	2.5	2.4	2.4	2.3	2.3	2.2
22	4.3	3.4	3.0	2.8	2.7	2.6	2.5	2.4	2.3	2.3	2.3	2.2
23	4.3	3.4	3.0	2.8	2.6	2.5	2.4	2.4	2.3	2.3	2.2	2.2
24	4.3	3.4	3.0	2.8	2.6	2.5	2.4	2.4	2.3	2.3	2.2	2.2
25	4.2	3.4	3.0	2.8	2.6	2.5	2.4	2.3	2.3	2.2	2.2	2.2
26	4.2	3.4	3.0	2.7	2.6	2.5	2.4	2.3	2.3	2.2	2.2	2.2
27	4.2	3.4	3.0	2.7	2.6	2.5	2.4	2.3	2.2	2.2	2.2	2.1
28	4.2	3.3	3.0	2.7	2.6	2.4	2.4	2.3	2.2	2.2	2.2	2.1
29	4.2	3.3	2.9	2.7	2.6	2.4	2.4	2.3	2.2	2.2	2.1	2.1
30	4.2	3.3	2.9	2.7	2.5	2.4	2.3	2.3	2.2	2.2	2.1	2.1
40	4.1	3.3	2.9	2.7	2.5	2.4	2.3	2.3	2.2	2.1	2.0	2.0
50	4.0	3.2	2.8	2.6	2.4	2.3	2.2	2.1	2.1	2.0	2.0	2.0
60	4.0	3.2	2.8	2.5	2.4	2.2	2.2	2.1	2.0	2.0	2.0	1.9
70	4.0	3.1	2.7	2.5	2.4	2.2	2.1	2.1	2.0	2.0	1.9	1.9
80	4.0	3.1	2.7	2.5	2.3	2.2	2.1	2.0	2.0	2.0	1.9	1.9
100	3.9	3.1	2.7	2.5	2.3	2.2	2.1	2.0	2.0	1.9	1.9	1.8
120	3.9	3.1	2.7	2.4	2.3	2.2	2.1	2.0	2.0	1.9	1.9	1.8
∞	3.8	3.0	2.6	2.4	2.2	2.1	2.0	1.9	1.9	1.8	1.8	1.8

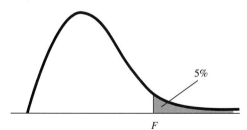

5%

F

From Leon F. Marzillier, *Elementary Statistics* ©1990 by Wm. C. Brown Publishers.

TABLE E

(continued)

$\alpha = .05$

df_1

df_2	15	20	24	30	40	50	60	75	100	120	∞
1	246	248	249	250	251	252	253	253	253	253	254
2	19.4	19.4	19.4	19.4	19.4	19.5	19.5	19.5	19.5	19.5	19.5
3	8.7	8.7	8.6	8.6	8.6	8.6	8.6	8.6	8.6	8.6	8.5
4	5.9	5.8	5.8	5.8	5.7	5.7	5.7	5.7	5.7	5.7	5.6
5	4.6	4.6	4.5	4.5	4.5	4.4	4.4	4.4	4.4	4.4	4.4
6	3.9	3.9	3.8	3.8	3.8	3.8	3.7	3.7	3.7	3.7	3.7
7	3.5	3.4	3.4	3.4	3.3	3.3	3.3	3.3	3.3	3.3	3.2
8	3.2	3.2	3.1	3.1	3.0	3.0	3.0	3.0	3.0	3.0	2.9
9	3.0	2.9	2.9	2.9	2.8	2.8	2.8	2.8	2.8	2.8	2.7
10	2.8	2.8	2.7	2.7	2.7	2.6	2.6	2.6	2.6	2.6	2.5
11	2.7	2.6	2.6	2.6	2.5	2.5	2.5	2.5	2.4	2.4	2.4
12	2.6	2.5	2.5	2.5	2.4	2.4	2.4	2.4	2.4	2.3	2.3
13	2.5	2.5	2.4	2.4	2.3	2.3	2.3	2.3	2.3	2.2	2.2
14	2.5	2.4	2.4	2.3	2.3	2.2	2.2	2.2	2.2	2.2	2.1
15	2.4	2.3	2.3	2.2	2.2	2.2	2.2	2.2	2.1	2.1	2.1
16	2.4	2.3	2.2	2.2	2.2	2.1	2.1	2.1	2.1	2.1	2.0
17	2.3	2.2	2.2	2.2	2.1	2.1	2.1	2.0	2.0	2.0	2.0
18	2.3	2.2	2.2	2.1	2.1	2.0	2.0	2.0	2.0	2.0	1.9
19	2.2	2.2	2.1	2.1	2.0	2.0	2.0	2.0	1.9	1.9	1.9
20	2.2	2.1	2.1	2.0	2.0	2.0	2.0	1.9	1.9	1.9	1.8
21	2.2	2.1	2.1	2.0	2.0	1.9	1.9	1.9	1.9	1.9	1.8
22	2.2	2.1	2.1	2.0	2.0	1.9	1.9	1.9	1.8	1.8	1.8
23	2.1	2.0	2.0	2.0	1.9	1.9	1.9	1.8	1.8	1.8	1.8
24	2.1	2.0	2.0	1.9	1.9	1.9	1.8	1.8	1.8	1.8	1.7
25	2.1	2.0	2.0	1.9	1.9	1.8	1.8	1.8	1.8	1.8	1.7
26	2.1	2.0	2.0	1.9	1.8	1.8	1.8	1.8	1.8	1.8	1.7
27	2.1	2.0	1.9	1.9	1.8	1.8	1.8	1.8	1.7	1.7	1.7
28	2.0	2.0	1.9	1.9	1.8	1.8	1.8	1.8	1.7	1.7	1.6
29	2.0	1.9	1.9	1.8	1.8	1.8	1.8	1.7	1.7	1.7	1.6
30	2.0	1.9	1.9	1.8	1.8	1.8	1.7	1.7	1.7	1.7	1.6
40	1.9	1.8	1.8	1.7	1.7	1.7	1.6	1.6	1.6	1.6	1.5
50	1.9	1.8	1.7	1.7	1.6	1.6	1.6	1.6	1.5	1.5	1.4
60	1.8	1.8	1.7	1.6	1.6	1.6	1.5	I.S	1.5	1.5	1.4
70	1.8	1.7	1.7	1.6	1.6	1.5	1.5	1.5	1.4	1.4	1.4
80	1.8	1.7	1.6	1.6	1.5	1.5	1.5	1.4	1.4	1.4	1.3
100	1.8	1.7	1.6	1.6	1.5	1.5	1.4	1.4	1.4	1.4	1.3
120	1.8	1.7	1.6	1.6	1.5	1.5	1.4	1.4	1.4	1.3	1.2
∞	1.7	1.6	1.5	1.5	1.4	1.4	1.3	1.3	1.2	1.2	1.0

TABLE E

(continued)

$$\alpha = 0.01$$

df_1

df_2	1	2	3	4	5	6	7	8	9	10	11	12
1	4052	4999	5403	5625	5764	5859	5928	5981	6022	6056	6082	6106
2	98.5	99.0	99.2	99.2	99.3	99.3	99.3	99.4	99.4	99.4	99.4	99.4
3	34.1	30.8	29.5	28.7	28.2	27.9	27.7	27.5	27.3	27.2	27.1	27.0
4	21.2	18.0	16.7	16.0	15.5	15.2	15.0	14.8	14.7	14.5	14.4	14.4
5	16.3	13.3	12.1	11.4	11.0	10.7	10.4	10.3	10.2	10.0	10.0	9.9
6	13.7	10.9	9.8	9.2	8.8	8.5	8.3	8.1	8.0	7.9	7.8	7.7
7	12.2	9.6	8.4	7.8	7.5	7.2	7.0	6.8	6.7	6.6	6.5	6.5
8	11.3	8.6	7.6	7.0	6.6	6.4	6.2	6.0	5.9	5.8	5.7	5.7
9	10.6	8.0	7.0	6.4	6.1	5.8	5.6	5.5	5.4	5.3	5.2	5.1
10	10.0	7.6	6.6	6.0	5.6	5.4	5.2	5.1	5.0	4.8	4.8	4.7
11	9.6	7.2	6.2	5.7	5.3	5.1	4.9	4.7	4.6	4.5	4.5	4.4
12	9.3	6.9	6.0	5.4	5.1	4.8	4.6	4.5	4.4	4.3	4.2	4.2
13	9.1	6.7	5.7	5.2	4.9	4.6	4.4	4.3	4.2	4.1	4.0	4.0
14	8.9	6.5	5.6	5.0	4.7	4.5	4.3	4.1	4.0	3.9	3.9	3.8
15	8.7	6.4	5.4	4.9	4.6	4.3	4.1	4.0	3.9	3.8	3.7	3.7
16	8.5	6.2	5.3	4.8	4.4	4.2	4.0	3.9	3.8	3.7	3.6	3.6
17	8.4	6.1	5.2	4.7	4.3	4.1	3.9	3.8	3.7	3.6	3.5	3.4
18	8.3	6.0	5.1	4.6	4.2	4.0	3.8	3.7	3.6	3.5	3.4	3.4
19	8.2	5.9	5.0	4.5	4.2	3.9	3.8	3.6	3.5	3.4	3.4	3.3
20	8.1	5.8	4.9	4.4	4.1	3.9	3.7	3.6	3.4	3.4	3.3	3.2
21	8.0	5.8	4.9	4.4	4.0	3.8	3.6	3.5	3.4	3.3	3.2	3.2
22	7.9	5.7	4.8	4.3	4.0	3.8	3.6	3.4	3.4	3.3	3.2	3.1
23	7.9	5.7	4.8	4.3	3.9	3.7	3.5	3.4	3.3	3.2	3.1	3.1
24	7.8	5.6	4.7	4.2	3.9	3.7	3.5	3.4	3.2	3.2	3.1	3.0
25	7.8	5.6	4.7	4.2	3.9	3.6	3.5	3.3	3.2	3.1	3.0	3.0
26	7.7	5.5	4.6	4.1	3.8	3.6	3.4	3.3	3.2	3.1	3.0	3.0
27	7.7	5.5	4.6	4.1	3.8	3.6	3.4	3.3	3.1	3.1	3.0	2.9
28	7.6	5.4	4.6	4.1	3.8	3.5	3.4	3.2	3.1	3.0	3.0	2.9
29	7.6	5.4	4.5	4.0	3.7	3.5	3.3	3.2	3.1	3.0	2.9	2.9
30	7.6	5.4	4.5	4.0	3.7	3.5	3.3	3.2	3.1	3.0	2.9	2.8
40	7.3	5.2	4.3	3.8	3.5	3.3	3.1	3.0	2.9	2.8	2.7	2.7
50	7.2	5.1	4.2	3.7	3.4	3.2	3.0	2.9	2.8	2.7	2.6	2.6
60	7.1	5.0	4.1	3.6	3.3	3.1	3.0	2.8	2.7	2.6	2.6	2.5
70	7.0	4.9	4.1	3.6	3.3	3.1	2.9	2.8	2.7	2.6	2.5	2.4
80	7.0	4.9	4.0	3.6	3.2	3.0	2.9	2.7	2.6	2.6	2.5	2.4
100	6.9	4.8	4.0	3.5	3.2	3.0	2.8	2.7	2.6	2.5	2.4	2.4
120	6.8	4.8	4.0	3.5	3.2	3.0	2.8	2.7	2.6	2.5	2.4	2.3
∞	6.6	4.6	3.8	3.3	3.0	2.8	2.6	2.5	2.4	2.3	2.2	2.2

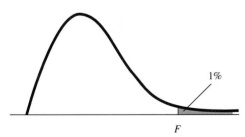

1%

F

From Leon F. Marzillier, *Elementary Statistics* ©1990 by Wm. C. Brown Publishers.

TABLE E

(continued)

$\alpha = 0.01$

df_1

df_2	15	20	24	30	40	50	60	75	100	120	∞
1	6157	6209	6235	6261	6287	6302	6313	6323	6334	6339	6366
2	99.4	99.4	99.5	99.5	99.5	99.5	99.5	99.5	99.5	99.5	99.5
3	26.9	26.7	26.6	26.5	26.4	26.3	26.2	26.3	26.2	26.2	26.1
4	14.2	14.0	13.9	13.8	13.8	13.7	13.6	13.6	13.6	13.6	13.5
5	9.7	9.6	9.5	9.4	9.3	9.2	9.2	9.2	9.1	9.1	9.0
6	7.6	7.4	7.3	7.2	7.1	7.1	7.1	7.0	7.0	7.0	6.9
7	6.3	6.2	6.1	6.0	5.9	5.8	5.8	5.8	5.8	5.7	5.6
8	5.5	5.4	5.3	5.2	5.1	5.1	5.0	5.0	5.0	5.0	4.9
9	5.0	4.8	4.7	4.6	4.6	4.5	4.4	4.4	4.4	4.4	4.3
10	4.6	4.4	4.3	4.2	4.2	4.1	4.1	4.0	4.0	4.0	3.9
11	4.2	4.1	4.0	3.9	3.9	3.8	3.8	3.7	3.7	3.7	3.6
12	4.0	3.9	3.8	3.7	3.6	3.6	3.5	3.5	3.5	3.4	3.4
13	3.8	3.7	3.6	3.5	3.4	3.4	3.3	3.3	3.3	3.2	3.2
14	3.7	3.5	3.4	3.4	3.3	3.2	3.2	3.1	3.1	3.1	3.0
15	3.5	3.4	3.3	3.2	3.1	3.1	3.0	3.0	3.0	3.0	2.9
16	3.4	3.3	3.2	3.1	3.0	3.0	2.9	2.9	2.9	2.8	2.8
17	3.3	3.2	3.1	3.0	2.9	2.9	2.8	2.8	2.8	2.8	2.7
18	3.2	3.1	3.0	2.9	2.8	2.8	2.8	2.7	2.7	2.7	2.6
19	3.2	3.0	2.9	2.8	2.8	2.7	2.7	2.6	2.6	2.6	2.5
20	3.1	2.9	2.9	2.8	2.7	2.6	2.6	2.6	2.5	2.5	2.4
21	3.0	2.9	2.8	2.7	2.6	2.6	2.6	2.5	2.5	2.5	2.4
22	3.0	2.8	2.8	2.7	2.6	2.5	2.5	2.5	2.4	2.4	2.3
23	2.9	2.8	2.7	2.6	2.5	2.5	2.4	2.4	2.4	2.4	2.3
24	2.9	2.7	2.7	2.6	2.5	2.4	2.4	2.4	2.3	2.3	2.2
25	2.8	2.7	2.6	2.5	2.4	2.4	2.4	2.3	2.3	2.3	2.2
26	2.8	2.7	2.6	2.5	2.4	2.4	2.3	2.3	2.2	2.2	2.1
27	2.8	2.6	2.6	2.5	2.4	2.3	2.3	2.2	2.2	2.2	2.1
28	2.8	2.6	2.5	2.4	2.4	2.3	2.3	2.2	2.2	2.2	2.1
29	2.7	2.6	2.5	2.4	2.3	2.3	2.2	2.2	2.2	2.1	2.0
30	2.7	2.6	2.5	2.4	2.3	2.2	2.2	2.2	2.1	2.1	2.0
40	2.5	2.4	2.3	2.2	2.1	2.0	2.0	2.0	1.9	1.9	1.8
50	2.4	2.3	2.2	2.1	2.0	1.9	1.9	1.9	1.8	1.8	1.7
60	2.4	2.2	2.1	2.0	1.9	1.9	1.8	1.8	1.7	1.7	1.6
70	2.3	2.2	2.1	2.0	1.9	1.8	1.8	1.7	1.7	1.7	1.5
80	2.3	2.1	2.0	1.9	1.8	1.8	1.8	1.7	1.6	1.6	1.5
100	2.2	2.1	2.0	1.9	1.8	1.7	1.7	1.6	1.6	1.5	1.4
120	2.2	2.0	2.0	1.9	1.8	1.7	1.6	1.6	1.5	1.5	1.4
∞	2.0	1.5	1.8	1.7	1.6	1.5	1.5	1.4	1.4	1.3	1.0

TABLE F

p-values for χ^2

χ^2	Degrees of freedom		
	2	**3**	**4**
3.2	2019		
3.3	1920		
3.4	1827		
3.5	1738		
3.6	1653		
3.7	1572		
3.8	1496		
3.9	1423		
4.0	1353		
4.1	1287		
4.2	1225		
4.3	1165		
4.4	1108		
4.5	1054		
4.6	1003	2035	
4.7	0954	1951	
4.8	0907	1870	
4.9	0863	1793	
5.0	0821	1718	
5.1	0781	1646	
5.2	0743	1577	
5.3	0707	1511	
5.4	0672	1447	
5.5	0639	1386	
5.6	0608	1328	
5.7	0578	1272	
5.8	0550	1218	
5.9	0523	1166	2067
6.0	0498	1116	1991
6.1	0474	1068	1918
6.2	0450	1023	1847
6.3	0429	0979	1778
6.4	0408	0937	1712
6.5	0388	0897	1648
6.6	0302	0858	1586
6.7	0351	0821	1526
6.8	0334	0786	1468
6.9	0317	0752	1413
7.0	0302	0719	1359
7.1	0287	0688	1307

χ^2	Degrees of freedom					
	2	**3**	**4**	**5**	**6**	**7**
7.2	0273	0658	1257	2062		
7.3	0260	0629	1209	1993		
7.4	0247	0602	1162	1926		
7.5	0235	0576	1117	1860		
7.6	0224	0550	1074	1797		
7.7	0213	0526	1032	1736		
7.8	0202	0503	0992	1676		
7.9	0193	0481	0953	1618		
8.0	0183	0460	0916	1562		
8.1	0174	0440	0880	1508		
8.2	0166	0421	0845	1456		
8.3	0158	0402	0812	1405		
8.4	0150	0384	0780	1355		
8.5	0143	0367	0749	1307	2037	
8.6	0136	0351	0719	1261	1974	
8.7	0129	0336	0691	1216	1912	
8.8	0123	0321	0663	1173	1851	
8.9	0117	0307	0636	1131	1793	
9.0	0111	0293	0611	1091	1736	
9.1	0106	0280	0586	1051	1680	
9.2	0101	0267	0563	1013	1626	
9.3	0096	0256	0540	0977	1574	
9.4	0091	0244	0518	0941	1523	
9.5	0087	0233	0497	0907	1473	
9.6	0082	0223	0477	0874	1425	
9.7	0078	0213	0456	0842	1379	
9.8	0074	0203	0439	0811	1333	2002
9.9	0071	0194	0421	0781	1289	1943
10	0067	0186	0404	0752	1247	1886

All missing entries on this page give p-values > 0.20.
See the next page for p-values corresponding to larger values of χ^2.

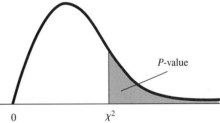

TABLE F

(continued)

χ^2	Degrees of freedom								
	2	**3**	**4**	**5**	**6**	**7**	**8**	**9**	**10**
11	0041	0117	0266	0514	0884	1386	2017		
12		0074	0174	0348	0620	1006	1512	2133	
13		0046	0113	0234	0430	0721	1118	1626	2237
14			0073	0156	0296	0512	0818	1223	1730
15			0047	0104	0203	0360	0591	0909	1321
16				0068	0138	0251	0424	0669	0996
17				0045	0093	0174	0301	0487	0744
18					0062	0120	0212	0352	0550
19					0042	0082	0149	0252	0403
20						0056	0103	0179	0293
21						0038	0071	0127	0211
22							0049	0089	0151
23								0062	0107
24								0043	0076
25									0053
26									0037

The missing entries in the top right-hand corner give p-value > 0.20. All other missing entries give p-value < 0.005.

From Leon F. Marzillier, *Elementary Statistics* ©1990 by Wm. C. Brown Publishers.

INDEX

absolute descriptive measures, 44
absolute dispersion, 56–58
accuracy, 21
actual (observed) frequency distribution, 154
agglomeration strategy of classification, 23
alternate (alternative) hypothesis, 117–118
analysis of variance (ANOVA), 146–149, 228
    boxed inset, 149
    geographic application, 150–153
area pattern analysis, 171–172, 182–189, 228
    geographic application, 185–189
arithmetic mean, 38–40
artificial sample, 126
association, 193–196
    direction of, 194–196
    linear, 212–214
    strength of, 195–196
autocorrelation, 172

between-group variability (in ANOVA), 148
bimodal distribution, 39–40
binomial distribution, 69–71
bivariate regression, 211–222, 228
    boxed inset, 222
    geographic application, 211–222
bivariate relationship, 32–33
boundary delineation, 46–47
bounds (of a confidence interval), 101

categorical difference tests, 154
categorical (qualitative) variable, 18
center of gravity, 53
central limit theorem, 98–100
central tendency measures, 37–40
chi-square test, 155–156, 166–170, 228
    boxed inset, 156
    geographic application (proportional, 160–163
    geographic application (uniform), 158–160
choropleth map correlation, 206–208
class midpoint, 39
classical hypothesis testing, 116–121
classification methods and strategies, 22–31
    agglomeration, 23
    equal intervals based on range, 23–24
    equal intervals not based on range, 23–24
    natural breaks, 24–25
    quantile breaks, 24–25
    subdivision (logical subdivision), 22–23
cluster sampling, 91–92, 94–95
clustered patterns, 171, 178–179

coefficient of determination, 217
coefficient of multiple determination, 224
coefficient of variation, 44, 49–50
combination (hybrid) sampling, 92, 94–96
complement (in probability), 67
confidence interval, 98, 100–107
    for the mean, 103–107
    for the proportion, 105, 107
    for the total, 104–105, 107
confidence level, 101–102
contingency analysis, 154, 166–170, 228
    boxed inset, 168
    geographic application, 168–170
continuous probability distribution, 69
correlation, 127–128, 193–209, 228
    of choropleth maps, 206–208
    of dot maps, 204–205
    of interval/ratio data, 196–201
    of isoline maps, 204–206
    issues regarding, 207–208
    of ordinal data, 201–204
covariation, 196–197
cross-tabulation, 167
crude mode, 38
cumulative frequency diagram (ogive), 31–32
cumulative normal value, 156–157
cumulative relative frequency, 31–32, 156–157
curvi-linear relationship, 211, 221

data, 13–14, 16
    categorical, 18
    continuous, 18
    discrete, 18
    explicitly spatial, 17
    implicitly spatial, 17
    individual, 17
    matrix, 13
    primary, 16–17
    secondary (archival), 17
    set, 13
    spatially aggregated, 17–18
    value, 13
degrees of freedom, 125
dependent variable (in regression), 211
descriptive statistics, 14, 37–51
deterministic processes, 65–66
deviation from mean, 41–42
difference of means tests (one sample), 118–119, 123–124
    boxed inset, 119
    geographic application, 118–124